Sustainable Brownfi

Due

Editors' Dedications

Tim Dixon would like to dedicate this book to Rachel and Sam for all their love and support during the course of editing this book.

Philip Catney would like to dedicate this book to Rachel for her love, understanding and support over the years.

Sustainable Brownfield Regeneration

Liveable Places from Problem Spaces

Edited by

Tim Dixon

Professor of Real Estate
Oxford Institute of Sustainable Development
Oxford Brookes University

Mike Raco

Senior Lecturer
Department of Geography
King's College London

Philip Catney

Research Associate
Department of Town & Country Planning
University of Sheffield

David N. Lerner

Professor of Environmental Engineering
Catchment Science Centre
University of Sheffield

Blackwell
Publishing

© 2007 by Blackwell Publishing Ltd

Blackwell Publishing editorial offices:
Blackwell Publishing Ltd, 9600 Garsington Road, Oxford OX4 2DQ, UK
    Tel: +44 (0)1865 776868
Blackwell Publishing Inc., 350 Main Street, Malden, MA 02148-5020, USA
    Tel: +1 781 388 8250
Blackwell Publishing Asia Pty Ltd, 550 Swanston Street, Carlton, Victoria 3053, Australia
    Tel: +61 (0)3 8359 1011

The right of the Authors to be identified as the Authors of this Work has been asserted in accordance with the Copyright, Designs and Patents Act 1988.

First published 2007 by Blackwell Publishing Ltd

ISBN-13: 978-1-4051-4403-2

Library of Congress Cataloging-in-Publication Data

Sustainable brownfield regeneration : liveable places from problem spaces / edited by Tim Dixon . . . [*et al.*].
    p. cm.
  Includes bibliographical references and index.
  ISBN-13: 978-1-4051-4403-2 (pbk. : alk. paper)
1. Soil remediation. 2. Brownfields. 3. Reclamation of land. 4. Sustainable buildings. 5. Building sites. I. Dixon, Timothy J., 1958–

  TD878.S872 2007
  333.77′137—dc22

                                                                    2007010208

A catalogue record for this title is available from the British Library

Set in 10/13pt Trump Mediaeval by Graphicraft Limited, Hong Kong
Printed and bound in Singapore by Utopia Press Pte Ltd

The publisher's policy is to use permanent paper from mills that operate a sustainable forestry policy, and which has been manufactured from pulp processed using acid-free and elementary chlorine-free practices. Furthermore, the publisher ensures that the text paper and cover board used have met acceptable environmental accreditation standards.

For further information on Blackwell Publishing, visit our website:
www.blackwellpublishing.com/construction

Contents

Notes on the Contributors

Editors

Tim Dixon is Professor of Real Estate and Co-Director of the Oxford Institute of Sustainable Development (OISD) based at Oxford Brookes University, UK. With more than 20 years' experience of research and education in the built environment, he is a member of SEEDA's South East Excellence Advisory Board and of the EPSRC Infrastructure and Environment Strategic Advisory Team as well as the editorial boards of five leading journals. He has worked on collaborative research projects with UK and overseas academics and practitioners, and co-led (with David Lerner) the successful 2003 bid to EPSRC for a £1.8 million four-year programme of research into brownfield issues (SUBR:IM – www.subrim.org.uk) in which Oxford Brookes was a key partner. Email: tdixon@brookes.ac.uk

Mike Raco is Senior Lecturer in Human Geography in the Department of Geography, King's College London. He has published widely on the topics of urban and regional development, local governance, urban regeneration and community development. His publications include *Building Sustainable Communities: Spatial Policy and Labour Mobility in Post-war Britain* (Policy Press, Bristol), *Urban Renaissance? New Labour, Community, and Urban Policy* (with Rob Imrie; Policy Press, Bristol). He formerly lectured at the universities of Reading and Glasgow. Email: mike.raco@kcl.ac.uk

Philip Catney is a political scientist. He is currently a research associate in the Department of Town and Regional Planning at the University of Sheffield, where he also teaches in the Department of Politics, and is the SUBR:IM Consortium Manager. He has published a number of articles on environmental, local and urban governance and the ideology of New Labour. Philip is also a member of the Sustainable Development Commission Forum, a stakeholder forum of the UK government's watchdog on sustainable development issues. Email: p.catney@sheffield.ac.uk

David N. Lerner was Director of the SUBR:IM Research Consortium with research interests in urban groundwater and contaminated land. He is Professor of Environmental Engineering, leader of the Groundwater Protection and Restoration Group and Director of the Catchment Science Centre at the University of Sheffield, UK. As well as researching on restoration of contaminated land and groundwater, he leads projects on integrated catchment modelling and management, especially on the use of systems analysis to build decision support tools. The interdisciplinary and multinational CatSci programme is researching a range of Water Framework Directive issues using the mixed urban–rural River Don catchment as a case study. Email: d.n.lerner@sheffield.ac.uk

Main authors

Abir Al-Tabbaa is Reader in Geotechnical Engineering at the University of Cambridge. She is involved in research in a range of areas in geotechnical and environmental engineering including ground improvement, contaminated land remediation, waste management and reuse and sustainable construction materials and products. Email: aa22@cam.ac.uk

Sophie Bowlby is a Senior Lecturer in Human Geography at the University of Reading. Her research focuses on issues of inclusion related to gender, race and class with particular reference to issues of mobility, access to employment and the geographies of informal care demands. Email: s.r.bowlby@rdg.ac.uk

Kate Burningham is Senior Lecturer in Sociology of the Environment in the Department of Sociology and the Centre for Environmental Strategy at the University of Surrey. Email: k.burningham@surrey.ac.uk

Philip Catney is a political scientist. He is currently a research associate in the Department of Town and Regional Planning at the University of Sheffield, where he also teaches in the Department of Politics, and is the SUBR:IM Consortium Manager.

Tim Dixon is Professor of Real Estate and Co-Director of the Oxford Institute of Sustainable Development (OISD) based at Oxford Brookes University, UK. Email: tdixon@brookes.ac.uk

Joe Doak is a Senior Lecturer in Planning and Urban Development in the Department of Real Estate and Planning at the University of Reading, and is Director of Postgraduate Planning Programmes. Email: a.j.doak@ reading.ac.uk

Uche Duru is a PhD student in Geoenvironmental Engineering at the University of Cambridge's Department of Engineering, working on the impact of climate change on chemical and biological properties of contaminated soils. Email: ued20@cam.ac.uk

Richard Eiser is Professor of Psychology at the University of Sheffield, a post he has held since 2000. From 1979 to 2000 he was Professor of Psychology at the University of Exeter. He is the author of several articles and books in social psychology with a special focus on attitudes and the judgement of health and environmental risks. Email: J.R.Eiser@sheffield.ac.uk

Chris Evans is Technical Director at Arcadis Geraghty and Miller International Limited, based in Newmarket, UK, specialising in contaminated land remediation and geotechnical engineering. Email: cevans@arcadisgmi.com

Stephen Garvin is a Director of BRE, based at BRE Scotland. He has undertaken research for central government and EPSRC over a 17-year period, producing guidance documents and research papers. He has a wide range of research interests that include land quality through to materials technology. Email: GarvinS@bre.co.uk

Michael Harbottle was formerly a Postdoctoral Research Associate at Cambridge University's Engineering Department and is now a Lecturer in Geoenvironmental Engineering at Cardiff University. Email: harbottlem@cardiff.ac.uk

Steven Henderson is a lecturer in Geography and Planning in the School of Applied Sciences at the University of Wolverhampton. Email: steven.henderson@wlv.ac.uk

John Henneberry is Professor of Property Development in the Department of Town and Regional Planning, University of Sheffield. His research focuses on the structure and behaviour of the property market and its relations with the wider economy and polity; particularly policy and decision-making with regard to the redevelopment of brownfield land. Email: J.Henneberry@sheffield.ac.uk

René van Herwijnen is Research Associate for Forest Research and the University of Surrey for whom he is doing research into new bioremediation technologies for contaminated land. Email: rene.vanherwijnen@forestry.gsi.gov.uk

Tony Hutchings is Head of the Land Regeneration and Urban Greening Group in Forest Research, the research agency of the Forestry Commission. Email: tony.hutchings@forestry.gsi.gov.uk

Srinath R. Iyengar is a PhD student in Geoenvironmental Engineering at the University of Cambridge's Department of Engineering. He is working on the development of sustainable contaminated land stabilisation/solidification remediation systems. Email: sri@cam.ac.uk

Michael Johns is a lecturer in Chemical Engineering at the University of Cambridge. He is working on developing magnetic resonance and other tomographic methods to better understand and control industrial process problems in addition to modelling the flow of pollutants in the subsurface. Email: mlj21@cam.ac.uk

Kalliope Pediaditi is Research Fellow at Mediterranean Agronomic Institute Chania (MAICh), Alsyllion Agrokepion, Crete, Greece. Email: pediaditi@maich.gr

Nikos Karadimitriou is a Lecturer in Land and Property Development at the Bartlett School of Planning, University College London. Email: n.karadimitriou@ucl.ac.uk

Nigel Lawson is a senior member of staff within the School of Environment and Development at the University of Manchester. Nigel has considerable experience in contaminated land and demolition waste research following a successful career in industry and business. Email: nigel.lawson @manchester.ac.uk

David N. Lerner was Director of the SUBR:IM Research Consortium with research interests in urban groundwater and contaminated land. He is Professor of Environmental Engineering, leader of the Groundwater Protection and Restoration Group and Director of the Catchment Science Centre at the University of Sheffield, UK. Email: d.n.lerner@sheffield.ac.uk

Andy Moffat is Head of Environmental and Human Sciences Division in Forest Research, the research agency of the Forestry Commission. Email: andy.moffat@forestry.gsi.gov.uk

Cécile De Munck is an Environmental Scientist and Modeller in the Land Regeneration and Urban Greening Group of Forest Research, the research agency of the Forestry Commission. Email: cecile.de_munck@forestry. gsi.gov.uk

Sabeha Ouki is a Reader in Pollution Control and Waste Management in the Centre for Environmental Health Engineering (CEHE) and Director of Research Studies for the School of Engineering, University of Surrey. Email: s.ouki@surrey.ac.uk

Mike Raco is Senior Lecturer in Human Geography in the Department of Geography, King's College London. Email: mike.raco@kcl.ac.uk

Colin Smith is a Senior Lecturer in Geotechnical Engineering in the Department of Civil and Structural Engineering at the University of Sheffield. Email: c.c.smith@sheffield.ac.uk

Sinéad Smith is a PhD student in Geoenvironmental Engineering at the University of Cambridge's Department of Engineering working on the impact of climate change on contaminated land and containment systems. Email: ses50@cam.ac.uk

Tom Stafford is a lecturer in the Department of Psychology at the University of Sheffield. He is lucky enough to have engaged in the study of decision-making at biological, psychological and social levels of analysis. Email: t.stafford@sheffield.ac.uk

Simon Talbot is the Director of GMGU, a part of Urban Vision Partnership Ltd, which supplies specialist land use and contaminated land consultancy services to the public and private sectors and provides teaching, training and research through its close links with the University of Manchester. Email: simon.talbot@gmgu.org.uk

Walter Wehrmeyer is Reader in Environmental Business Management at the Centre for Environmental Strategy of the University of Surrey. His research interests include organisational approaches to innovation and sustainable development, and participatory approaches to decision-making, including measurement systems to support progress evaluation in these areas. Email: w.wehrmeyer@surrey.ac.uk

Other contributors (by chapter)

Chapter 6

Steven Henderson, Geography, School of Applied Sciences, University of Wolverhampton, Wulfruna Street, Wolverhampton WV1 1SB. Email: steven.henderson@wlv.ac.uk

Chapter 11

Uche Duru is a PhD student in Geoenvironmental Engineering at the University of Cambridge's Department of Engineering working on the impact of climate change on chemical and biological properties of contaminated soils. Email: ued20@cam.ac.uk

Srinath Iyengar is a PhD student in Geoenvironmental Engineering at the University of Cambridge's Department of Engineering working on the development of sustainable contaminated land stabilisation/solidification remediation systems. Email: sri@cam.ac.uk

Julian Ridal is a Senior Consultant at BRE Scotland. He specialises in contaminated land, construction materials and waste, sustainability and health and safety. Email: ridalj@bre.co.uk

Acknowledgements

Our thanks go to the Engineering and Physical Science Research Council (EPSRC), which funded the research in SUBR:IM (Sustainable Urban Brownfield Regencration: Integrated Management) under Grant Number GR/S18809/01.

The editors and contributors would like to note the special contribution that Mike Brown (now based at the University of Dundee) made in making this volume possible. Mike's efficiency was essential in ensuring that SUBR:IM was successfully established.

The SUBR:IM website is at www.subrim.org.uk

Foreword

Ensuring the effective and efficient reuse of brownfield land is an essential part of the British Government's land use policies in support of sustainable communities. English Partnerships, as the Government's specialist advisor on brownfield land, has prepared new policy recommendations and over-arching principles to form a National Brownfield Strategy for England, aimed at stimulating the redevelopment and reuse of land by both the public and private sectors. The recommendations acknowledge the importance of reusing brownfield land for a full range of activities, including housing, employment, recreation and open space as well as increasing wildlife habitats and the prevention of flooding.

There is already a good track record in England for recycling brownfield land but there are many barriers that make the process less efficient and less attractive than Government would like. The aim is to increase the beneficial reuse of brownfield land and buildings, including tackling some of the more difficult, long-term derelict and vacant sites that often blight communities. The six 'over-arching principles' aim to ensure the country's 63 000 hectares of previously developed land is used to better effect. They include focusing on the widest possible range of uses, not just housing; concentrating efforts in areas where existing infrastructure has the capacity to support redevelopment; and ensuring that, where possible, future uses should help support families and assist in combating anti-social behaviour. The Strategy also recognises the need to improve the brownfield skills base and for Government departments to adopt a more 'joined up' approach to resolving brownfield problems.

The recommendations are the result of a three-year consultation programme, carried out in conjunction with DCLG (the Department for Communities and Local Government) and involving Defra (the Department for the Environment, Food and Rural Affairs), the Environment Agency and other Government departments, along with local authorities and private sector stakeholders. Throughout the process of developing the Strategy the work has been informed by the efforts of the SUBR:IM consortium.

SUBR:IM has played a very important role in raising the profile of brownfield issues and the impact they can have on all aspects of modern society. *Sustainable Brownfield Regeneration: Liveable Places from Problem Spaces* recognises that reusing brownfield land is not just about overcoming 'technical issues to remove contamination or other physical problems with the ground. It highlights the importance of engaging with the many different stakeholders whose opinions and concerns need to be taken into account if sustainable outcomes are to be achieved. The authors also recognise that brownfield land reuse is not just about building new homes or places of employment – the creation of new green spaces can be just as important.

Members of the research team should be congratulated on their contributions to the brownfield debate and I hope that they will continue their research interests in this challenging, and rewarding, area. I am therefore very pleased to welcome the publication of the research outputs from the project.

Professor Paul Syms
National Brownfield Advisor
English Partnerships
May 2007

Part 1
Introduction

1

Introduction

Tim Dixon and Mike Raco

1.1 Background

Brownfield regeneration has become a major policy driver in the UK and other developed countries. It is estimated that there are 64 000 hectares of brownfield land in England, much of which presents severe environmental challenges and lies alongside some of the most deprived communities in the country.

Brownfields have been defined by CABERNET[1] as sites that

- have been affected by former uses of the site or surrounding land
- are derelict or underused
- are mainly in fully or partly developed urban areas
- require intervention to bring them back to beneficial use, and
- may have real or perceived contamination problems

In the UK, land that has been 'previously developed' is commonly known as brownfield land. Previously developed land is defined in Planning Policy Guidance Document 3: *Housing* (2000) PPG3 as land that 'is or was occupied by a permanent structure (excluding agricultural or forestry buildings), and associated fixed surface infrastructure'. In practice, this means brownfield land comprises the following categories:

- Land type A – previously developed land now vacant
- Land type B – vacant buildings
- Land type C – derelict land and buildings
- Land type D – land or buildings currently in use and allocated in the local plan and/or having planning permission
- Land type E – land or buildings currently in use with redevelopment potential

Bringing such land back into active use has taken on a new urgency among policy makers, developers and other stakeholders in the development process. Frequently, however, policy thinking and practice has been underpinned by 'silo' mentalities, in which integrated and multidisciplinary approaches to problem-solving have been limited. Important issues such as the technical identification of forms of contamination and appropriate remediation techniques, the creation of interactive and inclusive systems of governance and policy-making, and mechanisms to encourage the mobilisation of important stakeholders such as the development industry and local communities have often been dealt with in isolation. As a consequence, those applying technologies in cleaning up contaminated brownfield land have often done so in a deterministic way, without fully incorporating an understanding of the impact on communities and other stakeholders. Similarly, some policy makers and local communities have tried to adopt ambitious approaches with little appreciation of the technical processes involved in site clearance, remediation and development.

To overcome these discontinuities, and to develop more integrated approaches, a new research consortium called SUBR:IM (Sustainable Brownfield Regeneration: Integrated Management) was formed in 2003. With more than £1.9 million of initial funding over four years, SUBR:IM (www.subrim.org.uk) brought together ten major research institutions to work on 18 inter-related projects.[2] SUBR:IM's researchers were drawn from across the science and social science disciplines and had experience of working in fields such as engineering, construction management, property and real estate, and development planning. The research also included support from stakeholders, including industry, civic associations, and national, regional and local government, and SUBR:IM's work focused on sites in Greater Manchester and the Thames Gateway, which have some of the largest concentrations of brownfield land and deprived communities in the UK.

This book systematically and comprehensively documents the core evidence and findings from SUBR:IM's research programme. It adopts an integrated approach to the subject by drawing on the lessons learned from the research, not only from the individual projects themselves, but also from the synergies established through the process of working together in a multidisciplinary team. It is intended to provide a highly original and innovative account of the processes and practices of brownfield regeneration in the UK and how different types of knowledge can be brought to bear to develop a more holistic, and ultimately effective, urban policy. It also establishes wider lessons for multidisciplinary and cross-subject research, something that is likely to become more and more significant across the social and technical sciences as the value of such work in tackling multiple urban problems becomes more apparent.

1.2 Aims and objectives

The book has two principal aims. The first is to examine the ways in which science and social science research disciplines can be brought together to help solve important brownfield regeneration issues, with a focus on the UK. The second is to assess the efficiency and effectiveness of different types of regeneration policy and practice, and to show how 'liveable spaces' can be produced from 'problem places'.

In order to address these aims the research projects within SUBR:IM have covered four principal themes (reflected diagrammatically in Figure 1.1):

(1) the property development and investment industries and their role in brownfield regeneration
(2) the processes of governance and multi-level decision-making relating to brownfield regeneration, including institutional structures and community engagement as well as risk, trust and systems of democratic representation
(3) the development of robust technical solutions to contamination and examination of the impact of climate change within this
(4) the ways in which integrated solutions to brownfield renewal can be developed, including how the greening of former brownfield spaces can open up new opportunities for urban regeneration

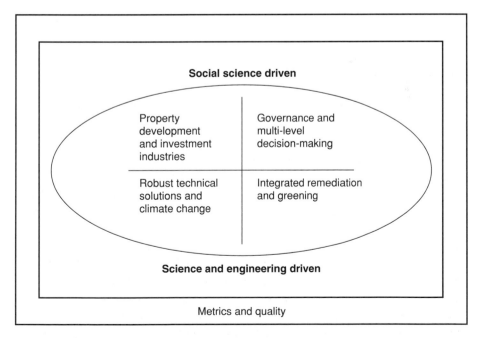

Figure 1.1   Conceptual diagram of SUBR:IM.

Themes (1) and (2) primarily related to the 'social science' disciplines, whereas themes (3) and (4) leaned more towards the environmental and engineering sciences (see Chapter 2 for a broader discussion).

All the SUBR:IM projects sought to develop cross-cutting methodologies and approaches. Some projects were also developed that were designed to weave together the project's broader research findings; for example, the 'metrics' project in SUBR:IM developed a more holistic view of the brownfield development process and questions of sustainability (see Chapter 12).

A large part of SUBR:IM's work has also focused on two major sub-regions where brownfield development is of particular significance: the Thames Gateway and Greater Manchester. This shared empirical focus was essential for the coherence of the research programme. The Thames Gateway in the south of England represented a site of core regional and national significance, and extends for approximately 60 kilometres along the River Thames from the London Docklands to Southend in Essex and Sheerness in Kent. Regional Planning Guidance for the South East (RPG 9) and the *Sustainable Communities Plan* had already identified it as a priority for regeneration and growth. It is also one of the four target areas for new housing in the South East.

Similarly in Greater Manchester, Manchester and Salford have both received an increased amount of government and media attention not only as a result of the Northern Way strategy and the broader sustainable communities agenda, but also because the 'Manchester model' of regeneration encapsulates much of what many commentators consider best practice in British post-war regeneration, particularly in relation to joint venture schemes for redevelopment.

1.3 Structure of the book

In order to address its aims the book is divided into four interrelated parts. These sections are designed to ensure that the contributions do not simply become a collection of research papers. Instead, they are designed to provide a framework within which the various chapters can be integrated and developed in a theoretically focused and robust manner.

Part 1 ('Introduction') comprises the current chapter (Chapter 1) and Chapter 2. It outlines the structures of the SUBR:IM consortium, the processes through which it emerged, and the problems and opportunities associated with these new, increasingly popular, academic working practices. In Chapter 2 Mike Raco and Tim Dixon examine broader conceptions of multidisciplinarity and how the SUBR:IM portfolio of work has fitted together. They assess the ways in which integrated, 'socio-technical' approaches were applied to the brownfield regeneration 'problem' under SUBR:IM and the lessons for future research in this field and beyond.

Part 2 deals with processes of *regeneration*, exploring socio-technical problems and solutions, and the role of actors/stakeholders in the regeneration process, together with governance issues. It consists of four chapters, each of which explores different dimensions associated with brownfield regeneration. In Chapter 3 Philip Catney *et al.* examine questions of 'Democracy, Trust and Risk Related to Contaminated Sites in the UK'. The authors examine the relationships between democracy, trust and risk and draw on two case studies of risk assessment and communication to explore how these issues relate to contaminated brownfield sites in England. The results show how different approaches to risk contamination affect the degree to which local residents trust their councils. Chapter 4 goes on to examine what Joe Doak and Nikos Karadimitriou term 'Actor Networks and the Brownfield Merry-Go-Round'. The chapter explores the relationships between property investors and high-profile development projects. The authors argue that we need to understand the development process through a form of complex and contextual network analysis. Examples from London and Manchester are used to examine this framework. In Chapter 5, entitled 'Heroes or Villains? The Role of the Development Industry in Brownfield Regeneration', Tim Dixon examines the nature and challenge of brownfield development in the UK, and the role of the development industry and its attitudes towards brownfield development through case study-based work in Thames Gateway and Greater Manchester. This chapter is closely integrated with Chapter 6, 'Delivering Brownfield Regeneration – Sustainable Community-Building in London and Manchester', in which Mike Raco *et al.* look at high-profile examples from London and Manchester and broader questions concerning brownfield regeneration processes and their wider sustainability. In both chapters the authors argue, from differing perspectives, that policy makers and others need to be more open to the possibility that brownfield development may represent only one part of a broader set of sustainable development policies.

Part 3 contains four chapters that focus on broader processes of *remediation* with an emphasis on the relationships between scientific 'problems' and 'solutions' in particular brownfields. Chapter 7, by Andy Moffat and Tony Hutchings, addresses the theme of 'Greening Brownfield Land' and examines the challenges of establishing and maintaining greenspace on brownfield land in a sustainable way. The chapter draws on work from the SUBR:IM projects and beyond. It is closely linked to Chapter 8 and its focus on 'Novel Special-Purpose Composts for Sustainable Remediation', where the researchers (Sabeha Ouki *et al.*) give an overview of the major results obtained during experimental investigations in testing composts combined with naturally occurring minerals (clays and zeolites) for their ability to reduce plant availability and leachability of heavy metals when applied to contaminated soils. Technical innovation and knowledge such as this is often essential to the remediation process. Other remediation techniques are examined in Chapter 9 in which Abir Al-Tabbaa *et al.* concentrate on the appropriateness of 'Robust Technical Solutions'. The discussion focuses on the assessment of sustainability in relation to containment and clean-up methods of remediation and how improvements can be made to specific technical solutions based on sustainability assessment and experimental investigations. Chapter 10 (Simon Talbot *et al.*) focuses on a particularly challenging example of brownfield remediation, that of 'acid tar lagoons'. It examines the scale of the problem, the particular difficulties associated with tar remediation, and provides examples of such remediation 'in practice'.

The concluding section, Part 4, is focused on what might be termed 'Joined-up solutions'. It draws on SUBR:IM projects that have adopted a cross-cutting methodology and provides conclusions to the book. In Chapter 11 Abir Al-Tabbaa *et al.* examine the theme of 'Climate Change, Pollutant Linkage and Brownfield Regeneration'. The research examines the impact that climate change could have on pollutant linkages in the soil and the extent to which stakeholders are developing adaptation strategies to account for this impact. Chapter 12 provides an overview of the work carried out by Walter Wehrmeyer and Kalliope Pediaditi on 'Evaluating the Sustainability of Brownfield Redevelopment Projects'. The chapter describes a particular framework, named the Redevelopment Assessment Framework (or RAF), used to assess and monitor the long-term sustainability of brownfield redevelopment projects. This is followed by a concluding chapter by Philip Catney, David Lerner, Tim Dixon and Mike Raco entitled 'Is Brown the New Green?', which summarises and synthesises the key messages and findings that have emerged from the SUBR:IM research and assesses their significance for wider debates over urban regeneration, environmental remediation and sustainable urban redevelopment. The chapter also highlights the wider lessons concerning the research process, the contested nature of the concept of sustainable development, and the value added by this type of multidisciplinary and integrated research.

Notes

1. CABERNET (Concerted Action on Brownfield and Economic Regeneration Network), is a multidisciplinary network comprising six expert Working Groups that aims to facilitate new practical solutions for urban brownfields (see www.cabernet.org.uk).
2. The ten institutions are the University of Sheffield, Oxford Brookes University, King's College London, Forest Research, the University of Reading, the University of Cambridge, the University of Surrey, the Building Research Establishment, the University of Manchester and the Greater Manchester Geological Unit. The website is at www.subrim.org.uk

2

Researching Sustainability: The Possibilities and Limitations of Cross-Cutting Research in the Urban Environment

Mike Raco and Tim Dixon

2.1 Introduction

> Science has spoken, with growing urgency and conviction, to society for more than half a millennium. Not only has it determined technical processes, economic systems and social structures, it has also shaped our everyday experience of the world, our conscious thoughts and even our unconscious feelings. Science and modernity have become inseparable. In the past half-century society has begun to speak back to science, with equal urgency and conviction. (Nowotny *et al.*, 2001)
>
> [I]n the academic world people fight constantly over the question of who, in this universe, is socially mandated, and authorised to tell the truth. (Bourdieu, 2003, pp. 70–71)

Across academia there are new and expanding pressures for researchers to engage in interdisciplinary (ID), multidisciplinary (MD) or transdisciplinary (TD) work. There are intellectual and more practical drivers underpinning this process. On the one hand there are new conceptual challenges in the twenty-first century that lend themselves to such study. Issues such as environmental change and sustainability, energy generation and habitat destruction represent expanding problems whose causes and potential solutions arguably require the creation and deployment of multiple knowledges. In parallel with this, in countries such as the UK, research environments are changing rapidly and encouraging academics from diverse disciplinary backgrounds to work together. The focus on 'evidence-based' policy, the New Public Management, and 'user-engagement' have increasingly made research funding conditional on the ability of researchers to work in new networks and partnerships and disseminate their knowledge to a wider public. Whatever the intellectual strengths and weaknesses of such work, it has become something of a conventional wisdom that mirrors broader shifts in the governance, management, and delivery of the public sector.

And yet, this shift towards non-disciplinary approaches raises a number of questions. What do such terms really mean, both conceptually and practically? What are their limitations and how do bureaucratic definitions and funding mechanisms help or hinder such work? Is such research practically feasible, and if so, what conditions are necessary for it be successful? Moreover, is such work desirable anyway and an effective use of increasingly scarce resources? This chapter (and see also Chapter 13) examines some of the factors that influenced the work of SUBR:IM and develops a critical reflexive approach (cf. Bourdieu, 2003) to the consortium and the broader multidisciplinary agenda that it is a part of. It begins by discussing disciplinarity and its significance for the research process before examining the experiences of SUBR:IM. It then concludes by highlighting some of the wider lessons that emerged from the research process for future research on urban sustainability.

2.2 Intellectual disciplines, interdisciplinarity and the construction of knowledge

2.2.1 The challenge to disciplinarity

The creation of knowledge is intimately linked to the formation and reproduction of academic *disciplines*. The *Collins English Dictionary* defines 'discipline' as 'systematic training in obedience' and 'training or conditions imposed for the improvement of physical powers, self-control, etc.'. To discipline individuals, therefore, involves the institutionalisation of common rules and ways of thinking and acting. In the field of knowledge production, disciplinary boundaries correspond to 'technical languages and university departments' (Fuller, 1993, p. 48), which have a social and a functional dimension through the provision of a shared language, sets of tools and resources. For Petts *et al.* (2005) these intellectual disciplines are constructs borne out of 'historical processes' and 'have survived for so long in the academic world because they serve a very useful function of constraining what the academic has to think about'. Disciplines provide rules for what may or may not provide a research problem or question, count as 'evidence' and constitute acceptable methods to generate, evaluate and transmit knowledge. As Fuller (1993, p. 33) argues, 'left to their own devices academic disciplines follow trajectories that isolate them from one another and from the most interesting intellectual and social issues of our time'.

And yet, this tradition of working to disciplinary boundaries has come under significant challenge both from within academia and outside of it. It is increasingly argued that disciplines are inherently limiting and restrict the types of problems that are addressed and examined. There has been a growing critique of the rationalities of intellectual separation and recognition that expert knowledge and expertise is becoming increasingly esoteric and isolated, thereby limiting its broader potential. During the period of modernity, knowledge growth has been subject to the fracturing and re-fracturing of disciplines into new 'specialities' (Scott, 1984). For Ostreng (2004) the scientific community has deliberately divided disciplines into smaller and more manageable entities, and further subdivided those. Specialisation has become self-amplifying. For example, by the year 1987 there were 8530 definable knowledge fields (Crane and Small, 1992, p. 197), fuelled by increased specialisation and interdisciplinary overlaps (Thompson Klein, 1996). By 1990 some 8000 research topics in science alone were being sustained by specialist networks. Thompson Klein (1996) sees this as evidence of the inner development of science posing even greater tasks leading to interconnections among natural, social and technical sciences. For example, the same object, an organism, can be viewed as a physical (atomic), chemical (molecular), biological (macromolecular), physiological, mental, social and cultural object. Just as cartography maps physical landscapes, so mapping knowledge is based on 'boundaries', which may be characterised by shared theories or ideologies, common techniques or socio-cultural characteristics.

The establishment of boundaries limits outlooks and perspectives and undermines the potential for researchers to develop broader and/or more imaginative perspectives and ways of viewing the world. As Fuller (1993) notes, the shift in thinking involves a growing recognition that:

> certain sorts of problems – increasingly those of general public interest – are not adequately addressed by the resources of particular disciplines, but rather require that practitioners of several such disciplines organise themselves in novel settings and adopt new ways of regarding their work and co-workers . . . ID responds to the failure of expertise to live up to its own hype. (p. 33)

Many research skills, it is argued, are common to several disciplines and are not confined to discrete units. The problem is often that 'disciplinary differences encourage inquirers to forgo points of contact and to concentrate, instead, on meeting local [disciplinary] standards of evaluation' (Fuller, 1993, p. 49). Commonalities are often stronger than imagined, particularly where disciplines 'overlap, touch, mix or merge . . . when it comes to concepts, methodological tools, insights and theories' (Ostreng, 2005).

In recent decades there has therefore been a greater emphasis on the power of approaches that move beyond single disciplinary approaches. As Ramadier (2004) shows, there has been growing interest in *multidisciplinary* (MD) and *interdisciplinary* (ID) working (these relationships are shown diagrammatically in Figure 2.1). In the case of MD work, the key aim is to juxtapose theoretical models from different disciplines, which are seen as being complementary in the understanding of problems and phenomena and which respect the plurality of points of view. In contrast, ID constructs a common model for the disciplines involved, based on dialogue between disciplines and the transfer of tools or models of working. The methodological implications of an ID approach are that researchers 'need to take the boundaries themselves as entry points for inquiry into the relations between science and power and to ask how they come about, and what functions they serve in channelling both knowledge and politics' (Jasanoff, 2003, p. 394). This offers the prospect of a new approach to scientific enquiry, one that is grounded in pluralist and inclusive viewpoints and is able to overcome some of the intellectual straightjackets of disciplinary thinking and working. In short, it is argued that a new set of more open approaches can allow researchers to 'call into question the differences between the disciplines involved and thereby serve as forums for the renegotiation of disciplinary boundaries' (Fuller, 1993, p. 33).

However, MD and ID ways of thinking do not necessarily challenge the primacy and value of the disciplinary model. By focusing on disciplines they may, ironically, reflect and reproduce existing disciplinary divisions and modes of working by taking their boundaries for granted. Authors such as Ramadier, therefore, propose the concept of *transdisciplinarity* (TD),

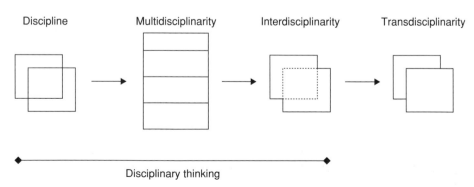

Figure 2.1   Types of cross-cutting research (after Ramadier, 2004).

which has as its core concern a break away from disciplinary thinking by not only unravelling complexities and confronting disjointed knowledge, but also by creating new knowledge paradigms for solving problems. The goal of TD is 'no longer the search for consensus but . . . the search for articulations . . . [and] to avoid reproducing fragmentary models' (Ramadier, 2004, p. 12). Or in Petts *et al.*'s terms (2005, p. 6), TD is a qualitatively different process from ID and MD as it is:

> something different, a practice that transcends, challenges or renegotiates traditional disciplinary boundaries and in some cases reconstructs them in new positions. Transdisciplinarity focuses on the organisation of knowledge around complex subjects, rather than around disciplines. Some argued that transdisciplinarity is the practice most likely to produce outcomes that amount to 'more than the sum of the parts'.

They go on to identify the factors that are necessary for success in TD research:

- mutual trust
- robust disciplinary science, combined with
- individuals' confidence in their own disciplines, and
- mutual respect (some participants added 'humility' and 'a sense of humour')
- space and time for sharing of knowledge, and exploration of differing constructions of problems and methods
- planned opportunities to negotiate at the boundaries, and
- agreement that 'the problem' is real and important

A number of authors also argue that the growing popularity of such approaches is in large part based on wider changes in the social, political and economic contexts of research. The nature of the research process is being transformed, and this transformation has many separate elements (Nowotny *et al.*, 2001). Scholars disagree about their respective novelty and intensity, but three trends appear to be significant – (a) the 'steering' of research priorities, (b) the commercialisation of research and (c) the accountability of science. These and other trends, or changes in practice, have given rise to new discourses of science and research. Nowotny *et al.* (2001) argue that there has also been a fundamental shift from considering science as an 'autonomous space' clearly demarcated from society, culture and the economy to a view that sees all these elements as being fundamentally linked. This is supported by Latour (1999), who suggests that science and society cannot be separated and depend on the same foundation. What has changed is the relationship with science, which is today viewed as 'internal' rather than 'external' to society. Scientific knowledge cannot be divorced from the contexts in which it is produced and which it, in turn, helps to shape. In Fuller's (1993) terms, 'the value of knowledge lies in the ability of its possessor to influence the subsequent course of its production' (p. 30).

Nowotny *et al.* (2001) also characterise current trends as consisting of a shift from Mode 1 knowledge to Mode 2 knowledge. Mode 1 knowledge is the production of 'disciplinary science' in which pure knowledge dominates, and where the goal is to produce theoretical knowledge of a physical and human nature. Mode 1 knowledge is conventionally organised along disciplinary lines in faculties and departments. In contrast Mode 2 knowledge is interdisciplinary and application-oriented knowledge production, where the emphasis is not so much on studying the laws of nature but rather on studying artefacts and the operation of complex systems. Nowotny also suggests Mode 2 knowledge production is characterised by

- applied nature
- multiple actors with expertise and skills
- transdisciplinarity

Expertise, therefore, needs to be conceptualised as

not merely something that is in the heads and hands of skilled persons . . . but is something acquired, and deployed, within particular historical, political, and cultural contexts . . . expertise relevant to public decisions responds to specific institutional imperatives that vary within and between nation states. (Jasanoff, 1990, p. 393)

The structuring of knowledge production and expertise is dependent on processes of governance and broader ways of thinking about how scientific knowledge is mobilised and produced and what 'effects' it has on broader decision-making processes. As Miller and Edwards (2001, p. 5) argue, 'science is . . . less an independent input to global governance than an integral part of it: a human institution deeply engaged in the practice of ordering social and political worlds'. Or in Latour's (1999, p. 80) terms, 'the notion of science isolated from the rest of society will become as meaningless as the idea of the system of arteries disconnected from the system of veins'. Moreover, 'maintaining the public authority of science requires trust not only amongst experts but also between experts and other social groups . . . that acquires its credibility and solidity in policy settings at least in part from socially embedded, culturally specific norms and practices for warranting public knowledge' (p. 15; see also Chapter 3).

2.2.2 Forced approaches, intellectual barriers and the (mis)translation of approaches

However, despite this growing consensus over the value of MD work, there are also clear, but often understated, limitations in the new approaches. One of the dangers with trying to institutionalise such changes is that it may encourage researchers to 'force' connections that may or may not have developed organically. Combining these approaches is not necessarily the most effective way of generating new methodologies and ways of working and thinking. There also remains the thorny issue of how and why it is that different disciplines emerged in the first place. Simply pretending that all disciplinary boundaries are procedural and instrumental and can be re-drawn to address any issue understates some of the fundamental differences that exist between different disciplines such as those between Chicago School economics and Marxist social science.

It is also often assumed within the new approaches that knowledge consists of 'packets' of intellectual capital of equal worth that can be moved around and traded between disciplines in an unproblematic manner (see Evans, 2006). Yet in reality, as we will see below, the processes involved in the *translation* and (re)*interpretation* of knowledge are extremely complex and diverse. Translation represents a 'way of knowing: a way of seeing one thing in terms of another' (White, 1990; quoted in Harvey, 2000: p. 245). As Harvey argues, translation between experts, and in their relations with policy makers and others, derives from 'dominant principles handed down to us [which] so limit our conceptions as to inhibit alternative visions of how the world might be' (p. 247). For example, all too often there is a simple elision between 'user-engagement' and the direct transfer and deployment of scientific knowledge across boundaries. The contexts within which research gets produced is simplified into a division between 'science' (facts) and 'politics' (values). It is this simple schism that provides the focus for funding and research in which a process of knowledge dissemination can bridge differences and allow the flow of facts into decision-making processes.

Furthermore, the process of translation involves two principal problems. First, as Latour (1999) argues, there is a tendency for experts to translate knowledge by *converting it into their (existing) own frames of reference*. Translation is always a reflexive process in which the translator mobilises their own knowledge base to interpret and 'make sense of' new information. In order for translation processes to operate effectively, scientists (and non-scientists) would need to develop a completely different set of skills, unrelated to others that they possess. Second, the ways in which scientific knowledge and 'facts' are constructed rely directly on the position, status and funding of scientific work. This in turn, depends upon political values, with particular 'problems' identified and mobilised in and through broader political processes. As Gieryn argues, in order to understand 'the pliability and suppleness of the cultural space "science" that accounts for its long running success – [we] need to look for local and episodic constructions of science' (1999, p. xi). Scientific knowledge claims do not exist in isolation. Instead, in Bourdieu's terms (2003, p. 236) 'each society at each moment elaborates a body of social problems taken to be legitimate, worthy of being debated, of being made public and sometimes officialised, and in a sense guaranteed by the state'. The role of research is to examine how scientists and professionals are 'constructed in ways that obliterate all differences across the field and to think about the work of aggregation and symbolic imposition that was necessary to produce it . . . and a structured space of social forces and struggles' (Bourdieu, 2003, p. 243).

In the practical world of politics and planning, it is difficult for academic researchers and practitioners to move beyond their own knowledge bases. For example, Evans's study (2006) of the working relationships between a group of academic scientists and local planners in Birmingham provides strong evidence of the ways in which 'epistemological closure and organisational differences between scientists and policymakers hindered the translation of local science into local governance by dislocating the two groups' (p. 527). The network was designed to provide interdisciplinary solutions to the increasingly complex environmental planning of the city. However, the research found that significant tensions developed over the perceived inability of the scientists to provide the definitive 'yes' or 'no' answers that planners and local politicians had expected and were looking for. The latter were also unwilling to change their long-established definitions of 'environmental corridors', even though local scientists were critical of their environmental value to the city. Policy makers expected scientific models to support their existing policies, rather than challenge them. But at the same time, the scientists also lacked awareness of the genuine constraints faced by policy makers and planners. All actors brought their own pre-existing knowledge and prejudices to the decision-making process and these could not be simply erased by the creation of a new scientific base or the formation of a new network. Similar examples can be found in other work on environmental management (see Eden and Tunstall, 2006), scientific advisory networks (Jasanoff, 1990), and sustainability indicators (Irwin, 1995).

2.2.3 Moving beyond disciplines, research funding and the new public management in the UK

These changes in the emphasis of research are being institutionalised by the Research Councils. Research Councils UK (RCUK), for example, describe research that crosses boundaries as 'research requiring knowledge and/or expertise from two or more discrete fields in order to successfully deliver the objectives of the project or field of research' (RCUK, 2005a). In its *Vision for Growth* document RCUK states that novel, MD approaches are needed to solve many if not all of the big research challenges over the next 10 to 20 years. The RCUK belief is that harnessing knowledge and skills across a wide range of disciplines and providing the necessary funding for infrastructure will support government aims to make the UK 'the most attractive location globally' for science and innovation (RCUK, 2005b). Indeed the government's support for MD research was outlined in its Science and Innovation Framework 2004–2014 (HM Treasury, 2004).

According to RCUK data, nearly £1 billion has been committed to MD research since 2001, and a range of initiatives have been used to support this through research, training and knowledge transfer initiatives. Outside the immediate urban environment, MD programmes straddle a number of science disciplines and include:

- energy research towards a sustainable energy economy programme (£28 million)
- e-science (£213 million)
- genomics/post-genomics (£246 million)
- stem cells (£40 million)
- basic technology (£104 million)
- rural economy and land use (£20 million)

RCUK see the use of MD approaches as providing the potential for 'innovative and paradigm-shifting research' (RCUK, 2005b). Generally, however, this relates to different sciences working together. As RCUK state, 'multidisciplinary research occurs optimally when the very best scientists from different disciplines are able to work together free from discipline or structural barriers'. Fostering an MD approach is at the heart of each research council's strategy; for example, ERSC Centre on Economic Learning and Social Evolution enabled collaboration between economists and psychologists. MD research is seen as a means to address the important research issues that are emerging and the creation of RCUK in 2002 added impetus to this activity.

However, despite this new emphasis, a recent evaluation by SDRN (2006) suggests that support for such research is continuing to 'fall through the gaps'. A number of factors are leading to this, which include:

- a mismatch in timescales between short-term political agendas and timescales and longer-term research horizons
- the continuity of programmes in which funding horizons lack transparency for policy makers and stakeholders

- the existing peer review processes
- an insufficient emphasis based on addressing the challenges of inter- and transdisciplinarity in the consortium-building phases
- the tendency for research councils to be risk averse and unwilling to fund innovative research
- the Research Assessment Exercise (RAE), which favours individual excellence/performance
- a lack of quality training and career development in such research

Some of these criticisms were echoed by RCUK (2005a), which saw the RAE as acting as a key barrier with their disciplinary panels and their emphasis on publications as an evaluation factor.

These changes to the management of scientific research in countries such as the UK have also become increasingly bound up with the modernisation of state (bureaucratic) systems and what Clarke and Vidler (2005) and Newman (2005) call the New Public Management (NPM). This NPM is premised on the principle that public sector organisations, including research councils, could and should act more like private sector businesses, and that this can be achieved through the implementation of new systems of targets, auditing, output-driven accountability and efficiency gains (see also Sennett, 2005). The consequence of this has been the re-institutionalisation of disciplinary boundaries with resources being allocated on a discipline-by-discipline basis, in direct contrast to the emphasis on MD research. In seeking to mimic global business markets there is a new emphasis on concentrating university research in so-called 'globally competitive' centres, with resources concentrated in a smaller number of centres and the work of a smaller number of 'world-class' researchers.

In many ways this reproduces inconsistencies in the rhetoric and realities of MD working and funding and indicates that it is impossible to divorce the process of knowledge production and transfer from these broader changes in the institutionalisation of research funding. Under this NPM the acquisition of intellectual authority, through disciplinary solidarity or what Bourdieu (2003) terms, 'intellectual capital' has become more and more important. Where MD work is funded, it is often driven by the requirement to meet new targets and agendas set from on high. It often goes hand in hand with a managed devolution of responsibilities that is implemented to save costs and ensure that researchers and universities fulfil the wider objectives of government departments, agencies and state policy agendas. The NPM can be used as a management tool to force academics and knowledge producers to conduct their work in particular ways. Such trends have been apparent in research on sustainability and sustainable cities and it is to this theme that the next section now turns.

2.3 The rise and rise of the sustainability agenda

Broader questions over the relationships between the creation of scientific knowledge, its dissemination and the role of science in public policy-making have been brought to the fore by the rise of sustainability discourses since the mid 1970s. For Meadowcroft (2000) the discourse of sustainability has come to represent a new meta-narrative that calls for the establishment of more equitable, accountable and future-oriented development agendas. Within such discourses 'the future' takes on the features of a *definable object*, that should be integrated into planning frameworks at all stages of their development and design. It becomes something tangible to be controlled, ordered and worked towards, with active citizens expected to consider their own actions in the context of how they might impact on the 'future'. As Kenny and Meadowcroft (1999, p. 4) note, 'nearly all the definitions [of sustainability] conceded that it involves the re-orientation of the meta objectives of a given society – by raising questions about different possible social trajectories through which the society may move'.

The rise of the sustainability agenda has close links to the broader movement towards ID and TD research. In particular it involves the following:

- The propagation of new *holistic* development agendas in which explicit connections are made between social, economic, and environmental processes. Experts working in these fields are increasingly expected to share their knowledge and change their ways of working with other experts, non-experts and the community.
- A new primacy to environmental science as a vehicle through which environmental changes can be monitored, managed, and mitigated. There is a certain irony in the lack of recognition given to science's role in facilitating environmental destruction and modernist development (Smith, 2001). Sustainability discourses highlight new ways in which science can be linked to broader systems of governance and decision-making in order to protect, rather than destroy environments.
- A breaking-open of decision-making processes and the knowledge that underpins them. This includes a challenge to the legitimacy of 'expert systems' and the ways in which experts create and disseminate their knowledge. This could translate into new ways of thinking about the relationships between knowledge production and the ways in which it can be used by other groups.
- A recognition that a plurality of knowledges exist beyond disciplinary boundaries and academia. This involves not merely new platforms of engagement between experts and lay communities but a breakdown of these boundaries and the mobilisation of different stakeholder groups.

In short, sustainability discourses open up new opportunities for scientists to engage with policy-making processes: in O'Riordan's terms (2004):

> [a] more engaged sustainability science, more willingness to work with new governing partnerships, a greater commitment to common endeavour, to delivery, and a fresh approach to research and teaching that prepares geographers and environmental scientists to be standard bearers for the tortuous but necessary transition to sustainability.

At present, it is argued, few environmental scientists are politically skilled enough to create a new, more politicised science. Within the cacophony of voices debating 'sustainability', too much environmental science is data-focused and modelling-based, making it too esoteric for broader dissemination and analysis. Environmental research is, therefore, 'more accurately described as multidisciplinary rather than interdisciplinary because of its multitude of mutually incommensurable methods and languages' (Eden, 2003, p. 243).

Such criticisms are especially relevant to recent research on sustainable cities such as *The Sustainable Cities Programme* in the UK in the 1990s. In recent years the 'urban environment' has become a coherent concept as well as a political issue and a research issue (Petts *et al.*, 2005). For Evans and Marvin (2004) the emergence of sustainability as a research and policy priority within the 'sustainable cities' agenda in the UK has been characterised as being a 'new and distinctive type of research problematic', underpinned by the interdisciplinarity of the research. Despite these aspirations Evans and Marvin suggest that the practical problems of ID research were more intractable than was assumed. These assumptions were based on the premise that new disciplinary collaborations were needed, with social science engaging in research on technological and/or ecological issues and science/engineering disciplines focusing on the social and political aspects of urban issues. However, for Evans and Marvin the common ground between different disciplines was 'far smaller and more antagonistic than was initially assumed'. The new research agenda also presented challenges for researchers because of the increased importance of users and policy makers in framing the research and throughout the research process itself. Finally, the 'social organisation' of research changed because research councils that represented different disciplines had to realign and reorganise their relationships in order to tackle research questions which were cross-cutting.

In examining the evolution of the 'sustainable cities' programme during the 1990s, Evans and Marvin also highlight how the differing disciplinary perspectives within a programme can colour and shape the perspective. Thus the EPSRC programme (*Sustainable Cities*, 1993), perhaps the most 'radical' of the ID programmes[1] examined under the sustainable city agenda, saw the city as a 'technology'. The challenge for SD was to break down the city into its constituent parts and thus find more efficient ways of performing the key functions. In this sense the engineering perspective is based on a systems approach with a city made up of subsystems such as transport, energy, water and building stock. As Evans and Marvin point out:

> By dividing these up into spatial units and mapping the flows of energy and waste between them, the engineering perspective is based on the assumption of a more or less static social structure being played out within a more dynamic urban infrastructure that can, with appropriate technologies, adapt to the demands made upon it. These changes might be one-off changes, such as the introduction of new bus lanes or a more dynamic process in which the city continually adjusts and changes, e.g. through flexible prices for road use or domestic energy.

But the programme did not see a city or society as something that could be changed. Society was therefore considered 'at the end of the process, as something that may need to be persuaded to change its ways, but it is neither an input nor an output of the process'.

For Evans and Marvin in each example, the councils recognised that the problems of the sustainable city crossed disciplinary boundaries but the joint outcome of their research has remained disciplinary in focus. This is also the result of the differing research council perspectives:

> Thus EPSRC see the 'sustainable city' mainly in terms of technological systems and fixes; NERC see it in terms of the flows and stocks of natural resources; ESRC sees it as a distinctive form of social organisation. Unfortunately, in setting the problem up in this way, what was originally a complex combination of science AND technology AND society has been reduced to science OR technology OR society. In other words, to the extent that interdisciplinary research occurred, then it was within research councils not between research councils. (Evans and Marvin, 2004, p. 2)

Evans and Marvin conclude that research needs a shared vision or understanding of the city, which ultimately is tantamount to searching for the Holy Grail. These criticisms are borne out by Petts *et al.* (2005), who suggest that research initiatives are often predicated on the basis of a 'technical-rational' model. But as Owens *et al.* (2006) suggest, this model assumes that a 'separation of powers' is deemed to exist between neutral, authoritative experts and the decision makers whom they advise, and that an objective assessment is assumed to lead to better decisions and in some cases more sustainable development.

The remainder of the chapter now turns to an assessment of the SUBR:IM consortium and the ways in which the tensions and dilemmas discussed above were reflected and reproduced in its research activities. The discussion begins by examining the emergence of the EPSRC's Sustainable Urban Environments programme, out of which SUBR:IM emerged. It then critically explores the strengths and weaknesses of the consortium and the SUE programme more generally.

2.4 The EPSRC's Sustainable Urban Environments programme and the emergence of the SUBR:IM consortium

The EPSRC's Sustainable Urban Environments (SUE) programme call was made in 2001. The aim was to sponsor a number of new MD research consortia to 'address key research issues in the development of a more sustainable urban environment' and ostensibly sought to address three interrelated objectives:

(1) improving the quality of life of the UK's citizens
(2) supporting the sustainable development of the UK economy and society
(3) meeting the needs of users of EPSRC-funded research in industry, commerce, government and the service sector

The call viewed the urban environment as a 'system', suggesting that

> [i]n order to meet the Brundtland Report requirements, future sustainable development of the urban environment must involve radically reduced usage of natural resources and energy . . . the design, construction and operation of the built infrastructure, urban spaces, transport systems and their related infrastructure must also take full account of the whole life cycle of the urban environment as a holistic system.

It therefore had a strong normative dimension and was grounded in a perspective that saw enhanced knowledge production as an essential element in the creation of sustainable (and practical) policy measures covering issues such as crime and pollution reduction and the accessibility of buildings and transport systems. To successfully undertake this 'highly multidisciplinary challenge' would require researchers drawn from engineering, physical science, environmental science and social science and to tackle five research themes: (i) towards a new infrastructure; (ii) the sustainable built environment; (iii) waste, pollution and urban land use; (iv) urban transport and urban design; and (v) social inclusion. In the first round of SUE, the EPSRC supported 12 consortia within four clusters:

- Cluster 1 – urban and built environment
- Cluster 2 – waste, water and land management
- Cluster 3 – transport
- Cluster 4 – metrics, knowledge management and decision-making

In addition, the SUE programme has supported related research projects, such as demonstrator studies of sustainable urban redevelopment and also a network to promote engagement and dialogue between the EPSRC research community and lay citizens.

The SUBR:IM consortium was funded under Cluster 2. Its core object-ives were very much a product of their time and the intellectual and policy contexts within which it was formed. During the 1990s there had been a growing concern with urban change, culminating in the Urban Task Force Report of 1999 and the Labour government's Urban White Paper of 2000. These policy blueprints argued that more resources should be made available for the regeneration of inner urban areas in order to facil-itate new forms of regeneration, while limiting the effects of urban sprawl on surrounding rural areas. The very practical problem of how to bring brownfields back into active use also connected with broader debates over so-called 'compact city' living and the ways in which cities could be made to be more sustainable (see Jenks *et al.*, 1996; Burton, 2000). In the UK, the new emphasis on brownfields became increasingly elided with a broadly defined urban sustainability agenda. Brownfield redevelopment involves the reuse of an existing resource and provided a platform for more integrated and energy-efficient urban living (see Greenstein and Sungu-Eryilmaz, 2004).

The focus on brownfield regeneration also lent itself to a cross-cutting research agenda. On the one hand, the brownfield 'problem' was conceptu-alised primarily in technical terms as something that could to be addressed through the application of technical, scientific solutions. Most urban brown-fields in British cities are former industrial sites containing harmful con-taminants and pollutants. Bringing such sites back into active use required the application of types of technical knowledge such as those involving land remediation technologies. On the other hand, however, such technical responses had to be put into the wider context of the brownfield regenera-tion process. The research sought to identify the ways in which brown-fields were defined in particular places, the influence of different groups in the governance of brownfield sites, the politics of redevelopment and the impacts of any (technical) changes for different groups. Collectively, the 18 projects funded under the SUBR:IM consortium were indicative of the multidimensional brownfield problem (see Chapter 1). They covered a range of different aspects of the brownfield development process and, ostensibly, provided the basis for a new MD approach to the problem. It was decided that the research teams should work in the common case study areas of Thames Gateway and Greater Manchester to provide a comparative dimension to their work and to tie the research in with the broader sustainable communities agendas of central government.

SUBR:IM's experiences reveal much about the relationships between knowledge production and existing funding and disciplinary regimes, and we divide the discussion into four interrelated themes: the establishment process; knowledge production; case studies and knowledge transferability; and the sustainability of the research consortium.

2.4.1 The establishment process: definitions, resources, and problem identification

SUBR:IM was formed through a process of top-down definition. Individual researchers responded to the EPSRC's original call with a brief statement of their research interests in the fields of environmental sustainability and their expertise and knowledge. EPSRC subsequently selected groups of researchers to participate at workshops where consortia formed around themes of common interest and consortium champions were nominated, all of whom had a track record of EPSRC funding and were, therefore, by default predominantly engineers and scientists. The selected groups of researchers often had little knowledge of each other's research backgrounds or their perspectives on brownfield regeneration processes. The advantages of the approach were supposed to be that as a new team, often working together for the first time, fresh perspectives and ways of working on common problems could be forged. Disciplinary boundaries, it was hoped, would be broken down by institutionalising MD working.

And yet within this radical new set-up disciplinary boundaries were not discarded altogether. Within the consortium resources were allocated to specific projects, devised and managed by individual researchers. Researchers were, therefore, encouraged to work on projects that reflected *their existing skills and expertise* and think about the ways in which their findings could then contribute to the wider outputs of the consortium. There was no requirement for ID, MD or TD methods to be included in the up-stream phases of project development and design. Instead, it was expected that researchers would develop common vocabularies and ways of thinking about common problems (such as brownfields) as they worked through a consortium structure. Integrated MD approaches would emerge during the down-stream phases of the research.

Moreover, as the funding was originating from the EPSRC, all the consortia would be judged and managed using EPSRC rules. Its procedures have been devised to meet the needs and expectations of the engineering and science disciplines, rather than social sciences where funding traditionally comes from other sources such as the Economic and Social Research Council and government departments. The overall thrust of the research, therefore, had to be dominated by science- and/or engineering-based disciplines. The ill-defined nature of the term 'social inclusion' in the original SUE call meant that social research was essentially a 'bolt-on' to the wider research aims of the scientists and engineers. Here we see the institutional pressures of the New Public Management and the central government push towards 'evidence-based' research having a significant impact on the types of research that are funded and the capacities of researchers to extend the scope and scale of their knowledge.

In addition, the consortium-forming process also involved a bureaucratic separation of research issues and themes and set down rules for who was 'allowed' to examine what. For example, any understanding of the broader brownfield development process has to involve some understanding of flooding processes and movement of water through the environment. This was particularly important for research in the Thames Gateway where fears of future flooding have been of concern to many investors and planners. However, in the early stages of SUE, flooding issues were being covered by a specific consortium, and others, such as SUBR:IM, were actively discouraged from addressing any of the issues relating to it. Similar separations were made over topics such as waste management and groundwater movement, all of which are important factors in the brownfield development process. In other words there was a significant mismatch between the objects and environments being studied and the bureaucratic funding boundaries established by the EPSRC. This also reflected territorial divisions within and between research councils and complex questions over divisions of responsibility and funding priorities.

All of these structural conditions had an impact on the research process – from the formulation of research problems, to the methods used to address them, and the types of data, findings and outputs that would be produced. Some of this related to the perception that it was not altogether clear what the EPSRC saw as the overall purpose of the SUE consortia. They were, of course, partly designed to produce new research. But they were also expected to act as vehicles for new forms of networking to be established between academics and to create obvious nodes through which future research money could be channelled with a minimum of costly bureaucratic and procedural selection processes. Inevitably, in the longer term it was expected that this would result in new (elite) research groups, with a proven track record of ID research.

2.4.2 Knowledge production, processes of translation and the capacities of researchers

As discussed earlier, one of the limitations of any MD work is the tendency for experts to translate other fields of knowledge into their own (Latour, 1999; 2004). Research objectives often become identified and re-packaged so that they become manageable *within the existing disciplinary boundaries that experts are working to*. This is a particular problem when the research object, in this case brownfield urban regeneration, is traditionally approached in different ways by different disciplines. This was compounded by the funding mechanisms for SUBR:IM and the other consortia in which funding was given to experienced researchers to expand their knowledge within disciplinary boundaries. The search for intellectual common ground and a common interpretation of the core research problems and focus took place *after* funding had been allocated by EPSRC to applicants based on their existing intellectual (and disciplinary) strengths.

In addition, the delivery of programmes such as SUBR:IM raises significant questions over broader questions of expertise and academic divisions of labour. The existing structure of disciplinary-led academic careers in the UK's university system has a major influence on the incentives and disincentives for the pursuit of MD work. Less experienced researchers are expected to simultaneously develop the disciplinary competences required for their career progression and take on a leading role in MD projects. Within the SUE consortia much of the practical, day-to-day research has been undertaken by dedicated, short-term researchers and in some cases PhD students. These individuals were employed, primarily, on a disciplinary basis and tasked to carry out the research for the individual, disciplinary-driven projects within SUBR:IM. Such researchers came under particular pressure as they were expected to expand their knowledge base by incorporating the methods and approaches of very different disciplines. Those trained in community-based, qualitative research, for example, had to develop a working knowledge and understanding of technical, quantitative research laboratory-based experimental techniques and vice versa. In practice, the researchers responded strongly to the particular challenges of SUBR:IM and most benefited significantly from the experience. However, without a wider recognition of the value of MD knowledge in the academic labour market, the incentives for researchers (at all levels) to undertake such work are limited, indeed in the current context there are many disincentives.

2.4.3 Case studies and knowledge transfer

One of the most significant difficulties encountered through the research process related to epistemological and ontological differences over the relative value of abstract and concrete research. There were frequent discussions within SUBR:IM over the differences between 'experiments' and 'case studies' and the value of both to the study of brownfield redevelopment. Many of the technical projects were 'problem-based' and sought to identify specific problems that are common to many (but by no means all) brownfields. Such research sought to abstract processes from contexts and aimed to identify generalisable findings primarily based on laboratory experiments. There was less interest in examining specific examples or case studies as the inclusion of 'contexts' would be seen to limit the scientific validity of any findings. Where case study work was undertaken it was based on specific, individual sites in which an identifiable 'technical problem' could be addressed with the application of particular techniques. These techniques, it was argued, should also be replicable in different physical and urban environments.

For most of the social science projects, however, research methodologies were very different. Much of the work was focused on the ways in which brownfield sites, in case study areas, were defined and the types of strategies that were used to bring them back into 'active' use. The focus was also on what the social, economic and environmental implications of redevelopment decisions were and the part that scientists, engineers and other technical groups played in the regeneration process. Brownfield development, it was argued, could only be understood in and through the contexts that exist in urban areas, and intensive case study work was required to draw out and develop and assess generalisable trends and concepts.

Attempts were frequently made to overcome these differences of approach, and it would be too simplistic to say that there were definite, intractable differences. Some of the projects within SUBR:IM, for example, focused on themes such as the value of sustainability 'metrics'. They were designed to act as umbrella projects that brought together the various approaches and findings. Within individual projects, too, efforts were made to link the findings from laboratory-based work directly with the experiences of practitioners (see, for example, Chapters 8 and 9). In addition, efforts were made to develop integrated outputs. Public- and practitioner-focused conferences and road shows were held and authors within SUBR:IM worked on joint publications (such as this book). Within SUBR:IM a management committee was also set up with the intention of coordinating the activities of the consortium and ensuring that its members were always clear about its broader objectives and purpose.

2.4.4 The sustainability of the research consortium

One of the ironies of the EPSRC's interdisciplinary SUE research programme is that the initial consortia, like SUBR:IM, are now being wound up and researchers are expected to develop new networks and research teams. This is not due to its lack of research outputs. Instead, it relates to two interconnected processes.

First, the wider policy contexts within which the research was conceived and carried out have moved on. For the Research Councils brownfields have become 'yesterday's problem' as debates over sustainability have shifted significantly during the 2000s. The new emphasis is on topics such as climate change, sustainable communities, sustainable urban infrastructure and waste management. Just as stocks of knowledge on brownfields from SUBR:IM and elsewhere are growing, so the focus of attention has moved on. Research funding priorities for academic and policy-related work shifts rapidly and all research programmes and agents find themselves caught on the tide of intellectual fashions and changes in social, economic and political priorities. In relation to British urban policy, for example, there has been a recent backlash against the practices of brownfield development in many towns and cities. In some cases critics have even challenged the principles that underpin such projects (see Hall, 2001). The focus of the sustainable urban environments agenda is, therefore, subject to rapid and ongoing change.

Second, research funding is guided by the wider principles of the New Public Management and the emphasis on modernisation and instability as a management tool. Change is seen as a valuable mechanism for breaking up established ways of doing things and developing new, innovative forms of working and knowledge production. This indicates the ways in which so-called 'objective' research is, in reality, circumscribed by the contexts within which it is funded, formed, and implemented.

In summary, then, the SUBR:IM consortium acted as an MD research team, in which the key focus has been to juxtapose theoretical models from different disciplines, which are seen as being complementary in understanding problems and phenomena and which respect the plurality of points of view (cf. Ramadier, 2004). It cannot claim to have been TD in nature, even though attempts have been made to develop more integrated ways of working and establish common platforms and/or frames of reference for analysis. As subsequent chapters will show, the SUBR:IM model has created a broad and deep range of research on a specified problem – that of urban brownfield development in the UK. This, in itself, is of great value.

2.5 Conclusions: SUBR:IM and new ways of working

It seems likely that MD research is here to stay. Its administrative attractiveness to policy makers and research funders has gone hand in hand with the emergence of new agendas, such as those surrounding urban sustainability, that seem to lend themselves naturally to MD approaches and ways of thinking. Its implications for social and scientific research are potentially enormous. This chapter has sought to further understanding of the intellectual and practical issues that surround such work and has used the example of the SUE programme and the SUBR:IM consortium to explore these issues. It has argued there is much potential in such agendas but that greater attention needs to be paid to the processes of research design and implementation. There are four core findings that have significance to other work in this field:

(1) The experience of working across disciplinary boundaries has undoubtedly encouraged many of the consortium's researchers to develop more *reflexive* approaches to their own, disciplinary-based work. In some cases SUBR:IM represented the first opportunity to engage in MD work and provided a framework in and through which alternative approaches to shared research problems could be translated and interpreted. The existence of such consortia has much potential in making researchers think actively about their own expertise, what it consists of and how it relates to other disciplines and forms of academic work. In Bourdieu's terms (2003), it could initiate new forms of 'critical reflexivity' or a process that can generate genuinely new insights and ways of interpreting, communicating and translating knowledge. It can broaden horizons and opens up new interpretations of what it is that constitutes a research 'problem' and what methods and practical initiatives can be used to address it. However, as noted above, such a process is not axiomatic. Without common agreement over the aims and objectives of the research there is a danger that an MD project could *reinforce* disciplinary boundaries, rather than dissolve them. It should also be noted that research that is not explicitly MD in nature should not be undervalued simply because it draws on the traditions and knowledge of an existing discipline. The integrity of all research should be treated on its own merits.

(2) One of the lessons of the SUBR:IM experience has been that MD research needs to be a key element in the *up-stream* phases of the research process if it is to develop into anything like an effective work programme and genuinely TD ways of working. Participants also have to work towards common *conceptions* of key terms, such as (in this case) 'sustainability' and 'brownfields', and develop complementary research around integrated case studies. If this is done successfully then there exists the potential for new types of research to be forged. A failure to confront such differences, however, may well undermine engagement efforts.

(3) The formation of effective user networks and dissemination strategies is a core part of any research programme and has particular value for MD work. Given the range of research skills and knowledge within the consortia, the opportunities for widespread user engagement are significant. However, this breadth of knowledge and expertise can also present particular challenges. There may, for example, be difficulties in sharing out information and key findings. For academic researchers there are particular challenges and opportunities in the creation of outputs that can be interpreted and translated by a range of different users in an efficient and effective manner. In addition it is important to recall that different types of users, such as planners and engineers, have been schooled in and through the disciplinary traditions of the academic, university system. There needs to be more emphasis on the ways in which those receiving the knowledge produced by researchers are themselves limited by their own disciplinary and technical boundaries. It is not always clear how multidisciplinary non-academic research users are in their own thinking.

(4) The shift towards new forms of MD research also raises questions over leadership and organisation and how the parameters and targets of the research should be defined. Consortium-building is a time-consuming process that on the one hand opens up new opportunities for working and communicating with others but on the other involves a significant commitment of increasingly scarce time and resources. In line with NPM thinking, the costs of administration and leadership have been devolved away from research councils to academics and their departments. No resources, for example, were available under the SUE programme for the initial period of consortia formation. There is no administrative support for the running of networks such as SUBR:IM and the costs of administering the programme fall, primarily, on the academic partners and their support staff. Much of the management process within such consortia inevitably becomes bound up with administrative questions of resource allocations, targets and budget rather than concentrating on the academic and intellectual questions raised by ID research. There are also questions over how leadership should be organised and whether the 'project leader' model of academic research funding is necessarily the most conducive to ID academic work.

Note

1. The other programmes examined by Evans and Marvin were NERC Urban Regeneration and the Environment and ESRC Cities and Competitiveness programme.

References

Bourdieu, P. (2003) *An Invitation to Reflexive Sociology*. Polity Press, Cambridge.

Burton, E. (2000) The compact city: Just or just compact? A preliminary analysis. *Urban Studies*, **37**, 1969–2006.

Clarke, J. and Vidler, E. (2005) Creating citizen-consumers: New Labour and the remaking of public services. *Public Policy and Administration*, **20**, 19–37.

Eden, S. (2003) People and the contemporary environment. In: *A Century of British Geography* (eds R. Johnston and M. Williams), pp. 213–46. Oxford University Press, Oxford.

Eden, S. and Tunstall, S. (2006) Ecological versus social restoration? How urban river restoration challenges but also fails to challenge the science–policy nexus in the United Kingdom. *Environment and Planning C: Government and Policy*, **24**, 661–80.

Evans, J. (2006) Lost in translation? Exploring the interface between local environmental research and policymaking. *Environment and Planning A*, **38**, 517–31.

Evans, R. and Marvin, S. (2004) *Disciplining the Sustainable City: Moving Beyond Science, Technology or Society*. Working paper 65, School of Social Sciences, Cardiff University.

Fuller, S. (1993) *Philosophy, Rhetoric and the End of Knowledge: The Coming of Science and Technology Studies*. The University of Wisconsin Press, Madison.

Gieryn, T. (1999) *Cultural Boundaries of Science: Credibility on the Line*. University of Chicago Press, Chicago.

Greenstein, R. and Sungu-Eryilmaz, Y. (eds) (2004) *Recycling the City: The Use and Re-use of Urban Land*. Lincoln Institute of Land Policy, Cambridge, MA.

Hall, P. (2001) Sustainable cities or town cramming? In: *Planning for a Sustainable Future* (eds A. Layard, S. Davoudi and S. Batty), pp. 101–14. Spon Press, London.

Harvey, D. (2000) *Spaces of Hope*. University of California Press, Berkeley.

HM Treasury (2004) *Science and Innovation Investment Framework: 2004–2014*. HM Treasury, London.

Irwin, A. (1995) *Citizen Science: A Study of People, Expertise and Sustainable Development*. Routledge, London.

Jasanoff, S. (1990) *The Fifth Branch: Science Advisers as Policy Makers*. Harvard University Press, Cambridge, MA.

Jasanoff, S. (2003) Breaking the waves in science studies. *Social Studies of Science*, **33** (3), 389–400.

Jenks, M., Burton, E. and Williams, K. (eds) (1996) *The Compact City: A Sustainable Urban Form?* E & F Spon, London.

Kenny, M. and Meadowcroft, J. (eds) (1999) *Planning for Sustainability*. Routledge, London.

Latour, B. (1999) *Pandora's Hope: Essays on the Reality of Science Studies*. Harvard University Press, London.

Latour, B. (2004) *Politics of Nature: How to Bring the Sciences into Democracy*. Harvard University Press, London.

Meadowcroft, J. (2000) Sustainable development: A new(ish) idea for a new century? *Political Studies*, **48**, 370–87.

Miller, C.A. and Edwards, P.N. (eds) (2001) *Changing the Atmosphere: Expert Knowledge and Environmental Governance*. MIT Press, Cambridge, MA.

Newman, J. (2005) *Remaking Governance: Peoples, Politics and the Public Sphere*. Policy Press, Bristol.

Nowotny, H., Scott, P. and Gibbons, M. (2001) *Re-Thinking Science: Knowledge and the Public in an Age of Uncertainty*. Polity Press, London.

O'Riordan, T. (2004) Environmental science, sustainability and politics. *Transactions of the Institute of British Geographers*, **29**, 234–47.

Ostreng, W. (ed.) (2004) *Convergence: Interdisciplinary Studies (2004–2005)*. Centre for Advanced Study at the Norwegian Academy of Science and Letters.

Owens, S., Petts, J. and Bulkeley, H. (2006) Boundary work: knowledge, policy and the urban environment. *Environment and Planning C*, **24**, 633–43.

Petts, J., Owens, S. and Bulkeley, H. (2005) Summary of discussions and recommendations, ESRC transdisciplinary seminar series. *Knowledge and Power: Exploring the Science/Society Interface in the Urban Environment Context*, www.cert.bham.ac.uk/urbanenvironment/Downloads/FinalDiscussionPaper.pdf

Ramadier, T. (2004) Transdisciplinarity and its challenges: the case of urban studies. *Futures*, **36**, 423–39.

Research Councils UK (2005a) Supplementary evidence by Research Councils UK. *Minutes of House of Lords Select Committee on Science and Technology, March*. HMSO, London.

Research Councils UK (2005b) *Vision for Growth*. Research Councils UK, Swindon.

Scott, P. (1984) *The Crisis of the University*. Croom Helm, London.

SDRN (2006) *Securing the Future: The Role of Research-Interdisciplinarity, Cross-cutting and Strategic Research Needs to Support the UK Sustainable Development Strategy*. Policy Studies Institute, London.

Sennett, R. (2005) *The Culture of the New Capitalism*. Yale University Press, London.

Smith, M. (2001) *An Ethics of Place: Radical Ecology, Postmodernity and Social Theory*. State University of New York Press, New York.

Thompson Klein, J. (1996) *Crossing Boundaries: Knowledge, Disciplinarities and Interdisciplinarities*. University Press of Virginia, Charlottesville, VA.

White, J. (1990) *Justice as Translation: An Essay in Cultural and Legal Criticism*. Chicago University Press, Chicago.

Part 2
Regeneration

3

Democracy, Trust and Risk Related to Contaminated Sites in the UK

Philip Catney, Dick Eiser, John Henneberry and Tom Stafford

3.1 Introduction

Within debates on environmental governance, there are growing pressures to refine the expert systems and technical–scientific analyses involved in risk management so that the process may accommodate a broader set of lay knowledge into decision-making. In this way, decisions will be more democratic, sustainable and relevant to local needs. This chapter examines the processes by which risks arising from contaminated brownfield sites are identified, defined and assessed and how lay communities engage with these actions and decisions.

Statute and policy form important parts of the context within which contaminated land is dealt with in the UK.[1] As we will show, this policy regime is distinguished by its pragmatism and flexibility (although international practice has shown a recent tendency to converge upon it). Of significance is the room for manoeuvre that is presented to local actors. There is little prescription. Consequently, the scope for judgement and negotiation is considerable, and the way in which they are exercised may have a considerable impact on the effective implementation of contaminated land policy.

Risk is not just a 'thing' that can be measured, but a value-laden social construction. Risk management is not just a technical exercise but a process in which the public may participate as an empowered partner. This raises issues much larger than just 'How can we get the public's views on risk to align more closely with the experts'?' Politics, or at least public acceptance, is important. Experts may be better at assessing risk than lay people, but they possess no equivalent superiority over the public when considering the implications of particular risks or the appropriate policy responses to them.

In these circumstances, open and democratic risk management processes are necessary because they generate higher levels of social trust than do closed processes. We explore the relations between democracy, trust and risk through two case studies of risk assessment and communication relating to contaminated brownfield sites in England. The case studies show how risk is defined, contested and incorporated into the policy agendas and preferences of actors in two local authorities. In addition, we describe a large-scale survey of residents living adjacent to contaminated sites in the case study areas. The results show how different approaches to risk communication affect the degree to which local residents trust their councils.

3.2 Contaminated land in the UK: context and policy

The policy regime and institutional mechanisms that have emerged in the UK to deal with the problem of contaminated land are structured around the dominant discourse of 'development managerialism' (Catney *et al.*, 2006). This reflects a politico-administrative perspective which (i) while recognising that contamination poses health and environmental problems, frames the issue primarily in economic terms, as an obstacle to economic progress and urban (re)development; and (ii) structures the palliative response primarily through the existing administrative apparatus of planning. The emphasis within the discourse is on minimising urban blight, protecting economic interests and harnessing market-led development processes to bring contaminated land back into productive use. Pragmatism and cost-effectiveness have been recurrent themes in how the UK government has sought to deal with the problem.

For most of the post-war period the issue of contaminated land was considered a technical matter – one that could be handled at site level by scientists and engineers, without the need for elaborate policy. The catalyst for greater central government intervention came initially from local authorities that wanted clearer guidance over the redevelopment of heavily polluted sites. This prompted the UK government to establish the Inter-departmental Committee on the Redevelopment of Contaminated Land (ICRCL) in 1976. The committee provided technical advice to local authorities on a case-by-case basis and produced some 'trigger values' (levels of contaminated materials on a site that suggested the need for action). Thus, as Walker (2002) noted, the UK response initially defined contaminated land as a material consideration[2] in the process of development control decision-making. The issue found a place on the UK policy agenda not because it was viewed as a critical environmental or health concern, but because it was considered a block on the redevelopment of brownfield land. It was, therefore, primarily a planning issue. This association of contaminated land with planning and development 'was to be a lasting one' (Walker, 2002, p. 99).

During the 1980s the British government made little attempt to extend its role further. Yet, as the decade advanced, the closure of many polluting industries made it increasingly difficult to ignore the scale of the problem. In 1989 two House of Commons Environment Committee reports – on contaminated land and on toxic waste – forced the government to concede the need for better tracking of potentially polluted sites (Walker, 2002, p. 86). As part of the major overhaul of environmental law introduced in the Environmental Protection Act 1990 (hereafter, 'EPA') the then Conservative government set out draft regulations for land registers (section 143), in which up to 100 000 sites would be listed as potentially dangerous because of past and present 'contaminative uses'.

The register met with fierce resistance from lobby groups that represented property interests. They feared significant market blight would occur if contaminated sites were identified on a public register. Subsequently, a wide-ranging review of the legal powers for public bodies to control and tackle contaminated land was undertaken (DoE, 1994; Tromans and Turrall-Clarke, 2000). This 'explicitly took a "market centred" approach' (Smith, 1999, p. 380) and resulted in the final demise of the register and the affirmation of the idea that contamination should generally be managed within the normal development cycle. These ideas formed the basis for section 57 of the Environment Act 1995, which inserted a new Part IIA into the EPA 1990[3] (Walker, 2002, p. 87; hereafter, 'Part IIA'). Registers of remediation action would be established under Part IIA, rather than registers of contaminated sites. Even so, this provision was not put in place until 2000.

The new Part IIA sought to give greater legal clarity and to regulate more systematically the activities of local actors who guided the redevelopment and remediation of contaminated sites. The government was particularly eager to ensure that the resultant system would contain costs. This was made clear in Circular 02/2000 *Contaminated Land* (DETR, 2000), which by then was couched in the language of sustainable development. The basic objectives of the regime were to:

- identify and remove unacceptable risks to human health and the environment
- seek to bring damaged land back into beneficial use
- seek to ensure that the cost burdens faced by individuals, companies and society as a whole are proportionate, manageable and economically sustainable

The extant UK policy regime clearly reflects some balance of these three elements. However, the pursuit of 'development managerialism' resulted in the weight given to these objectives being pretty much the reverse of their lexical ordering.

There are now two principal legislative controls that exist to regulate the redevelopment and remediation of contaminated sites in the UK: the Town and Country Planning Act 1990 and Part IIA of the EPA 1990 (Environment Agency, 2002, sec. 2.3). Primacy is given to planning. Remediation is generally to be achieved through the planning system as part of its role in the regulation of development undertaken predominantly by the private sector. However, the 'broad approach, concepts and principles of [Part IIA] . . . with respect to identifying risks from land contamination and dealing with them should be applied to plan-making and the determination of planning applications' (ODPM, 2004, p. 1, para. 2.2, square brackets added).

3.2.1 *Part IIA of the Environmental Protection Act 1990*

Within this context, contamination is specifically addressed by Part IIA of the Environmental Protection Act 1990 (Part IIA). It establishes the legislative framework for the identification and remediation of contaminated land, introducing for the first time a statutory definition of 'contaminated land'. Its main purpose is to address land which has been contaminated by past practices and which poses 'unacceptable' risks to human health or to the wider environment in the context of the current use of the land.[4] Under Part IIA, local authorities are charged with developing and implementing strategies for identifying contaminated land in their areas.

We now consider three of the main characteristics of the Part IIA provisions. These relate to:

- the definition of contaminated land
- the 'suitable for use' doctrine
- 'risk-based' remediation

3.2.1.1 Defining 'contaminated land'
A distinctive characteristic of the UK approach is the very specific way in which 'contaminated land' is defined as:

> any land which appears to the local authority in whose area it is situated to be in such a condition, by reason of substances in, on or under the land, that:
> (a) significant harm is being caused or there is a significant possibility of such harm being caused; or

(b) pollution of controlled waters is being, or is likely to be, caused.
(Section 78A(2) of Annex A to Part IIA)

Thus, there is a distinction between a physical and an administrative/legal conception of 'contaminated land'. Contaminants may be physically present at a site, but the land is only 'contaminated land' in a legal sense if these substances can be demonstrated to be causing (or risk causing) 'significant harm'. The new regime emphasises the importance of risk assessment in the identification and remediation of contaminated sites. Indeed, as the guidance (DETR, 2000) states, the 'definition of contaminated land is based upon the principles of risk assessment'. This definition relies on:

[t]he relationship between a contaminant, a pathway and a receptor . . . unless all three elements of a pollutant linkage are identified in respect of a piece of land, that land should not be identified as 'contaminated land'. (DETR, 2000, p. 76)

Part IIA stipulates that harm to health and the environment arises not from the mere presence of contaminating substances but from 'Significant Pollutant Linkages' (SPLs). It is only where an SPL can be identified and where land meets this statutory definition that a local authority can formally determine the site to be 'contaminated land' (Syms, 2002, p. 107).

3.2.1.2 'Suitable for use'
UK policy emphasises 'suitability for use'. No absolute standards are laid down for remediation. Instead, the rule is that the standard applied to a site should be specific to the current or proposed land use. Central to this understanding of 'suitability for use' is a process of risk assessment. This involves risk estimation by surveying and risk evaluation, where risks are interpreted by experts, using different guidelines for different land uses (see Petts *et al.*, 1997); for example, the 'suitable for use' approach means that more stringent standards of remediation are applied if the end use of a site is housing or agriculture, rather than amenity or hard-surface uses. As with other aspects of the policy regime, the main advantage here is cost-effectiveness (see, for example, Defra, 2004).

The UK government's 'suitable for use' approach consists of three elements (DETR, 2000, sec. 10):

(1) ensuring that land is suitable for its current use
(2) ensuring that the land is made suitable for any new use, as planning permission is given for that new use
(3) limiting requirements for remediation to the work necessary to prevent unacceptable risks to human health or the environment in relation to

the current use or future use of the land for which planning permission is being sought

3.2.1.3 Risk assessment

Another aspect of the UK approach, which links to the last characteristic, is the critical role it assigns to formal risk assessment. De Sousa (2001, p. 138) outlines two types of criteria for evaluating the extent of soil pollution and the clean-up goals that are currently being used internationally. The first is generic numeric soil quality criteria,[5] and the second is site-specific risk assessment/risk management. The site-specific risk assessment consists of procedures that develop soil and groundwater criteria that consider tolerance and risk levels associated with a specific site (De Sousa, 2001, p. 138). Generic numeric soil quality criteria are not used in the UK as a trigger mechanism, but a number of such criteria have been developed by the Environment Agency through its Contaminated Land Exposure Assessment (CLEA) model. They indicate (in)tolerable levels of risk for a range of substances and assist in the definition of what constitutes 'significant risk' of 'significant harm' as outlined in the statutory guidance. The site-specific risk assessment process guides the 'suitable for use' approach. Indeed, it is a necessary corollary to such an approach.

Risk assessment is done by surveying the site, estimating the risks and then evaluating them by comparing the findings with the guidance as to what is deemed 'acceptable'. Central to the risk assessment process is establishing whether there is a 'significant possibility of significant harm'. But this is an ambiguous yardstick. The statutory guidance for Part IIA (DETR, 2000) sought to clarify matters by setting out two qualitative tables: Table A 'Categories of Significant Harm' and Table B 'Significant Possibility of Significant Harm'. However, as Walker (2002, p. 260) observed, 'both sets of criteria of harm and significance depend heavily on the availability and usage of what the guidance terms 'relevant information'.

3.2.2 An arena for negotiating risk

Under the Part IIA regime, risk is a contingency. There are no statutorily defined standards of concentration of contaminants that will trigger immediate remediation. The significance of any harm, or of the possibility of its occurrence, is dependent upon local circumstances. These circumstances relate to the character of the site and the contaminants therein; to the nature of the surrounding area; and to the existing and future uses of the site. Consequently, arguments over 'significance' may relate as much to what may be done to the site as to any contamination of it. The matter will turn upon the interrelations between three factors: (i) the nature and severity of the contamination and the cost of its remediation; (ii) the value of the proposed use relative to the current use and its ability to absorb remediation costs; and (iii) the standard of remediation required, given the varying sensitivity of the receptors associated with different land uses (Environment Agency, 2002, p. 10; ODPM, 2002, p. 12, para. 22).

Such relations may produce acute tensions. For example, the only way to deal with a badly contaminated site may be through its redevelopment for housing, because of the high development values that result; but local residents may be strongly opposed to new development and may value the site as an open space, yet still expect it to be treated. The resolution of such competing solutions poses a major challenge to all those involved, particularly to policy implementers in government. The UK approach to dealing with contaminated land ensures that the nature of risks and their significance are bound up with wider issues over which there is considerable scope for negotiation and debate. Not only does this result in risk being a fluid, contested concept; it also means that risk management is as much a social and political process as it is a technical, scientific one.

3.3 Democracy, trust and risk in environmental governance

3.3.1 Debating risk

Modern (post-)industrial societies are beset by a broad range of environmental and technological hazards that are a result of the first wave of modernity (see Beck, 1992). The response of governments to ameliorating and regulating a range of complex and controversial environmental hazards has been to rely increasingly upon the methodology of risk management and risk assessment, which is used to predict the potential a hazard has to harm an affected or exposed population. For government, the task for risk assessment is, therefore, to develop rigorous, comparable and durable scientific analysis which can be used to generate probabilities of risk. These probabilities can then be communicated to 'lay' audiences, such as affected residents.[6]

One of the prime difficulties of the utilisation of risk assessment in environmental governance has been, at one level, how to define risk. Most definitions of risk refer, in one way or another, to uncertainty over the incidence and/or level of potential harm that may arise from current processes or future events (Beck *et al.*, 2005, p. 397). For Furedi (2002, p. 7), risk is a protean concept, varying through space and time. Responses to a particular hazard or risk are 'shaped not so much by the disaster itself, as by a deeper consciousness which prevails in society as a whole at that moment'. Various environmental hazards, technological developments or personal activities are now considered more risky than hitherto. This is not simply because the nature of the threat has changed, but because society as a whole has become more sensitive to the potential harm posed in the light of more scientific data and/or media attention. In addition, an increasingly litigious culture compels public and private authorities alike to take more precautions to ensure that their decisions, and the effects of their actions, do not impose unacceptable costs upon groups within society.

In political debates about risk, the wider application of risk assessment and risk management is often criticised by some because it has been applied in too many areas and with too much vigour. For example, Furedi (2002, p. vii) argued that experts 'no longer simply dwell on risks – they are also busy evaluating *theoretical risks*. And, since theoretically anything can happen, there is an infinite variety of theoretical risks.' Indeed, Furedi sees this as one source of the creation of a 'culture of fear' and an obsessive preoccupation with safety. This preoccupation stems from the widespread adoption of the 'Precautionary Principle' in many industrial countries (Furedi, 2002, p. 8). The Precautionary Principle asserts that where the possibility of serious harm arises because of an environmental or technological risk, palliative responses should be enacted before the existence of scientific proof of harm (see O'Riorden and Cameron, 1994; Harding and Fisher, 1999). This principle requires decision makers to be more cautious. As a result, risk assessment is utilised more widely and is skewed towards the protection of human health as a core value. In this sense, the Precautionary Principle and risk assessment are intimately bound together.

However, risk assessment is criticised by others for serving the interests of the private sector and cost-conscious government (McLaren, 2000) at the expense of the wider public interest (interpreted in terms of ecological considerations). Indeed, in the United States it has been argued that the Bush administration has deliberately sought to undermine the science behind risk assessment to serve such interests (Kennedy, 2005; Macilwain, 2006). It is for this reason that Smith (2004, p. 329) argues that the processes of risk management, of which risk assessment is a key part, are implicitly bound up with the notion of power:

> Government's concern with risk cannot be divorced from a power/ knowledge framework. Governments are using the science of risk assessment to legitimise particular forms of behaviour. Risk assessments are seen as scientific but because they deal in probabilities and averages they are essentially ambiguous. The chances of pylons causing cancer are small or insignificant and therefore the government does not have to act and it is difficult for non-specialists to object. On the other hand, if government wants to legitimise other forms of behaviour it can do so. Risk is an excellent form of knowledge because it is scientific and ambiguous at the same time.

In Fischer's view (1995), the emphasis placed on expertise and expert judgement can be seen as another form of political manipulation because the nature of cause and effect between a technology and an environmental consequence can be difficult to prove scientifically. Consequently, 'the question of risk always remains open to interpretation' (Fischer, 1995, p. 185). Experts are hence considered to wield considerable influence over the judgement of what is defined as risky and what is not. For this reason, expert judgement in the risk assessment process is said to reduce the scope for wider forms of 'lay' public participation (Fischer, 2000).

3.3.2 Realist and social constructivist approaches

In academic debates on risk, two broad schools can be discerned: realist and social constructivist approaches. The realist approach suggests that risk is a measurable object which is independent of social and cultural processes, although it may be distorted through frameworks of interpretation. Within the realist paradigm, the emphasis of debate is on the accuracy of the methods employed to measure and calculate risk (Lupton, 1999, p. 18). An important distinction in the risk literature is made between objective risk and perceived risk (see Royal Society, 1983; Slovic *et al.*, 1982; Slovic, 1987; Royal Society, 1992; Adams, 1995, p. 7). This is generally considered to be a key cleavage between scientific (expert) interpretations of risk from any given hazard, and that of the general public (Margolis, 1998). In the realist tradition, an implied elitism exists whereby the public needs to be educated by the experts. The public's reaction to some risks is often claimed to be disproportionate with respect to their response to other comparable risks, with risks from a particular hazard being either over-estimated or underestimated. As the emphasis in risk assessment has been placed on 'solid science', the scope for the incorporation of the public's perception of risk into assessments is limited. Lay audiences are considered to have little to contribute to the process. Consequently, their role is usually relegated to the receiving of information in the risk communication part of the analysis.

Social constructivists identify another level of contestation of risk assessment from that offered in the realist paradigm. Here, the debate is centred less around the mechanics of risk assessment and more on the values implied by and embodied in current approaches. For Kunreuther and Slovic (1996), risk assessment is conceptualised as a game in which the rules must be socially negotiated within the context of a specific problem. This socially constructed or contextualised approach places emphasis on the institutional, procedural and societal processes involved with managing risk rather than on quantitative risk assessment. Treating risk as a game and a negotiation empowers affected citizens. They become legitimate partners in developing a response to a particular problem. As Skorupinski (2002, p. 97) observes, 'Talking about decisions under risk – which in most cases includes decisions made under conditions of uncertainty – is not possible without ethical considerations.' Managing risk requires the application of norms and values. Hence the incorporation of the views of affected citizens is necessary for risk decisions to be considered accountable, legitimate and democratic.

3.3.3 Beyond public–expert discrepancies

In contrast to the expert–lay dichotomy advanced in much of the extant literature, we do not presume, in our analysis, any fundamental difference between how the general public and experts arrive at their respective opinions about any given risk. It is not that such differences cannot be found, but rather that they are a red herring restricting understanding of the underlying processes. Despite this, it is the views of the *public* about risk, rather than those of experts, that have been and continue to be of most concern to (many) policy makers. Why is this?

One practical consideration is that the desirability of winning public acceptance can be a major constraint on how, or even whether, particular policies can be enacted. This is evident in the objections of local residents and environmental groups to new industrial or infrastructure developments (for example, wind farms, airport runways, waste incinerators, mobile phone masts, flood defence schemes). Processes of public consultation can be costly and time-consuming, even where the eventual outcome is favourable from the perspective of the policy makers. Perception of risk in environmental governance has a special place in such debates. It introduces an ethical dimension (see above). It is one thing to acknowledge that a development will produce economic losers as well as winners, and therefore that some deal needs to be done. It is quite another matter to maintain that an increase in the incidence of childhood cancer, or the destruction of a rare wildlife habitat, is a price worth paying for economic progress. In short, the discourse of risk is very powerful, whatever lies behind it.

Along the same lines, risk management decisions may be influenced by public perceptions of risk in ways that may sometimes distort priorities away from actual risk reductions. Policy makers may feel the need to be *seen to be doing something* about particular risks, even where these risks are relatively small and where the actions undertaken are more visible than effective (see Rothstein *et al.*, 2005). The other side of the story is that the general public can sometimes appear frustratingly complacent about the seriousness of other kinds of risks, and so be resistant to policy initiatives, and/or recommendations for individual lifestyle change, that could lead to risk reduction.

There is hence a clear need for policy makers to develop a greater understanding of how the public perceives risk in general, and specific risks in particular. However, it is dangerous to assume that there is something fundamentally different about the way in which risks are perceived by 'the public' on the one hand and by policy makers or 'experts' on the other, with the latter being taken as a benchmark of objectivity to which the former is contrasted. Depending on the circumstances, we *all* assume multiple roles at different times: of ordinary citizens, of local residents, of 'experts' with special experience, of decision makers needing to balance benefits and costs for ourselves and for those dependent on us.

There is increasing recognition, therefore, of the need to move away from the traditional research agenda. This focuses on questions such as how public perceptions differ from expert opinions, on potential reasons for such differences, and on what can be done – for instance, by better risk communication – to reduce this discrepancy (typically by shifting public perceptions towards those of the experts). Not only does this agenda pose the wrong questions: it makes many researchers wary of how their answers may be interpreted. A bleakly cynical view could be that much risk perception research has been a thinly disguised exercise in social control, directed towards manipulating public opinion so that it is brought into line with the wishes of the powers that be (Smith, 2004).

We argue that experts and the public can, and often do, come to very different conclusions about the seriousness of any risk. In arguing that the *processes* whereby they come to their respective conclusions are essentially the same, we are *not* suggesting that there is nothing to choose between the conclusions, in terms of correspondence with objective evidence. It is absurd to suggest that experts are omniscient, and never make mistakes. It is no less absurd, however, to suggest that there is nothing to choose between the views of experts and non-experts in terms of likely accuracy, so long as we acknowledge that all expertise is relative and not absolute. Of course, the business of identifying who is and who is not an expert is often controversial, but this only amounts to saying that some people who are conventionally regarded as experts may be no such thing. This is a far cry from saying that there is, in principle, no such thing as comparative expertise, or denying that some people truly are more expert and knowledgeable than others (even though their superior expertise may be imperfectly acknowledged). So arguing over whether the risk assessments of experts are 'better' than those of non-experts is specious. Either this must be trivially true (since making good assessments is part of what it means to be an expert) or else the argument is about something quite different – the validity of the criteria according to which we attribute 'expertise'.

Furthermore, there is a vital distinction here between prediction and prescription. Experts should be better at predicting what will happen than non-experts, but recommending what action should be taken on the basis of such predictions necessarily rests on some view of what outcomes are most desirable. However, if there is no agreement about the goals – and the different parties to an encounter espouse different values and priorities, as in choices between economic growth and environmental conservation – the prescriptions of experts may be rejected, however convincing their predictions. Put differently, predictions of consequences imply prescriptions for action only when we attach value to such consequences, and (not always, but sometimes) when taking action can be controversial.

3.3.4 'The virtuous circle' of democratic practice, trust and risk?

There are essentially two broad sets of factors that influence how people perceive risks: psychological and socio-cultural (Fiorino, 1995, p. 105). Psychological factors explore what goes on 'inside the heads' of *individuals*. The emphasis here is placed on *processing information*, in order to understand what individuals do and feel because of how our minds make sense of the information available to us. In general, these psychological approaches seek to provide 'lawful' accounts of human behaviour at a quite general level. To do this, much of psychology depends on analysing and interpreting data derived from rather abstract experimental or laboratory settings. Generalisation to 'real life' relies not on any surface similarity between the experimental conditions and an actual social setting, but on the argument that the processes that account for how individuals make sense of experimental tasks also account for everyday behaviour.

However, as Fiorino (1995, p. 106) observes, 'people do not react to risk in a state of social and cultural isolation. Shared values and a sense of community come into play.' There is an interactive relationship between individuals and the broader social structures within which they are located. Indeed, psychological approaches acknowledge the importance of both individual and situational differences in behaviour. Yet while psychological approaches have much to say about persons, they have a far looser grip (at least outside strict experimental settings) on distinguishing the critical elements of different environments. When the environments of interest are complex social and political settings, this difficulty becomes acute. For this reason it is important to appreciate the significance of the socio-cultural context of risk processes, particularly for risk managers in government as 'people's views about the acceptability of various kinds of risks reflect their judgements about the institutions that manage risks in society' (Fiorino, 1995, p. 106).

However, before considering this issue, it is important to examine at a more conceptual level how risk perception, attitudes and trust might relate to each other in terms of their underlying psychological processes. A difficulty with the phrase 'risk perception' is that it can lead us to think of risks as objects 'out there' to be 'perceived' rather than as complex interactions between physical hazards and human decisions. Consequently, the literature tends to emphasise those factors that result in over- or under-estimation of specific event probabilities, while work on the human decision-making processes that can lead to the occurrence of such events and/or to the production of adverse consequences is relatively limited. This is particularly true of hazardous substances, such as those found on contaminated sites. The main reason that hydrocarbons, asbestos, heavy metals and so on may constitute a risk is because, if improperly managed, they may come into contact with people who may inhale or ingest them. The risk arises at least as much from any improper management as from the physical properties of the contaminant itself. The implication of this is simple, but important: if risk arises in large part from human decision-making, risk *perception* involves, explicitly or implicitly, an evaluation of the quality of such decision-making.

In attempting to define the quality of decision-making, we draw on a classic psychological theory, originally developed to account for perceptual discriminations under conditions of 'noise' or uncertainty, known as Signal Detection Theory (SDT: Swets, 1973). In the present context, the question is how well a decision maker could tell whether a possibly (or somewhat) contaminated site is dangerous or safe. The theory distinguishes two parameters of decision quality: sensitivity (in this context the ability to differentiate greater from lesser degrees of risk) and response bias, or criterion (the level of risk at which one decides that a hazard is dangerous rather than safe). The first parameter reflects the decision maker's knowledge, competence or expertise, whereas the second can reflect the relative costs and benefits of making different correct or incorrect choices, and so may be affected by the decision maker's personal interests or motives.

Our interest here is less with the quality of the *actual* performance of decision makers than with how their performance is perceived; that is, how, say, a council's performance is *perceived* by local residents (for fuller discussions of the applicability of SDT concepts to risk perception and trust, see Eiser, 1990; White and Eiser, 2006). *Trust* should be higher if the council is seen as making better-quality decisions; that is, decisions characterised by higher expertise and by a lack of inappropriate bias in interpreting evidence as indicative of danger or safety. However, decision makers who are expert and use an appropriate cut-off point for *interpreting* signs of danger for themselves may not necessarily be trusted when *communicating* risk. Some might be (perceived as) being either too ready or too reluctant to tell the public about possible dangers, regardless of their own interpretation of the level of risk. In other words, trust could be undermined by a *communication bias*, over and above a bias in *interpretation* of the evidence. More generally, trust is likely to be undermined by a perceived lack of *openness* in risk communication (see also Mayer *et al.*, 1995; Peters *et al.*, 1997).

Whereas these aspects of trust deal very much with perceptions of risk communicators (here, a council) as sources of *information*, issues of bias and openness may be seen as tied into a perception of *motives.* Other research indicates that decision makers and communicators are trusted more to the extent that perceivers see them as sharing their values, that is, essentially being 'on the same side' (Earle and Cvetkovich, 1995). However knowledgeable decision makers or communicators are seen to be, if they are thought to be serving their own interests their motives will be suspect; any communication from them may be discounted as designed to serve such interests rather than to convey the truth. For this reason, one might expect that another major predictor of trust will be the extent to which a council is perceived to have residents' interests at heart.

The level of social trust within populations is thus claimed to be a critical factor in determining the success of risk communication and wider risk management strategies. Put simply, in places where levels of social trust in institutions involved in risk management are low, it is likely that populations will be more reluctant to accept the claims made about risk by these organisations. Indeed, within the same area there is often significant variation between the *types* of organisations that are trusted. For example, public organisations, such as local authorities, tend to be trusted more than private interests, such as private developers.

In turn, the willingness of institutions involved with risk management to engage in open and democratic risk discourse is often related to how these organisations perceive the likely reaction of the public to particular risks. Where trust is low, risk communication strategies are likely to place less emphasis on the active involvement of the public. In contrast, in places where trust between the public and risk managing institutions is high, it is more likely that participatory approaches will be promoted. Indeed, these processes are self-reinforcing whether the respective initial returns are negative or positive. In effect, there can be either a 'spiral of distrust', where risk communication is kept to a minimum, or a 'virtuous circle', where trust is high and democratic processes are promoted.

To what extent do these principles hold in the real world? In order to examine the interrelations between democracy, trust and risk we examined the risk management strategies pursued by two local authorities. Both had to deal with issues posed by contaminated sites in their areas, but they adopted quite different approaches to the problem. The authorities could be distinguished most clearly by the way they (did not) communicate with local residents about risk.

3.4 Case studies

3.4.1 *Communication strategies in Area A and Area B*

3.4.1.1 'Area A'

Area A's approach to communication was heavily coloured by what happened at the first site they dealt with under the Part IIA system. The council held a meeting between environmental officers and councillors and it was decided that they wanted to be 'as open as possible with residents and the wider public' (interview, communications officer, 2005: Area A). This view was heavily influenced by a presentation from the communications officer from Area B, which was described as 'helpful and . . . an influence on our decision to be open' (interview, communications officer, 2005: Area A). The local authority for Area A produced leaflets for residents and, at the same time, talked directly to the local press, issuing a press release. This was the first information that some people had received on the contamination problems on the site.

After the council had sent out the press release, a meeting was held between the two local newspapers and the Lead Member for the Environment. While it was felt that the 'regional' newspaper ran a 'fair' account of the troubles on the site, the local newspaper for the contaminated area ran the story on its front page with what was described as the 'wrong sort of tone . . . If the same story had been run on page 5 then that would have been fairer' (interview, communications officer, 2005: Area A).

Various actors inside the local authority noted the lack of trust between the local authority and the local press. The residents of the houses on the site also expressed anger about the tone that the local press had taken. They were concerned that the council had – apparently – informed the media without consulting them first. One of the key factors behind this anger was the effect residents felt that the story would have on local house prices. Property blight was an issue that was identified by several interviewees as a key concern of the local authority.

Approximately three weeks after the original notification letters went out to residents, the local authority held a meeting with them. Many of the people who attended the meeting described the residents' anger as 'obvious', with media relations being the main source of concern. Interviewees' opinions diverged over the cause of the tension in the meeting. Some suggested that the Lead Member for the Environment, who chaired the public meeting, was an inappropriate person to undertake this delicate function. Others suggested that the member was 'constructive and listened to the advice of officers' but was in an impossible position (interview, communications officer, 2005: Area A).

However, as the interviewees pointed out, knowledge of what happened with regard to this particular site is crucial to an understanding of the approach the council adopted at other contaminated sites within its area. Various local authority officers stated that relations with the (local) press had deteriorated to such a degree that it was impossible to trust them to run a 'balanced' story on any site. This led some to question the extent to which it was possible to communicate a message on contaminated sites that would not be reported in sensational terms: 'It is very hard to tell people that a site is contaminated with asbestos but that there is possibly no risk to them' (interview, communications officer, 2005: Area A).

As a result, a consensus was reached in the local authority that, in future, people or the press would be given information about contaminated land *if they requested it.* Affected local residents would still be informed by post of what was happening, but the local authority would not release media reports. Instead, a 'reactive' statement would be prepared by the council for use in response to any inquiry (interview, communications officer, 2005: Area A). In addition, council officers stated that they would not be holding any open public meetings. They preferred speaking to local residents on a one-to-one basis. They also stated that they would provide a 24-hour hotline for concerned local residents to call. When asked if a reactive approach might perhaps produce greater tensions between the press and the council and give the impression that the council had something to hide, the communications officer stated:

> There is a lack of trust in this area between the press and the council. It will take a long time before that trust is restored. It is very hard to persuade other officers and councillors to be open with the local press again. To be fair to them, however, one of the sites we have might one day become a live story and then they have a duty to report it how they like. However, it only takes one resident to complain about a cover-up for the story to blow. (Interview, communications officer, 2005: Area A)

The contaminated land officer for this site suggested that one of the key blocks on the local authority's communication strategy was the lack of resources:

> I think that going public without the necessary manpower is difficult. If there were more officers then the site could have been dealt with much sooner. We are hoping that two officers will be appointed in the near future. It's all about resources really. We don't have enough to do as much as we would like. In theory you should be doing X, Y and Z but limited resources mean that priorities must change. (Interview, contaminated land officer, 2005: Area A)

The head of the Environmental Services directorate at this particular council stated that it has been held up as a beacon for community consultation and:

[i]t has established a system of Community Committees, Community Strategies, etc. In terms of contaminated land, we learned a lesson from [the first site]. We were told by other local authorities to be open with the community and the press but we learned difficult lessons from being open and honest with the local community. The community did not like the way everyone in the wider area was told about the problems with the site. They especially didn't like the way that it was blasted across the local press. We felt that the [regional newspaper] was fair and balanced but that the tone of the local newspaper was unfortunate. Now we work on the basis of public reassurance as a core element of our communication strategy. We are careful about where the information goes to outside local residents. (Interview, Environmental Service directorate officer, 2005: Area A)

At a more general level, the risk assessment for the site – which concluded that remediation was necessary – was contested by officers within the local authority of Area A. When asked about it, the contaminated land officer for the site (interview, 2005) stated:

It's not completely up-to-date and some of the values put in have been adjusted to reflect the CLEA model. They have supported this with a study of the site's use. It was probably the best they could have come up with, with the information they had. I remember that in the conclusion it presented three possible scenarios for the contamination on the site, with each reducing the assumption of contamination and risk on the site to a point where in the final one there were only one or two chemicals that were problematic and it set very low exposure levels. The highest was the most unrealistic with the lowest closest to reality . . . If I was a site user and I picked it up then I would be concerned. I think the conclusion is fair relating to the knowledge of that time. The assumptions built into its first stage were too conservative . . . The conclusion suggests that action be undertaken immediately. However, we are going to be forced to delay again by about a month as I'm currently tied-up with the public relations work on [other sites].

However, following this the same officer stated:

If we find that with further testing . . . the problems were not as severe as the risk assessment suggested then I have the power to de-designate the site. I've checked the guidance and it doesn't say that I don't have the power to do this . . . Hopefully I will get some information from Defra on the benzo(a)pyrene. Hopefully this will be lower than the currently assumed levels of toxicity that [have] been assumed in the risk assessment. We are going to determine [the site] because [the local authority] and the Environment Agency agree on the problems of the site, but we might de-designate at a later date . . . At the moment I'm not tempted to remediate. I think risk-management would more prudent. It would be irresponsible to do something and not to do it properly because Defra will not fund anything that is seen as temporary.

3.4.1.2 'Area B'

The local authority for Area B has considerable experience of dealing with contaminated sites. On one of their sites, two deaths have been attributed to historical contamination and exposure to asbestos. In this case, the insurance company for the local authority decided to make a settlement with the families of those who died.

The communications officer for Area B stated that the council's actions on the site were driven not by economic considerations, but by ethical and political considerations. However, the officer did concede: 'Reputation is fundamental to issues like regeneration' (interview, communications officer, 2005: Area B). The officer highlighted the importance of the Freedom of Information Act (FOI) in determining the amount and type of information that the local authority gives to the community:

When the story broke we called a meeting that included cabinet members for [Environment] Housing and Health and the ward members. They were briefed on what had happened and we agreed with them a briefing that would be circulated around the rest of the members of the council. The next stage was to hold a meeting between regulators and councillors. This allowed members a chance to ask the regulators any questions they had. After this we went to the residents and asked residents to select representatives who could help the community manage risks themselves (interview, communications officer, 2005: Area B)

In terms of informing local residents, Area B does not hold open public meetings: only residents of the affected areas are invited (though it is made clear that the national press could attend if it wished). The communications officer felt that the local press reported contamination issues in a 'fair and fairly accurate manner' (interview, communications officer, 2005: Area B).

On another site, where conditions had not really made much media impact, a local valuer mistakenly told the banks that the area had been designated as 'contaminated land'. On this basis the banks stopped lending money for mortgages. The local authority wrote a letter of complaint to the banks. While it was accepted that there was no such designation on the site, the banks refused to rescind the block on lending. In response, the council organised a meeting with local valuers to make clear that the site had not been determined as contaminated. Next, a round-table meeting was held between the local authority, the main banks and the local valuers at which they were asked to help to resolve the issue and to adopt a more 'community-oriented approach'. Subsequently, the banks agreed to begin lending again in the area, re-establishing the housing market there.

What of the contamination that had given rise to these events? The council wanted to meet with residents to tell them what was going to be done on the site, but decided not to speak to them until more evidence was available: 'We have to been seen to be knowing what we are talking about' (interview, communications officer, 2005: Area B). The meeting was thus delayed for more than two months, in which time the community had noticed that they could not get mortgages in the area. The officer suggested that the residents were not really bothered about the lack of news on the contamination inspection, but were much more concerned about their property values. One critical factor on the site was the issue of liability. It was clear that the local authority was the responsible party and would bear the burden of costs. The communications officer felt that this reassured residents and lenders and diffused a good deal of tension.

The contaminated land officer for Area B stated that when the council first received reports back on the potential contamination of the site they created a focus group, and held a meeting with residents to inform them of the findings and of their intention to carry out onsite investigations:

> There was a group of residents who were actually much more vocal than the rest and we said we'd like you to be a focus group so we can come back to you. Their opinions weren't necessarily representative of the people on the estate. So we had regular meetings with them and we also sent out a message to residents saying this is the stage we're at, and letting them know of any developments before the local press gets hold of it. It's worked quite well because they can see that we're being up-front and straight forward, and also trying to listen to what their concerns are. One example is, 'is my dog okay?' None of us had thought 'are dogs okay, because they're digging around in the garden?' We had a chat and said that what we perceived from the survey is that dogs aren't affected. But it was one of those things, we have our preconceived ideas of what they think, and when you actually speak to them they usually come back and you think, right that is actually the issue. We just say to them 'this is what we're planning to do, does that make sense to you?' We try to get them to understand the ideas of risk assessment, and that's a very difficult thing to do. It's not really my area of expertise, the risk communication. (Interview, contaminated land officer, 2005: Area B)

The contaminated land officer also suggested that residents only really respond to the familiar-sounding contaminants:

> There was an asbestos issue and there's one thing that the residents know and it's that asbestos equals death. Arsenic was used in Sherlock Holmes and things like that, but lead and nickel they don't care. Hydro-carbon, no issue. And so you have to start saying okay, we can be up-front because they don't care. We can't say there's no asbestos because we've only sampled a small proportion of the estate but as far I'm concerned there isn't an issue. But we could have lead, 2000 milligrams a kilogram. I find that kind of thing interesting. We try to make it clear to them that we might have to do something because there's an issue with nickel. Nickel? Nobody gets it. (Interview, contaminated land officer, 2005: Area B)

3.4.2 A survey of the residents of Area A and Area B

It is clear that the residents and councils in Area A and Area B have different perceptions of one another's views and actions in relation to contaminated land. The two councils have developed distinct approaches to risk communication and management. In Area A processes tend to be cautious, reactive and 'closed'. In comparison, in Area B processes are more proactive and 'open' and involve local residents more substantively. How may these different approaches have affected the relations between democracy, trust and risk in the two areas?

To explore this, a postal survey of residents in Area A and Area B was conducted. The survey was entitled 'Redeveloping Urban Land: Tell Us What You Think' and covered a variety of topics, including attitudes to brownfield redevelopment, preferences for different forms of redevelopment (for example, housing, recreation), and perceptions of the impact of new housing developments on their area. We here report the findings of a subset of questions relating specifically to satisfaction with, and trust in, communication by the local council, and to perceptions of risk from contaminated land.

A total of 8378 copies of the questionnaire were sent out with freepost reply envelopes to addresses in selected wards within Area A (3603) and Area B (4775). The surveyed areas included communities adjacent too, and more distant from, contaminated land. Depending on the size of the ward, either all or alternate households were included in the sample. A total of 747 questionnaires were returned (407 from Area A, 340 from Area B). Although low, the response rate (8.9%) is not out of line with similar unsolicited mail surveys. There was no evidence that respondents differed demographically from other households in the sampled areas, although we can expect a self-selection bias towards individuals with greater interest in the issue. Of those responding, 48.6% were male, 65.6% owned their own home, 61.4% were employed or self-employed, with 7.8% seeking work and the remaining 30.8% being homemakers, retired or in education. Their average age was 51.0 years (SD = 16.3). The response rates and general demographic make-up did not differ notably between Area A and Area B.

3.4.2.1 Basic findings

The questionnaire began by asking residents, 'Thinking of the country as a whole, should most new homes be built on brownfield land?' and 'Thinking of your local area, should most new homes be built on brownfield land?' Responses to both questions were recorded on a five-point Likert scale (definitely no; no; not sure; yes; definitely yes; scored 1–5).

Respondents in both areas were, on average, broadly in favour of building new homes on brownfield land (Figure 3.1). On average, respondents in Area A were more positive than respondents in Area B to the idea of building new houses on brownfield land, both nationally and in their local area.

Three items on the questionnaire assessed satisfaction with the council in the context of housing and redevelopment. Respondents were asked how satisfied they were that the local council had (a) kept residents informed, (b) sought residents' views and (c) taken residents' views into account. Responses were again recorded on a five-point Likert scale (1 = definitely no; 5 = definitely yes).

The response to these questions indicated that respondents in both areas were, on average, moderately dissatisfied with the council. Respondents in Area B were, however, less dissatisfied than respondents in Area A.

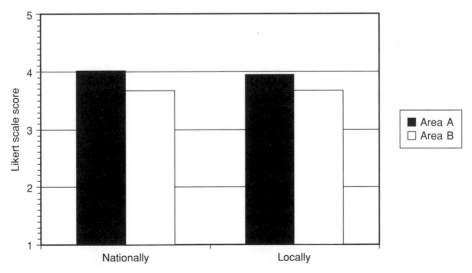

Figure 3.1 Support for the redevelopment of brownfield land for new houses (a) nationally and (b) locally.

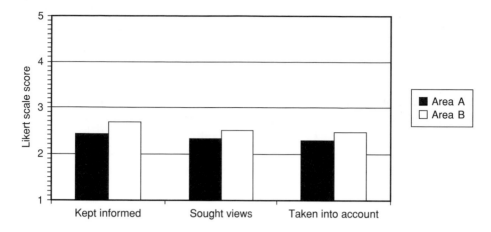

Figure 3.2   Satisfaction that local council had (a) kept residents informed, (b) sought residents' views and (c) taken residents' views into account over housing and brownfield redevelopment, by area.

This can be seen in Figure 3.2. Because individuals' responses to these three items were highly consistent with each other (Cronbach's $\alpha = 0.91$), they were averaged to yield a single score ('satisfaction') for subsequent analyses.

The section of the questionnaire that dealt with residents' perceptions of the risk of contamination on brownfield land was introduced so as to avoid implying to residents that they had been selected because their own home was at risk. The text began:

> It is always possible that brownfield land is contaminated because of the way it has been used in the past. For example the soil may contain residues from factory processes, or left-over materials that might be toxic. Because of this one has to be careful how the land is developed.

Respondents were then asked the following questions to assess their perceived *vulnerability*:

- Do you think any brownfield land in your local area might be contaminated? (five-point scale: definitely no = 1 to definitely yes = 5)
- Compared with other urban areas in the UK do you think there is more or less contaminated land in your neighbourhood? (five-point scale: lots less; less; not sure/about average; more; lots more; scored 1-5)
- Compared with other homes in your neighbourhood, do you think there is more or less contaminated land near your own home? (five-point scale as above)

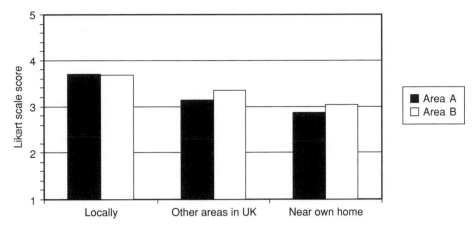

Figure 3.3 Perceptions of risk of contamination in the local area, compared to other urban areas in the UK, compared to other homes near to own home, by area.

The responses to these questions are shown in Figure 3.3. It is clear from responses to the first question that there is no significant difference between Area A and Area B in the respondents' belief that there was contaminated land in their local area. The combined mean response (3.70) significantly exceeded the scale midpoint of 3 ($t = 20.11$, $p < .001$), indicating that both groups considered their local area to be affected. Likewise, the combined sample tended overall to believe that there was more contaminated land in their neighbourhood than in other urban areas ($M = 3.23$, $t = 7.53$, $p < .001$), but in this case the two groups differed significantly, with residents of Area B believing they were relatively more affected. When asked about risk of contamination near their own home compared with the rest of their neighbourhood, Area B residents again considered themselves more at risk than those in Area A. A striking aspect of these data, however, is the comparative reluctance of respondents, at least in Area A, to believe their own home was at risk, even when acknowledging the presence of contamination in their neighbourhood, as shown by a highly reliable mean difference between these latter two items for the sample as a whole ($t = 9.23$, $p < .001$). Thus, even against a background of generally heightened risk perceptions, residents saw their own home as no more at risk (Area B) or even less at risk (Area A) than other homes in their neighbourhood, a finding reminiscent of an effect known as 'unrealistic' optimism or optimistic bias (Weinstein, 1987; 1989).

These three items measuring vulnerability were significantly ($p < .001$) intercorrelated. For the sake of simplicity, therefore, we averaged them to yield a single score (*vulnerability*) for use in subsequent analyses (Cronbach's $\alpha = 0.68$).

Concern with the consequences of contamination was measured by asking respondents how concerned they would be, if they learned they lived near contaminated land, about 11 potential effects (including effects on their own health and that of friends, children, pets; on wildlife, house prices and mortgages; on local recreation and bathing; and the implications of eating locally caught fish or locally grown vegetables). Items were rated on a five-point scale (not at all concerned = 1 to extremely concerned = 5). A twelfth question asked how concern over contamination compared with that over other urban risks (crime, air pollution and traffic accidents). The average responses for these items are shown in Figures 3.4 and 3.5.

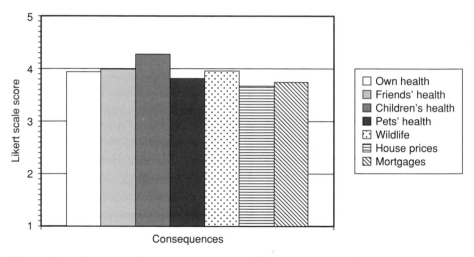

Figure 3.4   Concerns about consequences of living near contaminated land.

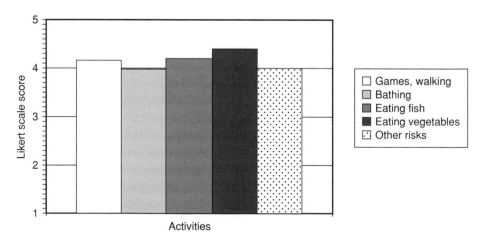

Figure 3.5   Concerns about loss of opportunities for different activities which might result from living near contaminated land.

The only items on which there were significant differences between Area A and Area B were those concerning property prices and the ease of selling and/or mortgaging respondents' houses. Residents in Area B were significantly less concerned than those in Area A about this, despite the temporary difficulty the former had had in relation to securing mortgages. In view of their high internal consistency (Cronbach's $\alpha = 0.91$), these 12 items were averaged to yield a single measure of *concern*.

A major part of our survey concerned residents' trust in their local council, and relevant aspects of council decision-making and communication. Respondents rated their general trust in their local council in the context of decision-making about contaminated land (wouldn't trust at all = 1; would trust completely = 5), together with five aspects of decision-making and communication that might contribute to such trust:

(a) *expertise*: not at all able to judge how safe or dangerous it was = 1; extremely able to judge = 5
(b) *interpretation bias*: would definitely see the risk as safer than it really was = 1; would definitely see the risk as more dangerous than it really was = 5
(c) *communication bias*: would definitely underplay the risks when communicating to the public = 1: would definitely exaggerate the risks when communicating to the public = 5
(d) *openness*: not at all prepared to tell what they know = 1; extremely prepared to tell = 5
(e) *shared interests*: definitely hasn't got my interests at heart = 1; definitely has got my interests at heart = 5

The mean scores given by residents of Area A and Area B for these aspects of trust are shown in Figure 3.6. There were no significant differences between the areas with respect to their views on these matters (but see below for the analysis of differences between the areas in how aspects of trust related to other variables).

3.4.2.2 Interrelation between variables
The mean scores for the two areas on the principal variables are shown in Table 3.1, together with the results of *t*-tests for the univariate differences. Residents of both Area A and Area B said they would be quite concerned if they learnt about local contamination, but did not differ from each other in this regard. Concern and vulnerability were only weakly associated $(r = 0.08, p < .05)$.

Residents of both areas were rather dissatisfied with their local council in terms of their perceived style of communication, the overall M (2.44) being significantly to the negative side of neutral $(t = 15.40, p < .001)$.

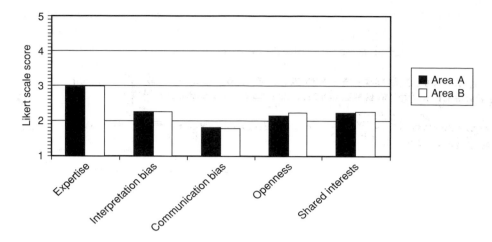

Figure 3.6   Residents' trust in aspects of their local council's decision-making and communication.

Table 3.1   Mean scores for the two areas on the principal variables.

Area:	A	B	*t*	*p*
Vulnerability to contamination	3.71	3.69	0.26	ns.
Concern about contamination	4.02	3.97	0.97	ns.
Satisfaction with council	2.35	2.55	2.80	<.005
General *trust* in council	2.05	2.23	1.73	<.09

Such dissatisfaction, however, was significantly stronger in Area A than in Area B. Satisfaction was negatively associated with perceived vulnerability $(r = -.21, p < .001)$ and, more weakly, with concern $(r = -.08, p < .05)$. When we performed an analysis of covariance to look at area differences in satisfaction, controlling for vulnerability, the A vs B comparison was even clearer (Adjusted Ms = 2.33, 2.58, $F (1,727) = 11.80, p < .001)$. The effect of vulnerability was also highly significant $(\beta = -.22, F (1,727) = 37.00, p < .001)$.

3.4.2.2.1 General trust
Overall, respondents were quite distrustful of how their local council would deal with issues of contaminated land $(M = 2.13;$ difference from midpoint, $t = 17.38, p < .001)$. This was slightly more so for Area A than for Area B. Controlling for vulnerability, concern and satisfaction rendered the area difference in general trust clearly non-significant $(F (1,699) = 1.45, ns.)$. The effects of vulnerability $(\beta = -.11, F (1,699) = 9.45, p < .002)$ and especially

satisfaction (β = .35, F (1,699) = 93.92, p < .001) were highly significant, but not that of concern (β = –.05, F (1,699) = 2.40, *ns.*). In other words, irrespective of area, residents trusted their council far more if they were satisfied with its record on communication, but somewhat less if they perceived themselves to be at relatively greater risk from contamination.

3.4.2.2.2 *Aspects of trust*

Finally, we performed a multiple regression to examine how general trust might be predicted from other aspects of trust. Since there were no area differences on these items, the analysis was performed on the total sample. The items measuring interpretation and communication bias were each recoded from 1 to 3 (i.e. 1,5 = 1; 2,4 = 2; 3 = 3) so that 1 represented maximum bias either in the direction of underplaying or exaggerating the extent of risk, and 3 represented a lack of bias. After such recoding, the five aspects together accounted for 59.7% of the variance in general trust (F (5,651) = 193.20, p < .001). The two most important predictors were openness (β = .38, t = 11.60, p < .001) and shared interests (β = .37, t = 11.15, p < .001), followed by a lack of communication bias (β = .13, t = 4.03, p < .001). The effects of a lack of interpretation bias (β = .05, t = 1.76, p < .08) and expertise (β = .04, t = 1.41, *ns.*) were not statistically reliable. In other words, trust in the council was only weakly related to perceptions of the quality of their decisions as such, but strongly related to perceptions of their openness and lack of bias as communicators, and to their perceived motives; that is, whether they had residents' interests at heart. This is illustrated by Figure 3.7.

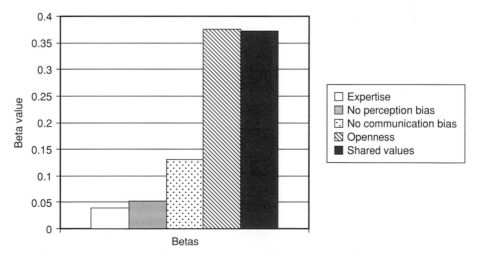

Figure 3.7   Regression of trust in council onto predictors of trust.

3.5 Conclusions

3.5.1 *Is there a good way to communicate risk?*

These findings provide both discouraging and encouraging news for those charged with looking after the interests of local communities and, in particular, protecting residents from the potentially harmful consequences of environmental contamination. On the one hand, both local councils were held in somewhat low regard, attracting ratings of dissatisfaction and distrust overall, rather than satisfaction and trust. Such negative evaluations were strongest among residents who perceived themselves to be relatively more vulnerable to the risks associated with contaminated land. On the other hand, where council officials appeared to have made more effort to be open with local residents and to engage them in discussion about the relevant issues (and/or were perceived to have done so), such distrust and dissatisfaction were significantly reduced.

It is difficult within the limitations of the design of this study to attribute these differences between the two areas to any single incident or example of good (or bad) practice by any particular local government officer(s). Our background interviews and observations suggested that it was differences in organisational culture that mattered, rather than merely the attitudes of specific individuals. Our data are likewise silent on how far, if at all, such differences may have spread to other aspects of the councils' activities. Nonetheless, our characterisations of the communication strategies of the two councils was broadly borne out by residents' responses to our questionnaire. Extra confidence can be placed in these findings because our postal questionnaire method lessened any chance that they could be the result of demand characteristics. Direct contact between researchers and respondents was avoided and the questionnaire itself made no mention of our intention to use the data to draw comparisons between different areas. Furthermore, the area differences in residents' satisfaction cannot be simply put down to differences in the (perceived) extent of contamination per se. Although greater perceived vulnerability predicted more dissatisfaction at an individual level, residents of Area B were *less* dissatisfied with their local council while seeing themselves as *more* affected by contamination than residents of Area A.

While commending the more open approach adopted by the council in Area B, what we observed amounted merely to a preparedness to adhere to principles that, from the perspective of the broader literature on risk communication (Kasperson and Stallen, 1991; Calman, 2002), one might have hoped to be more widely acknowledged and adopted. In contrast, the reluctance of some council officers in Area A proactively to engage with residents when aware of possible contamination risks is of concern. It flies in the face of the principle that the withholding of information can lead to a loss of trust that may be very difficult to recover. A more worrying possibility is that such examples of poorer practice, and possibly a lack of professional training in risk communication, may not be uncommon within local government or other relevant agencies. As mentioned, the council for Area B had built up more experience of dealing with serious contamination issues and therefore had developed a more thought-through strategy for communicating with local residents. This could imply that many councils or other agencies with less experience of such issues may rarely have a prepared risk communication strategy in place before being confronted with particular incidents. Under such circumstances, their response may be largely determined by short-term considerations and/or the intuitions, good or bad, of individual officers. None of this suggested a systematic dissemination of evidence-based good practice.

At a more conceptual level, our findings reinforce the message that risk perceptions, attitudes and trust are closely interconnected, as noted in previous research on other forms of risk (e.g. Eiser *et al.*, 2002). Those residents who perceived their neighbourhood and/or home to be more vulnerable to the effects of contaminated land were more dissatisfied with and distrustful of their council. Furthermore, when considering different predictors of trust, residents attach greatest weight to aspects that reflect their perceptions of the council's motives (having residents' interests at heart) and their openness in communication. These aspects take precedence over those aspects more central to the judgement of danger or safety itself (expertise and interpretation bias). In other words, it helps a little (in residents' minds) if the council appears to know what it is doing, but it helps build trust even more if it is seen to be acting honestly, and for the right reasons. Such findings suggest that ordinary citizens, even though they often lack the expertise to interpret technical information concerning levels of specific contaminants, may yet rely on their everyday knowledge about *people* and their motives when forming judgements of risk and trust. In so doing, they recognise, explicitly or implicitly, the extent to which risk is a product not merely of physical hazards but of human behaviour.

Acknowledgements

This paper is based on work undertaken for a collaborative research programme on 'Sustainable Urban Brownfield Regeneration: Integrated Management' (SUBR:IM) funded by the Engineering and Physical Sciences Research Council (grant number GR/S148809/01). The authors are grateful for EPSRC's support. We also specifically thank Nigel Lawson, School of Environment and Development, University of Manchester, for his collaboration over the background interviews. The views presented are those of the authors and cannot be taken as indicative in any way of the position of SUBR:IM colleagues or of EPSRC on the subject. All errors are similarly those of the authors alone.

Notes

1. It should be noted that this paper refers to UK policy but essentially deals with the institutions and policies that apply to England. The policy approaches adopted in Scotland, Wales and Northern Ireland are similar to the one that operates in England, although there are differences in the actors involved and their relative competencies.
2. A 'material consideration' is a matter that must be taken into account when determining a planning application. The range of issues that can be material considerations is wide and can include representations made by the public, infringements on privacy, design issues, draft plans, development impacts and comments made by statutory and non-statutory consultees. However, issues that are not 'material considerations' and that cannot be used as reasons to refuse a planning application include the loss of a view, commercial competition and the effect the proposal might have on property values and ownership. It is for planning authorities to decide what weight to attach to each material consideration.
3. The terms 'Part IIA' and 'statutory guidance' are used here interchangeably throughout to refer to the overall policy regime for dealing with contaminated land (as defined by the guidance).
4. So far as land which was affected by contamination, and has been developed (where that development incorporates treatment specified by conditions attached to a planning permission) is concerned, once the development is complete and has started to be used the land should not be capable of being determined as 'contaminated land' under Part IIA.
5. These are numerical indices that can be used for both assessment and clean-up activities derived from toxicological studies that identify levels according to a tolerable health risk.
6. For this reason, an important subdivision of the risk assessment methodology is the emphasis placed on risk communication (Fischer, 1995, p. 168).

References

Adams, J. (1995) *Risk*. UCL Press, London.

Beck, M., Asenova, D. and Dickson, G. (2005) Public administration, science and risk assessment: a case study of the UK spongiform encephalopathy crisis. *Public Administration Review*, **65** (4), 396–408.

Beck, U. (1992) *Risk Society: Towards a New Modernity*. Sage, London.

Calman, K.C. (2002) Communication of risk: choice, consent and trust. *Lancet*, **360**, 166–8.

Catney, P., Henneberry, J., Meadowcroft, J. and Eiser, J.R. (2006) Dealing with contaminated land in the UK through 'development managerialism'. *Journal of Environmental Policy and Planning*, **8** (4), 331–56.

De Sousa, C. (2001) Contaminated sites: the Canadian situation in an international context. *Journal of Environmental Management*, **62** (2), 131–54.

Defra (2004) *Remediation of Land Affected by Contamination: A Regulatory Summary.* Department of the Environment, Food and Rural Affairs, London.

DoE (1994) *Paying for our Past: the Arrangements for Controlling Contaminated Land and Meeting the Cost of Remediating Damage to the Environment.* Consultation Paper, Department of the Environment, London.

DETR (2000) *Environmental Protection Act 1990: Part IIA Contaminated Land.* Circular 02/2000. Department of the Environment, Transport and the Regions, London.

Earle, T.C. and Cvetkovich, G.T. (1995) *Social Trust: Toward a Cosmopolitan Society.* Praeger, Westport, CT.

Eiser, J.R. (1990) *Social Judgment.* Open University Press, Buckingham.

Eiser, J.R., Miles, S. and Frewer, L.J. (2002) Trust, perceived risk, and attitudes toward food technologies. *Journal of Applied Social Psychology*, **32**, 2423–33.

Environment Agency (2002) *Dealing with Contaminated Land in England.* Environment Agency, Bristol.

Fiorino, D.J. (1995) *Making Environmental Policy.* University of California Press, California.

Fischer, F. (1995) Hazardous waste policy, community movements and the politics of Nimby: participatory risk assessment in the USA and Canada. In: *Greening Environmental Policy: The Politics of a Sustainable Future* (eds F. Fischer and M. Black), pp. 165–82. St Martin's Press, New York.

Fischer, F. (2000) *Citizens, Experts and the Environment: The Politics of Local Knowledge.* Duke University Press, London.

Furedi, F. (2002) *Culture of Fear: Risk Taking and the Morality of Low Expectation.* Continuum, London.

Harding, R. and Fisher, E. (1999) *Perspectives on the Precautionary Principle.* Federation Press, Sydney.

Kasperson, R.E. and Stallen, P.J.M. (eds) (1991) *Communicating Risks to the Public.* Kluwer, The Hague.

Kennedy, R.F. (2005) *Crimes Against Nature.* Penguin, Harmsworth.

Kunreuther, H. and Slovic, P. (1996) Science, values and risk. *Annals of the American Academy of Political and Social Science*, **545**, 116–25.

Lupton, D. (1999) *Risk.* Routledge, London.

Macilwain, C. (2006) Safe and sound? *Nature*, **442** (20 July), 242–3.

McLaren, D. (2000) 'Risky Business: An Environmentalist Viewpoint.' Presentation to Swiss Re Annual Client Conference, 13 October 2000. Available at: http://www.foe.co.uk/resource/presentations/riskybusiness.html

Margolis, H. (1998) *Dealing with Risk: Why the Public and the Experts Disagree on Environmental Issues.* University of Chicago Press, Chicago.

Mayer, R.C., Davis, J.H. and Schoorman, F.D. (1995) An integrative model of organizational trust. *Academy of Management Review*, **20**, 709–34.

ODPM (2002) *Development on Land Affected by Contamination.* Consultation paper on draft planning technical advice. Office of the Deputy Prime Minister, London.

ODPM (2004) *Planning Policy Statement 23: Planning and Pollution Control, Annex 2: Development on Land Affected by Contamination.* HMSO, Norwich.

O'Riordan, T. and Cameron, J. (eds) (1994) *Interpreting the Precautionary Principle.* Earthscan, London.

Peters, R.G., Covello, V.T. and McCallum, D.B. (1997) The determinants of trust and credibility in environmental risk communication: an empirical study. *Risk Analysis*, **17**, 43–54.

Petts, J., Cairney and Smith, M. (1997) *Risk-Based Contaminated Land Investigation and Assessment*. John Wiley & Sons, Chichester.

Rothstein, H., Irving, P., Walden, T. and Yearsley, R. (2005) *The risks of risk-based regulation*. Paper presented at the Centre for the Analysis of Risk and Regulation, London School of Economics, June.

Royal Society (1983) *Risk Assessment: a Study Group Report*. Royal Society, London.

Royal Society (1992) *Risk: Analysis, Perception and Management*. Royal Society, London.

Skorupinski, B. (2002) Putting precaution to debate: about the precautionary principle and participatory technology assessment. *Journal of Agricultural and Environmental Ethics*, **15**, 87–102.

Slovic, P. (1987) Perception of risk. *Science*, **236** (17 April), 280–85.

Slovic, P., Fischhoff, B. and Lichtenstein, S. (1982) Why study risk perception? *Risk Analysis*, **2**, 83–93.

Smith, M.J. (2004) Mad cows and mad money: problems of risk in the making and understanding of policy. *British Journal of Politics and International Relations*, **36** (3), 312–32.

Smith, R. (1999) Opposition to environmental policies: contaminated land in Britain. *Local Environment*, **4** (3), 377–383.

Swets, J.A. (1973) The receiver operating characteristic in psychology. *Science*, **182**, 990–1000.

Syms, P. (2002) *Land, Development and Design*. Blackwell Science Ltd, Oxford.

Tromans and Turrall-Clarke (2000) *Contaminated Land : The New Regime: Part IIa of the Environmental Protection Act 1990*. Sweet & Maxwell, London.

Walker, S. (2002) *The politics of contaminated land: a political history of UK contaminated land policy 1975–2002*. PhD thesis, University of Newcastle.

Weinstein, N.D. (1987) Unrealistic optimism about susceptibility to health problems: conclusions from a community-wide sample. *Journal of Behavioral Medicine*, **10**, 481–500.

Weinstein, N.D. (1989) Optimistic biases about personal risks. *Science*, **246**, 1223–32.

White, M.P. and Eiser, J.R. (2006) Marginal trust in risk managers: building and losing trust following decisions under uncertainty. *Risk Analysis*, **26**, 1187–203.

4

Actor Networks: The Brownfield Merry-Go-Round

Joe Doak and Nikos Karadimitriou

4.1 Introduction

The range of actors involved in brownfield regeneration fluctuates with the scale and complexity of individual sites or regeneration areas, but it is often substantial. They 'come to the table' with an equally disparate range of goals, demands, perceptions, requirements, resources, strategies and constraints. Apart from the landowners, developers and supporting specialists, there is a range of governmental and community agencies and interests who have a 'stake' and specific roles in the regeneration process. Many of the generic interests need to be deconstructed further. Developers, for example, come forward as traders and/or investors, and operate at different spatial scales – international, national, regional or local. Government agencies include English Partnerships, Regional Development Agencies, Urban Development Corporations, Urban Regeneration Companies, the Environment Agency, English Heritage, and the Commission for Architecture and the Built Environment. These organisations and interests regularly form coalitions or partnerships of some kind, or bring other actors into the process such as engineers, and others might draw on the knowledge and expertise of scientists, while local environmental groups can utilise the experience and advice of national and international networks. This formal or informal 'networking' will cement some actors into 'actor networks', and these can be significant in taking brownfield regeneration forward in certain ways. However, such partnerships and networks have the potential to exclude other actors and to be selective in the objectives and agreements that are forged through them.

These network interrelationships and processes do not take place in a vacuum. There are a range of 'forces' structuring their formation and operation. For example, most of the actors involved in brownfield governance are well aware of the economic imperatives that shape the options that are available, and governmental policy and legislation provides constraints, requirements and opportunities which can be significant factors in the network-building and regeneration process. However, this 'structuring process' takes place as part of an iterative set of actions in which such forces influence actor behaviour, but actor decisions (collectively) also shape the structures that impinge on the regeneration process. For instance, masterplans for sites or areas are formulated with certain 'assumptions' built into them (such as implicit or explicit calculations of market demand or landowner 'rights') but these selfsame plans are then likely to (re)structure markets (e.g. change the balance of supply/demand) and redistribute landownerships (e.g. through compulsory purchase, partnerships or planning gain).

A further element to emphasise within these processes is the important role that communication, understanding and trust play in 'cementing' some of the key objectives and approaches to brownfield regeneration. In exploring this, academics have focused on the creation and role of 'discourses' within the regeneration process (Atkinson, 1999; Hastings, 1999; Murdoch and Abram, 2002). Such discourses involve the construction of sets of ideas, meanings and material supports (e.g. planning policies, grant regimes, scientific techniques, management guides, performance indicators) that direct the regeneration process in certain ways, often quite subtly but sometimes more forcefully. As an example, Murdoch (1999) focuses on the evolution of the brownfield debate and:

> looks in some detail at the activities of a 'discourse coalition' in the UK planning sector and seeks to show how this coalition works to 'frame' local decision making through the use of diverse sets of resources. (Murdoch, 1999, p. 51)

Biddulph *et al.* (2003) concentrate on similar processes of discourse construction and reconstruction in their research on the urban village approach to brownfield regeneration. Central to their work is:

> an understanding of how multifarious strands of thinking have become fixed into a seemingly homogeneous concept, how and why this thinking has then been progressively unfixed as actors have tried to transform the concept into both paper planning schemes and built products and finally how the concept has been confronted by contradictory lived experiences. (Biddulph *et al.*, 2003, p. 189)

The important aspect about this emphasis on discourse construction within a web of social (and other) relations is that these constructs provide the framework within which action takes place and these relations are structured and restructured. In this way, 'discourse produces its own "regime of truth" in which knowledge and power are inextricably bound together' (Atkinson, 1999, p. 60) and this framing is then 'materialised' through the application of various resources and technologies (Murdoch, 1999, p. 51). Murdoch goes on to link this 'materiality of discourse' to the assembling of 'heterogeneous resources' through 'actor-network processes', which is similar to the approach developed in the next section of this chapter.

Table 4.1 maps out the range of discourses that permeate the history of brownfield policy-making and implementation. They are packages of ideas and material technologies that have shaped the debates and, more

Table 4.1 Brownfield discourses in UK policy and practice.

Sustainable brownfield development: the sustainability discourse
Key ideas
Environmental efficiency; resource conservation; futurity; quality of life; social equity; participation

Relevant technologies
Urban capacity studies; EIA; SEA; sustainability indicators; BREEAM; life cycle analysis

Market-generated brownfield development: the neoliberal discourse
Key ideas
Competition; economic utility; private sector involvement; polluter pays principle; market demand; owner-occupied housing; cost minimisation

Relevant technologies
Housing land availability studies; development appraisals/valuations; market analysis; privatisation; best value

Negotiated (or equitable) brownfield development: the participatory discourse
Key ideas
Community involvement; empowerment; social need; active citizenship; partnership; lay knowledge

Relevant technologies
Planning for real; forums; partnerships; community surveys; social need assessments; groundwork and other environmental activities

Organisationally efficient brownfield development: the managerial discourse
Key ideas
Good practice; measurement; efficiency; effectiveness; technical guidance.

Relevant technologies
Best value; Public Service Agreements; grant regimes; targets; indicators; good practice guides; development briefs; partnerships

Technically efficient brownfield development: the scientific discourse
Key ideas
Risk assessment; remediation techniques; land use efficiency; expert knowledge-systems

Relevant technologies
Risk assessments; good practice guides; evidenced-based policy; technical specifications and guidance; urban capacity studies; scientist/expert involvement; research papers and projects.

importantly, practices of brownfield regeneration for a number of decades. Each one has evolved, ebbed and flowed in terms of influencing the agenda, and increasingly they have become mixed in different ways in the construction of governmental policy statements and in the regeneration of particular areas and sites.

Another insight from academic research is provided by the renewed interest in the role of the physical environment (and 'things' in more general terms) in shaping human activities. This is explicit in Table 4.1, which places emphasis on a range of technologies influencing how brownfield regeneration is progressed. It also comes to mind when we think of how the 'forces of nature', such as flooding and other extreme weather events, have been influential in raising the profile of climate change and stimulating policy guidance and adaptation measures (see Chapter 11). Indeed, 'contaminants' take on a life of their own within this perspective and interact with humans in complex and unpredictable ways.

So in researching actors and their network relations in the brownfield regeneration process we should take into account a range of insights about how these actor networks are constituted and how they 'work'. In this project we sought to develop this conceptual approach and use it to gain insights into development processes. Our empirical research was concentrated in two case study brownfield sites – Paddington Waterside in London and New Islington, Manchester.

4.2 Actors and their roles

As stated above, the redevelopment of any significant brownfield site or area is likely to involve a constellation of agencies and impact on an equally wide range of interests. Figure 4.1 illustrates the diverse range of potential actors in brownfield regeneration. This highlights the range of actors and interests that the SUBR:IM teams encountered in their research, and illustrates the potential issues of coordination and conflict management within the process (see Chapters 3 and 5).

Some academics have researched (selected) actors in the brownfield development process, David Adams (Adams *et al.*, 2002) and Paul Syms (1999) being the most active in this area. Adams has used eclectic theoretical ideas to explore landownership constraints involved in brownfield development and started to unpack some of the forces that impinge on their room for manoeuvre, while Syms has employed a rather more simplistic actor perspective to identify the range of decision-making factors relevant to development actors but not really unpacking them in relation to structural context or the social relations of actors (i.e. without 'politics'!). Although their findings illustrate the complexity of the interests,

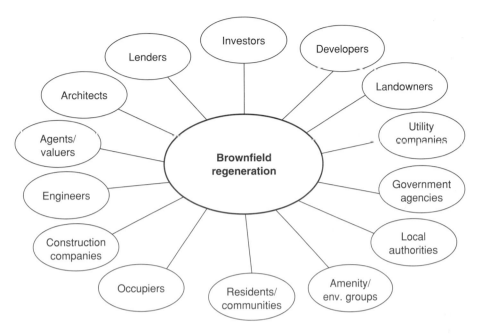

Figure 4.1   The (simplified!) network of actors around brownfield regeneration.

strategies and conflicts involved, they do not fully explore the relational dynamics involved, in which actors socially construct (and re-construct) these various factors and constraints, producing different regeneration strategies and redevelopment 'scripts'.

We can build on Adams's analysis to begin to map out the wider 'driving forces' that provide an important (often determining) context for the strategies and activities of actors. Figure 4.2 illustrates the main drivers that structure the action of actors.

Again this is a relatively simplistic picture, which is explored in different ways by the authors of other chapters. The diagram emphasises a range of drivers. In terms of economic forces, we have market demand/supply influences, creating a frame for financial commitments from market actors (including major investors and development companies but also individual investors or consumers). The market itself is heavily 'organised' into organisations and institutions that regulate and shape the processes involved (not as neo-classical theory would have it, as a set of individual sellers and buyers). The profit-making imperative is close to the hearts and minds of all market actors, but it is played out in a range of different strategies and operationalised though a set of networked interactions as transactions and 'deals' are negotiated and agreed. Given the significance of the extra works usually associated with remediation, development costs and

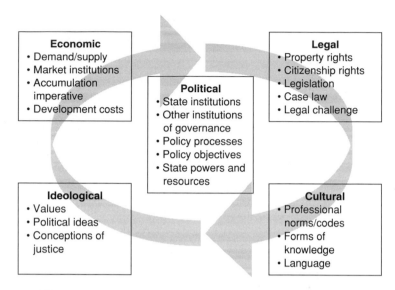

Figure 4.2   The main driving forces structuring the brownfield regeneration process.

risks often weigh heavily in the calculations and strategies of market (and other) actors involved in brownfield regeneration.

However, economic drivers are inherently linked to the other factors shown in the diagram and the circling arrow emphasises this. It suggests that these structural forces are often mutually reinforcing (e.g. state/legal support for private property rights and ideology help facilitate market institutions and relationships; the political culture and challenge of the environmental movement has led to take-up of sustainable development policy by political parties and state institutions, and contributed to emergence of the idea of environmental justice and an increased role for 'corporate social responsibility' in market processes). These forces or drivers don't just 'sit above' the actors in some abstract way, but are made and remade on a daily basis in all the actions of actors. In that sense, these drivers are socially constructed by the actors (produced, reproduced and/or changed) as the brownfield regeneration process unfolds.

As demonstrated through this chapter, effective understanding of these two major processes requires ideas and frameworks which can transcend the duality of 'structure' and 'agency'. Figure 4.3 brings the two aspects together in diagrammatic form, and begins to illustrate the kind of approach that is developed in the next section. A number of models exist that can help in this, such as the kind of discourse analysis alluded to earlier, or the ideas of 'structuration' developed by Giddens (1993) and used in relation to property research by authors such as Healey (1992) and Guy and Henneberry (2000). Similar conceptions pervade the actor-network approach (Law and Hassard,

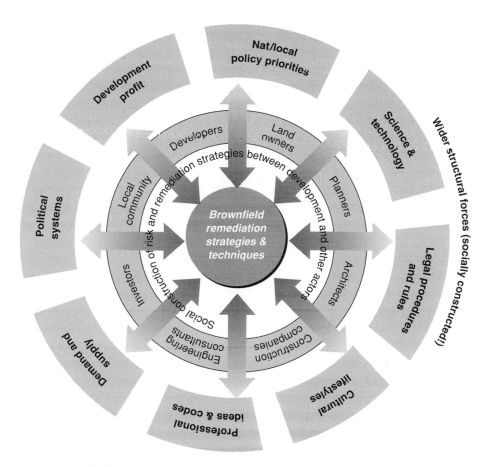

Figure 4.3 Exploring the role of 'agency' and 'structure' in the brownfield regeneration process.

1999), socio-technical systems (Rohracher, 2001) and ideas about 'communicative action', as applied to planning by Healey (1997) and others. They also relate very closely to the emergent set of perspectives provided by complexity theory which are increasingly being applied to the interactions between physical and social phenomena (by writers such as Holland, 1998; Capra, 2002; Urry, 2003).

So we are dealing with a complex set of (network) relationships with a range of inputs and outputs. How can the ideas outlined above help us to conceptualise these relationships which involve combinations of human and material objects, which are both (at the same time) structured and structuring? Our answer is to carefully draw on the bodies of work alluded to above and, in particular, to judiciously combine some of the key ideas from complexity and actor-network thinking.

4.3 Networks and their construction

The holistic concerns of sustainable brownfield regeneration require a commensurably holistic conceptual framework. Complexity theory and the 'Complex Adaptive Systems' approach provide a useful starting point. Although we have expanded on this elsewhere (Doak and Karadimitriou, 2007), we can outline the main components of the approach here.

Complexity thinking specifically considers the diversity of actors and forces operating in brownfield regeneration, and the way in which change emerges out of the dynamic relationship between them. According to Taylor (2001), complex systems comprise many different parts, connected in multiple ways, interacting both serially and in parallel. They display spontaneous self-organisation which gives rise to structures which 'are not necessarily reducible to the interactivity of the components' (Taylor, 2001, p. 142). The system is characterised by emergent qualities which tend to cover the whole system despite the fact that they are generated from local interactions.

Furthermore, complex systems evolve as a result of their openness and adaptiveness. They exist in the fluctuating zone between chaos and order. The 'system' (and the actors who are part of it) survives by both 'imposing' order – to maintain and operate the system – and yet also seeking out new avenues and opportunities, so as to increase the chances of survival into the future. Using the example of institutional investment companies, we can see that certain *schemata* (Gell Mann, 1994) or *standard operating procedures* (Holland, 1998) are developed within these bodies in the form of norms, rules, perceptions or routines in order to structure investment decision-making. These schemata secure ongoing investment returns but in doing so they can blinker investors towards certain new or alternative types of investment opportunity. It will usually require a jolt to the system for these schemata to be changed and then new operating procedures might be constructed and put into place as part of the process of adaptation. Those jolts can come from sources internal or external to the particular 'actor' or 'node' concerned.

The recent shift towards brownfield regeneration is a good example of how a combination of internal jolts (or in complexity terms, 'fluctuations') and external changes ('perturbations') can affect the webs of production and consumption within the built environment. Restrictive planning policies, which have reduced the amount of greenfield land coming forward for development, have affected the resource input into those organisational networks that specialised in such development and depended on the regular release of that type of land in order to make profits. Therefore restrictions in the inflow of greenfield land will lead many of these networks to extinction, while others will have to reorganise their activities and schemata in order to survive. Inevitably, these new emergent networks will need new inflows (of money, materials and human effort) to reach a new state in order to survive. This process, however, can also affect the web in its totality and in relatively unpredictable ways. As suggested, brownfield redevelopment requires the reorganisation of the ways in which actors operate, and this can lead to further changes in the wider webs of production and consumption for the built environment.

This emphasis on 'systems' can appear rather mechanical and 'determinist' but a number of writers have consistently emphasised the fact that complex systems are made up of human (and non-human) actors and other physical/chemical artefacts or materials. Capra (2002, p. 64) for instance has stressed that '[a] full understanding of social phenomena . . . must involve the integration of four perspectives – form, matter, process and meaning'. In our view, 'networks (form) and interaction (process) are the cause and "glue" that give rise to and sustain phenomena, "generating" meaning which is embodied into matter' (Doak and Karadimitriou, 2007, p. 2). A key mechanism for generating and communicating meaning in a complex system is through the construction of discourses which are circulated through social intercourse and material objects (such as conceptual frameworks, diagrams, plans and actual buildings).

In taking forward this focus on the social dimensions of complex systems and the interaction between human and non-human actors, we have also drawn on ideas about actor-network processes (Law and Mol, 2002; Urry, 2003), which are closely aligned to aspects of complexity theory. Actor network theory provides a number of interesting perspectives which shed further light on our understanding of brownfield regeneration. These insights include:

- the 'symmetry principle', which considers the role and effect of network actors or 'actants', irrespective of whether they are human or non-human (e.g. animals or things, such as great crested newts or contaminants)
- treating things as relationally fluid so that they are often combined in the same event or process (i.e. breaking down various dualities, such as local/global, centre/periphery, order/disorder, structure/agency and brownfield/greenfield)
- mapping the process of network-building (or 'translation') using four main stages (problematisation, interessment, enrolment and mobilisation), taking network actors/actants from initial problem identification to network stabilisation
- how networks begin to have enduring effects on the world (Latour, 1993, p. 792) by 'inscribing' the aims of the network into social relations and material things, but with variable levels of success (Akrich, 1992)

- the important role of 'intermediaries' ('anything passing between actors that defines the relationship between them': Callon, 1991, p. 134) in stabilising networks and extending them through the enrolment of other actors
- an emphasis on the relative view of space in which networks are always both local and global, transcending and incorporating global and local space (therefore being inherently 'glocal' in nature)

Combining these perspectives leads us to view brownfield regeneration in a different way. All the actors, material resources, technologies and 'structural' forces are now joined up and interrelated. 'Brownfield' sites themselves are part and parcel of past and existing networks that have created them and, indeed, may still use them for various purposes (such as storing building materials, recycling waste, carrying out motor repairs, walking the dog and even for living accommodation).

The regeneration process begins to build a new set of network relations and draws in other actors, in a way that can be mapped by actor network theory. The key network builders lead the process but they often compromise in various ways as other actors are enrolled and shape its objectives and 'inscriptions'. The new network of actors, money, materials and machines either becomes tied to the existing networks or obliterates them (or reaches some point in between). The new actors themselves are, of course, tied to other networks and these shape their ideas and schemata which will facilitate, shape and constrain the regeneration process in various ways. However, the redevelopment process is rarely a straightforward one; actors will join, others will leave or be excluded, and changing conditions locally and globally will affect possible outcomes.

Although some actors, for instance financial institutions, may not be directly involved in negotiating the redevelopment scheme, their 'enrolment' could be fundamental and their influence on the 'inscription' could be significant, showing the 'strength of weak ties' (Granovetter, 1973). At other times they will be non-existent and other actors will provide the financial resources for regeneration. Here we might expect different outcomes from the redevelopment process as the network of actors seeks alternative objectives based on their particular mix of interests and understandings.

The redevelopment itself will serve a range of functions that reflect the interests, objectives and perspectives forged in the network-building process. This might be in the form of multinational business districts with institutionally funded office complexes, corporate leisure facilities and 'status-enhancing' architecture, or it could involve a different mix of uses aimed at a different type of user applying and 'testing' new ideas or 'schemata'. The outcomes will 'feedback' into the complex system that these networks are part of and send messages to the actors involved and to others. These outcomes will confirm or challenge the schemata that have developed and possibly lead to changes that could spread throughout the system. The research undertaken and described below elaborates on these network processes, and illustrates how this method could be applied and what implications arise from it.

4.4 Network processes in brownfield regeneration

The two case studies reported here involve (to some extent) the two types of network suggested above: one constructed within the ambit of major institutional investors, and the other located in the 'shadow-lands' of corporate capital, even though both networks are linked in 'multiple and serial ways'. They combine elements of description (the nature and operation of network relations) and analysis (the critical evaluation of such processes and the outcomes they deliver). They also provide specific, if tentative, conclusions about the nature and role of investors in that process.

4.4.1 Paddington

Paddington Waterside is located in inner west London, on the edge of the central area and adjacent to Paddington Station. It is made up of a collection of sites with various owners and users (see Figure 4.4). Despite its location, excellent accessibility and some effort on the part of the local authority, earlier redevelopment proposals had failed to get past the 'problematisation' stage of the network-building process. The arrival of the Heathrow Express to Paddington Station (site 9 in Figure 4.4) provided a high-speed link to the airport and essentially brought the brownfield land around the station within easy reach of international business passengers. This, not surprisingly, kick-started a new initiative for the area.

At Paddington Canal Basin (site 2), an architect and an entrepreneurial developer (whom we shall call developer A), in cooperation with a long-established team of consultants, introduced the idea of developing the Basin to a major corporate developer (developer B). The latter acquired the site in a joint venture with A and enrolled new actors into the evolving network (banks, planning consultants, architects, urban designers, etc.) based on a masterplanning proposal created by the architect who helped 're-problematise' the future use of the site, based on the new-found accessibility. At about the same time, Network Rail were making plans for the redevelopment of the train station (site 8) and the National Health Service (NHS) was planning to redevelop St Mary's Hospital (site 7). These adjacent developments were of crucial significance to the quality of the environment

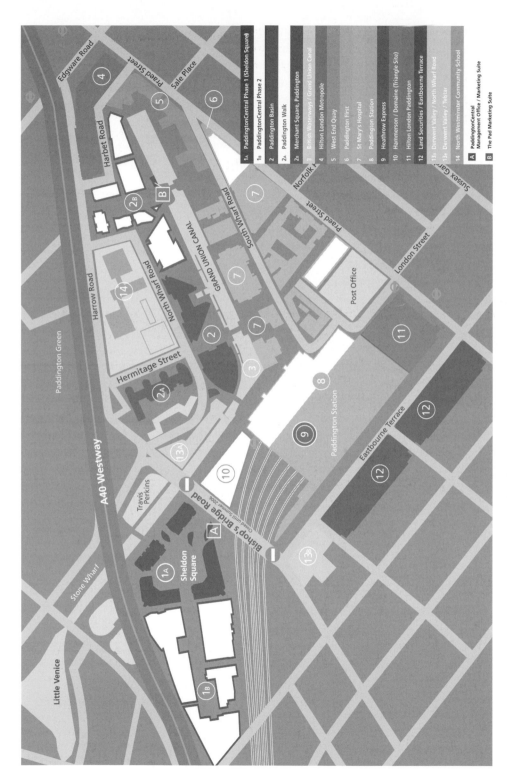

Figure 4.4 Paddington Waterside Development Sites. Source: Paddington Waterside Partnership, 2004, pp. 8–9.

1A PaddingtonCentral Phase 1 (Sheldon Square)
1B PaddingtonCentral Phase 2
2 Paddington Basin
2A Paddington Walk
2B Merchant Square, Paddington
3 British Waterways / Grand Union Canal
4 Hilton London Metropole
5 West End Quay
6 Paddington First
7 St Mary's Hospital
8 Paddington Station
9 Heathrow Express
10 Hammerson / Domaine (Triangle Site)
11 Hilton London Paddington
12 Land Securities / Eastbourne Terrace
13A Derwent Valley / North Wharf Road
13B Derwent Valley / Toktar
14 North Westminster Community School

A PaddingtonCentral
Management Office / Marketing Suite
B The Pad Marketing Suite

and the accessibility and permeability of the Waterside redevelopment scheme. Yet more actors were brought in via the involvement of the NHS and Network Rail in a parallel process of network-building.

In a neighbouring site, PaddingtonCentral (site 1), another major developer (developer C) was promoting its own vision, and bringing in a different set of actors with few direct linkages to the actor network being put together by developers A and B. Senior officers in the local planning authority and the Government Office for London were keen to bring these proposals together in the context of the development plan policies for the area and more site-specific development briefs. Working through a number of forum arrangements, the developers were brought together with other interests and a partnership was brokered, which allowed the differing visions and interests to be aired, negotiated and taken forward. This fragile convergence of interests developed some basic guidelines to shape site development, including a relatively strategic design guide and some minimum provisions for traffic and pedestrian movement. However, the members of the network(s) were not so well enrolled into the new partnership that a common masterplan could be agreed. This resulted in a more 'federal' structure in which each developer prepared its development schemes according to a set of evolving blueprints, albeit in consultation with the other.

This loose network of interests was, and still is, in constant flux. Once the plan/scheme had progressed to a point where a sizable uplift in land values could be realised, developer A sold its site and moved on. Other developers 'enrolled' into the scheme and its evolving vision by buying some of plots and/or undertaking developments, or were introduced to achieve specific tasks, such as initiating a housing element during the early stages. Slowly this evolving network of development (and, indeed, other) actors began to design and implement the scheme according to their variegated visions. These visions were structured by 'messages' and discourses that ran through the planning and development 'system', such as the government's policies supporting high-density brownfield regeneration, particularly around transport nodes, and the perceptions and 'schemata' of those (institutional) investors seeking to maximise the new opportunities for international business space in this 'new' extension of London's West End office market.

Site 5 (on the eastern end of the development) provides an example of how the network begins to inscribe these messages/schemata into the built environment and how these inscriptions start to close down the options for the rest of the area. A 'critical' decision was taken to use this land for luxury apartments and retail, which – in the light of the contemporary patterns of market demand – meant that the development would comprise a disproportionate amount of high-income individuals without ties to the area, many of them living abroad. This injection of a new set of actors into

Figure 4.5   Paddington Waterside.

the social mix of the area had a number of implications and outcomes. The needs and demands of the new residents (even their *assumed* needs) affected the type of retail establishments that could be attracted to the development and excluded serious consideration of other than 'up-market'/aesthetically pleasing uses for the canal (i.e. the transportation of waste and building materials had been suggested by some local people but never taken forward). Barges serving leisure and educational uses were installed along the sides of the canal, replacing existing residential barges and houseboats (Figure 4.5). At the same time, 'necessary' changes to the canal's layout aiming at its aesthetic and functional transformation limited the options for its future use as a commercial waterway.

At Paddington particular meanings and interpretations of 'brownfield redevelopment' were circulating through the networks that came to be involved in the regeneration process. Investor communities stayed at arm's length from this regeneration network and had few direct links with it. Institutional investors and banks plugged their money into it through the 'entrepreneurial expertise' of the developers and did not *directly* influence the process. However, the more indirect influence of the sector was clear and substantial in the way the visions and related 'inscriptions' came to be realised. The investors were drawn in through a series of personal, often imperceptible, ties with the developers and yet they provided a set of invest-ment criteria that pervaded the development process. In some respects they were instrumental in creating an emergent office market beyond the normal confines of the West End. However, systemic fluctuation was facilitated by a mix of 'external' planning and economic/transport policies supporting the growth of Heathrow Airport.

Measuring the 'form' and 'process' of network-building (of problematisation, interessment, enrolment and mobilisation) and exploring its reciprocal effects on material reality (e.g. luxury block of flats) and meaning (e.g. water as something nice to look at) helps us to link up all four elements that Capra (2002) refers to. The network can be strategically 'mapped' at various instances and its composition and structural characteristics can be compared against material reality, as expressed in blueprints for each development, proposed masterplans for the whole area and actual buildings constructed.

Whereas the Paddington Waterside redevelopment was clearly in the ambit and influence of the institutional investment market, the same cannot be said about the second case study. Here we explore a different network-building process, tied to a different set of forces and relations but also influenced by the existence and operations of national/global investment and other processes.

4.4.2 New Islington

The New Islington scheme in East Manchester involves the redevelopment of the former Cardroom Estate, a run-down local authority estate, itself a redevelopment scheme built in the late 1970s. By the 1990s the local community was suffering the effects of depopulation, poor services and high levels of crime. Only half the 204 homes were occupied when the site was identified for development as a Millennium Community in 2001. The site covers 12.5 ha (29 acres) and was previously bordered by two canals (the Ashton and Rochdale canals), some of which had been filled in as part of the earlier Cardroom Estate scheme (see Figure 4.6). The land around the Estate was contaminated in various places.

The proposal to regenerate the Cardroom Estate through a redevelopment scheme emerged from work by Manchester City Council (operating through an urban village company/UVC) in the neighbouring Ancoats area. A leading person in the UVC (the initial 'problemiser') planted the idea of regenerating Cardroom using a similar central government initiative with key members of the Urban Task Force, the city council, English Partnership and local residents (the latter, through a simple 'attitudinal survey'). After some sporadic CPO activity by English Partnerships, the government's Millennium Communities initiative gave the stimulus for more concerted network-building. A successful bid was made and this brought together the

Figure 4.6 New Islington development sites. (Urban Splash).

city council, English Partnerships and local residents (via the establishment of a residents' steering group).

With little or no interest from the major insurance companies, pension funds and investment banks, the challenge was to provide a platform for other forms of funding. However, in many respects (and certainly for most of the actors involved) the lack of interest from the major institutions opened up a more diverse range of opportunities for innovative forms of development and a mixture of activities. Indeed, this was reflected in the process by which the lead developer was 'enrolled' into the network. Based on previous regeneration experience in Manchester and elsewhere, the initial 'problemiser' encouraged a process in which the vision for the area was kept open, and the partners concentrated more on the cultural and other attributes of the competing developers. This led to the choice of a Manchester-based 'progressive' development company that was committed to an open and innovative style of place-making that suited both the experiential learning of the problemiser, and also the public policy goals of English Partnerships and the interests of the local residents.

Significantly, English Partnerships agreed to fund the necessary land purchase, remediation and infrastructural work as part of the national Millennium Communities programme. This was a fundamental underwriting of development risk and secured the necessary involvement and commitment of the developers and other interests, including the residents, who were inherently suspicious of the plans for the area given past promises and proposals that had floundered.

The process of network-building continued and the lead developer took centre stage in the process of actor enrolment and mobilisation. The 'cultural requirement' for openness and innovation was maintained in the selection of the lead architect/designers and site engineers, and as actors joined they were moved into a strong communicative relationship with 'The Client Group' who led the direction and work of the network. This group was made up of the lead developer, English Partnerships and the city council, represented through their arm's-length New East Manchester Urban Regeneration Company. Meanwhile the (locally elected) residents' steering group became a key sounding-board for the ideas and proposals being developed for the area. It took on an increasingly prominent role in the regeneration process when it was given the responsibility of selecting and briefing the social housing developer required to build the replacement accommodation for the Cardroom Estate.

A key event in the network process was the agreement of a 'strategic planning framework' for the area in 2001. The lead architect was central to this, although the ideas were worked out and tested through a wide-ranging stakeholder group, which included the residents' representatives. The process was characterised by innovation and 'blue skies thinking', which was

Figure 4.7   New Islington. (Urban Splash).

facilitated by a number of things: the Millennium Communities policies/ guidelines; the culture and experience of the lead developer and its design team; the loose and non-constraining funding schemata (from English Partnerships and the lead developer, who was putting together a package of funds from banks and private investors); and the relative openness of the stakeholder process.

An example of the ability of this network to accommodate challenging viewpoints is provided by the influence of the existing residents of the Cardroom Estate who selected and worked closely with the social housing developer. They rejected the high-density vision of brownfield housing development propagated by government policies (a key component of the sustainability discourse outlined in Table 4.1) and demanded individual 'houses' at much lower densities than 'required' under the Millennium Communities guidance. They also insisted that the social housing should be built on some of the 'prime' residential plots in the scheme, and again this was accommodated after some difficult debates (Figure 4.7).

Subsequent 'inscriptions' of this network process have included a general, unquantified, package of mixed tenure housing; high-quality public space; an (expensive) redesign of the canal network; an innovative onsite remediation solution which involves some elements of bioremediation; the imposition of traffic calming/reduction measures using an extensive system of shared road space; and the provision of local employment opportunities. The 'sustainability discourse' is also being taken forward into built form through the use of combined heat and power (CHP), high standards of energy efficiency, reduction of water consumption, green specification of materials and reduction of construction waste, design for life cycle adaptability and increased onsite biodiversity. When the redevelopment is complete it is intended that there will be up to 1734 new homes (including a range of social and affordable housing) alongside commercial space for new shops, pubs, restaurants, cafes, bars and a number of facilities, which will include a new canal and water park, a primary school and a state-of-the-art primary health clinic.

What is also significant about this 'local' example of 'sustainable property development' is that the site is being scrutinised closely by the wider property world, with institutional funders beginning to take an interest in the scheme. Indeed, a specialist regeneration fund has now invested money in the development, and others could follow. Similarly the private investor market has been quite active in helping to kick-start demand in New Islington. The lack of awareness (on the part of investors) and the suspicions (on the part of the development partners) appear to be receding somewhat as the investment opportunities become clearer. However, this is only the niche end of the investor network – those companies and individuals who have a different set of investment schemata (inclined towards riskier direct investments) from those of the major players.

There are, of course, dangers to the sustainability of this particular scheme as it unfolds into built forms that are structured by market institutions and forces. Area-wide regeneration often results in spatially variable gentrification, as other Manchester examples such as Salford Quays have shown (see Chapter 6). Such 'structuring forces' could sweep away the delicate social mix that is currently being implemented at New Islington. The danger is already apparent in house price inflation, and the increasing private investor interest in capital speculation rather than in securing steady returns through residential letting. This illustrates that wider actors and forces are relevant to the outcomes at New Islington, but also that New Islington will affect the wider webs of production and consumption, maybe in quite significant ways.

4.5 Conclusions

The 'complex adaptive network-building' approach to brownfield regeneration expounded in the above case studies shows some of the important dynamics involved in conceptualising and implementing such projects. It places the actors and their interrelations at the centre of the dynamic but does not deny their relations with wider networks, which interact in multiple and serial ways. These actors communicate and negotiate the meanings that go to construct their plans and the physical and semiotic outputs of the networks they form. They draw on and reconstruct broader discourses of brownfield regeneration, depending on their histories and interests. Sometimes, in schemes such as Paddington, the visions and discourses prioritise a more free-market set of ideas and technologies, although there was evidence of other participatory and managerial discourses influencing the process. In other cases, such as at New Islington, the nature of the network and its wider set of relations allows for the construction of a set of principles, ideas and technologies that are more akin to the 'sustainability discourse' outlined in Table 4.1. Again, this is not a pure form and other discourses continually jostle for influence through the different actors and forces that make up the complex web of relations.

In studying these cases we have seen how network schemata (or decision-making rules and techniques) are applied, based on past experience, but also how those 'operating procedures' are modified or challenged by network actors or changed circumstances (network fluctuations or perturbations). In 'agreeing', changing and using these schemata the network puts its visions into practice and 'inscribes' its agreements into material reality, by remediating land (in certain ways), building housing (of certain types) and providing business and other commercial space (to supply certain demands/needs). All this output feeds back into the system in multiple ways and contributes to the next round of activity, carried through the network (imperfectly) on the back of corporate accounts, individuals' lived experiences, professional meetings, conference presentations, web content, academic case studies and so forth.

Although the two case studies illustrate *particular* networks and outcomes within a wide spectrum, they cast light on the important role of financial actors in structuring the network negotiations and agreements that take place. Despite their limited direct involvement in the two cases reported here, their participation or non-participation is likely to have a number of implications. These relate to a set of generally risk-averse schemata (operationalised in investment strategies, allocation rules, appraisal software, reporting requirements, etc.) that pervade institutional investment decision-making, schemata that 'make sense' given the pressures those organisations are under to protect the returns of their policy- or shareholders, and that have stood them in good stead in the past. Despite recent initiatives such as the Igloo Urban Regeneration Fund, it has been difficult to attract institutional investment to brownfield locations in the UK. Few institutions have invested in physical regeneration projects in inner city areas because of the perceived range of uncertainties and the related schemata mentioned above.

However, the 'rules of the game' are changing in a number of ways and our interviews with investors and other actors suggest that there is an 'emergent' brownfield development network that is tentatively expanding on the back of these changes. A select but growing band of developers and 'specialist' investment funds have now worked together on a number of brownfield projects; they have developed experience and expertise in this area of practice and are reaping significant financial returns. Some of the major financial institutions are also becoming aware of the range of potential 'returns' from this kind of development and/or investment. Financially these sites and areas may be able to help diversify property portfolios, provide higher returns than less risky investments, create new areas of demand (i.e. create new markets), and support the investors' growing commitment to 'socially responsible investment'.

Researchers at the University of Ulster (Adair *et al.*, 2006) have proposed that there is a need to develop new investment vehicles to draw institutional investors into the brownfield redevelopment process. Although this addresses an obvious 'strategic gap' in the process, the research presented here (and also confirmed by Adair and his colleagues) reveals that investment fund managers do not particularly differentiate between brownfield and non-brownfield assets but treat each according to its merits (in relation to portfolio strategies and schemata). Indeed, as we have seen, investors have placed funds into some brownfield redevelopment sites and the existing 'brownfield development network' is sending its messages, techniques and built forms through the wider investment community, as was illustrated in both case studies. The network is being consolidated, and expanded, through ongoing joint working, personal friendships and social gatherings as well as joint memberships of organisations and attendance at regeneration conferences and workshops. These personalise the messages, and glue the network together using that very important social entity of trust.

These findings place some emphasis on the 'soft infrastructure' of investment processes in which personal interrelationships and actor networking are important determinants of the direction and scale of financial investment into brownfield regeneration areas. They sit alongside (and within) the more technical and strategic requirements for appropriate investment vehicles and the economic and legal drivers for increasing the flow of money into such redevelopments.

References

Adair, A.S., Berry, J.N., McGreal, W.S., Hutchinson, N. and Allan, S. (2006) *Institutional Investment in Regeneration: Necessary Conditions for Effective Funding.* IPD, London.

Adams, D., Disberry, A., Hutchinson, N. and Munjoma, T. (2002) Brownfield land: owner characteristics, attitudes and networks. In: *Planning in the UK: Agendas for the New Millennium* (eds Y. Rydin and A. Thornley), pp. 317–36. Ashgate, Aldershot.

Akrich, M. (1992) The description of technical objects. In: *Shaping Technology – Building Society: Studies in Socio-technical Change* (eds W.E. Bijker and J. Law). MIT Press, Cambridge, MA.

Atkinson, R. (1999) Discourses of partnership and empowerment in contemporary British urban regeneration. *Urban Studies*, **36** (1), 59–72.

Biddulph, M.J., Franklin, B. and Tait, M. (2003) From concept to completion: a critical analysis of the urban village. *Town Planning Review*, **74** (2), 165–93.

Callon, M. (1991) Techno-economic networks and irreversibility. In *A Sociology of Monsters, Essays on Power, Technology and Domination* (ed. J. Law), pp. 132–61. Routledge, London.

Capra, F. (2002) *The Hidden Connections: A Science for Sustainable Living*. Flamingo, London.

Doak, J. and Karadimitriou, N. (2007) (Re)development, complexity and networks: a framework for research. *Urban Studies*, **44** (2), 1–22.

Gell Mann, M. (1994) *The Quark and the Jaguar: Adventures in the Simple and the Complex*. W. H. Freeman, New York.

Giddens, A. (1993) *New Rules of Sociological Method*, 2nd edition. Polity Press, Cambridge.

Granovetter, M. (1973) The strength of weak ties. *American Journal of Sociology*, **78**, 1360–80.

Guy, S. and Henneberry, J. (2000) Understanding urban development processes: integrating the economic and the social in property research. *Urban Studies*, **37** (13), 2399–416.

Hastings, A. (1999) Analysing power relations in partnerships: is there a role for discourse analysis?' *Urban Studies*, **36** (1), 91–106.

Healey, P. (1992) An institutional model of the development process. *Journal of Property Research*, **9**, 33–44.

Healey, P. (1997) *Collaborative Planning: Shaping Places in Fragmented Societies*. Macmillan, Basingstoke.

Holland, J. (1998) *Emergence: From Chaos to Order*. Oxford University Press, Oxford.

Latour, B. (1993) *We Have Never Been Modern*. Harvester Wheatsheaf, Hemel Hempstead.

Law, J. and Hassard, J. (eds) (1999) *Actor Network Theory and After*. Blackwell, Oxford.

Law, J. and Mol, A. (eds) (2002) *Complexities: Social Studies of Knowledge Practices*. Duke University Press, Durham, NC.

Murdoch, J. (2001) Putting discourse in its place: planning, sustainability and the urban capacity study. *Area* (2004) **36.1**, 50–8.

Murdoch, J. and Abram, S. (2002) *Rationalities of Planning: Development Versus Environment in Planning for Housing*. Ashgate, Basingstoke.

Paddington Waterside Partnership (2004) *Introducing Paddington Waterside*, London, PWP.

Rohracher, H. (2001) Managing the technological transition to sustainable construction of buildings: a socio-technical perspective. *Technology Analysis & Strategic Management*, **13** (1), 137–50.

Syms, P. (1999) Redeveloping brownfield land: the decision-making process. *Journal of Property Investment and Finance*, **17** (5), 481–500.

Taylor, M. (2001) *The Moment of Complexity: Emerging Network Culture*. University of Chicago Press, Chicago.

Urry, J. (2003) *Global Complexity*. Polity Press, Cambridge.

5

Heroes or Villains? The Role of the UK Property Development Industry in Sustainable Urban Brownfield Regeneration

Tim Dixon

5.1 Introduction

The UK property development industry is a key player in regeneration projects, which conventionally have been focused around area-based initiatives. Emerging policy has attempted to interlink a 'sustainable development' agenda with a 'sustainable brownfield' agenda. These twin policy agendas, based on further normative policy aspirations, have found an additional focus through the UK government's 'Sustainable Communities Plan', predicated on the drive to provide additional housing (underpinned by the Barker Review (Barker, 2003; 2004)), and engender market renewal in key areas in England.

Given the importance of property's dual role in the economy, not only as a means of production and physical regeneration but also as a means of wealth ownership, the UK property development sector, comprising financial institutions such as pension funds, insurance companies and property companies (including investor/developers and housebuilding companies) has the power and capacity to influence patterns of economic activity as well as affect wealth and income distribution through engagement in urban regeneration. Examining the role of the development industry and its interaction with other key stakeholders in the brownfield regeneration process is therefore vital to an understanding of how the dynamics of brownfield recycling and regeneration works in practice.

This chapter examines the emergence of these agendas and related policies, and analyses the role of the property development industry nationally within the UK in terms of its engagement with the brownfield regeneration and sustainability agendas and also sub-regionally in the regeneration of six key brownfield sites, based in Thames Gateway and Greater Manchester. The chapter also investigates the perceived 'sustainability' of these projects, in terms of key aspects of the 'triple bottom line' approach and the development of a related conceptual model, and highlights key lessons emerging from the experience of developers, and their engagement and dialogue with other stakeholders in the brownfield regeneration process on these case study sites.

A key message emerging from the research is that although the property development industry is coming to terms with brownfield risks, including contamination, major policy barriers are hampering effective regeneration. Yet all stakeholders have a role to play in the development process, and a key issue is the extent to which developers really are engaging with the sustainability agenda. To simplify and characterise the debate, are developers 'heroes' or 'villains'?

In summary this chapter focuses on:

- the nature and challenge of brownfield development
- the role of the development industry and its attitudes towards brownfield regeneration
- how we can best learn from current practice in two key areas of brownfield regeneration (Thames Gateway and Greater Manchester)

5.2 The nature and challenge of brownfield development

The UK Labour government has placed a strong emphasis on brownfield[1] recycling as a foundation of urban regeneration, linked strongly with the concept of 'sustainable development'. This approach highlights the importance of reusing and recycling brownfield land, not only to improve urban environments but also to relieve development pressures in the countryside. The twin policy mantras of *sustainable development* and *brownfield regeneration* have therefore dominated the debate on urban redevelopment since the mid 1990s.

Traditionally, regeneration in the UK has been characterised by area-based initiatives driven largely by the property development industry, but often in close partnership with the public sector. The redevelopment of brownfield sites has been seen as a 'good' thing, preventing urban sprawl, keeping cities compact and reducing out-migration. This has led to a marrying of the brownfield and sustainability concepts to underpin a vision of *sustainable brownfield regeneration*.

Within the UK, the role of brownfield regeneration continues to be important and has been given a new resonance because of the focus of government policy on sustainable communities (through the Sustainable Communities Plan) and Kate Barker's review of the UK housebuilding industry (Barker, 2003; 2004). Williams and Dair (2005) highlight the evolution of brownfield policy in England. This first found a focus through Planning for the Communities of the Future (DETR, 1998), and was further developed through the Government's Urban White Paper (DETR, 2000, p. 29), which stated that it aimed to:

> accommodate the new homes we need . . . through a strategy that uses the available land, including, in particular, brownfield land and existing buildings in urban areas.

The brownfield agenda has also been underpinned through the Planning Directorate of the Office of the Deputy Prime Minister, which seeks 'to promote a sustainable pattern of physical development and land and property use in cities, towns and the countryside' (ODPM, 2001), and furthermore through planning policy guidance (PPG3, ODPM, 2000) and more recently PPS3), which has also reinforced the message on brownfield recycling, together with the key quality of life indicator, relating to land reuse (H25, HM Government, 2005).

As a result of the emergence of the sustainable development and brownfield regeneration agendas in the UK, there has been increased debate over the concept of *sustainable brownfield regeneration*. Inevitably this concept is founded on the three pillars model of sustainable development. RESCUE (2003) provide a helpful EU-wide definition of sustainable brownfield regeneration in this respect, which sets brownfields within a 'triple bottom line' framework:

> The management, rehabilitation and return to beneficial use of brownfields in such a manner as to ensure the attainment and continued satisfaction of human needs for present and future generations in environmentally sensitive, economically viable, institutionally robust and socially acceptable ways within the particular regional context.

Similarly, Williams and Dair (2005) suggest a sustainable brownfield development is a:

> development that has been produced in a sustainable way (e.g. in terms of design, construction and participation processes) and enables people and organisations involved in the end use of the site to act in a sustainable way.

Table 5.1 The benefits of brownfield regeneration.

Economic	Social	Environmental
Creation and retention of employment opportunities	Improved quality of life in neighbourhoods	Reduced urban sprawl pressures on greenfield sites
Increased competitiveness for cities	Removal of threats to human health and safety	Restoration of environmental quality
Increased export potential for clean-up technologies	Access to affordable housing	Improved air quality and reduced greenhouse gas emissions
Increased tax base		

Source: adapted from National Round Table on Environment and Economy (2003)

This has also found resonance at an international level. As Table 5.1 points out, experience from Canada, the USA and Europe suggests that, while specific circumstances vary, the key benefits of brownfield regeneration within a 'triple bottom line' model share common features.

However, Pahlen and Franz (2005) also highlight the fact that sustainability is neither static in time nor does it imply a fixed spatial perspective, in that it has to balance short- and long-term effects over generations and also has political, administrative and functional impacts at a local, regional, national and global level.

It is also a widely held view that the property and construction industry has been slow to react to the challenges of sustainability. A workshop for the DTI (Davis Langdon Consultancy, 2003) highlighted key findings from the Sustainable Construction Taskforce (2001) report, and found that although the social and environmental benefits of sustainability had been highlighted, not enough had been done to demonstrate the economic benefits, especially from the property investment point of view. Moreover, many initiatives had focused on 'pushing through' sustainable development, although the 'pull through' by property investors is currently limited. This was highlighted as a 'circle of blame', whereby investors claimed they would fund more sustainable developments if the market asked for them, but the construction industry said they are not asked by the development industry to build sustainable developments. The housebuilding industry has faced particular criticism regarding its effectiveness in mainstreaming sustainable development policies. The industry is an important part of the development and construction sector in the UK, and the majority of new and existing housing in the UK has significant environmental impacts (WWF, 2003; Environmental Audit Committee, 2005). Housing also contributes about 27% of carbon dioxide emissions in the UK with commensurate implications for climate change, and has significant impact on timber and water consumption (WWF, 2003; Entec, 2004).

5.3 The role of the UK property development industry in brownfield regeneration

On the face of it, the property development industry seems to have come to terms with brownfield regeneration. Current statistics show that the brownfield land total is about 64 000 ha in England, with some 16 500 ha comprising 'hardcore'[2] sites (English Partnerships, 2003), some of which may have contamination issues associated with them from an industrial legacy. The government's national target is that by 2008, 60% of new dwellings should be provided on 'previously developed' land (or brownfield land), and through the conversion of existing buildings. In 2005, some 74% of new dwellings were built on previously developed land (including conversions),[3] compared with 56% in 1993 (ODPM, 2005). As Figure 5.1 shows, the total number of new dwellings completed on brownfield sites was relatively stable between 1997 and 2001, although the absolute total appears to have increased more recently, with a bottoming out of 'greenfield' completions.

Currently the 60% target applies only to England and Northern Ireland (Department for Regional Development, 2001), and although planning policy in Scotland and Wales, for example, 'prefers' the recycling of vacant or derelict land to greenfield development, no equivalent targets currently exist in these constituent countries.

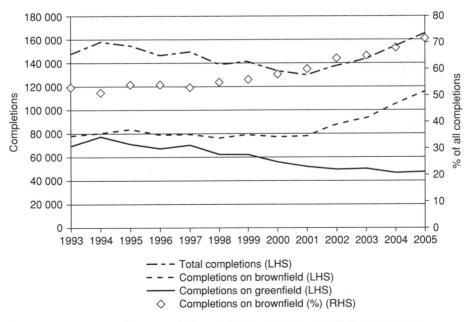

Figure 5.1 New dwelling completions in England (1993–2004) (ODPM, 2005).

However, as Williams and Dair (2005) point out, such targets (which focus on the quantity of land retained) have also been underpinned by increasing attention to the quality of brownfield developments. An example is the new PPS1 guidance in England, which now stresses that all developments must fully address sustainable development principles. On the other hand, as the same authors also suggest, for those wishing to deliver sustainable development, translating policy objectives into practice at a site level can be difficult because guidance and policy is often imprecise and open to interpretation, and sustainability is often not defined in precise terms. Moreover, it is not always clear which elements of a development need to be sustainable: for example, is it the land remediation process, the planning process, or the buildings themselves?

Figures such as those cited above seem to suggest targets are being met, but how are developers reacting on the ground to the challenges of brownfield development? This was the subject of a major SUBR:IM survey of commercial and residential developers carried out in mid 2004 (158 respondents in the stage 1 survey and 94 in the stage 2 survey), with follow-up interviews with 11 national developers during 2004–2005 (Shephard and Dixon, 2004; Dixon *et al.*, 2005a). Further details on the methodology are given in Appendix 1 of this chapter.

5.4 Survey and interview findings

The survey results confirmed that brownfield development is now wide-spread throughout the housebuilding industry. For example, more than 80% of the sample of developers developed entirely on brownfield sites (Figure 5.2). It was already apparent that brownfield development was no longer the preserve of specialists and had been adopted by volume house-builders (i.e. larger housebuilders were building some 50–74% of their units on brownfields). Findings from the survey also show that smaller and medium-sized operators have clearly shifted their output towards brownfield.

Given the policy emphasis on brownfield development, it is not sur-prising that housebuilders of all sizes are undertaking schemes on previ-ously developed land, to a greater or lesser degree. Maintaining output on greenfield sites has become increasingly difficult in the recent planning climate. Indeed, the 'availability of land' or 'government policy' (which underpins the former) were the key reasons given by the majority of devel-opers for increasing their output on brownfield over recent years. How-ever, the move towards brownfield development has not been solely policy-driven; a significant proportion of developers – both commercial and residential – viewed it as an opportunity for profitable development in what has been a relatively buoyant property market.

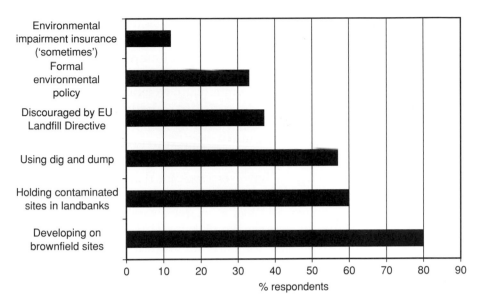

Figure 5.2   Key findings from survey of developers (based on Shephard and Dixon, 2004).

At present, there appears to be a clear intention among developers to continue to increase the amount of brownfield development they are undertaking, and for housebuilders this was supported by the composition of their land banks in which brownfield accounted for, on average, 70% of total plots.

Developing on sites with contamination is likely to become increasingly important if the brownfield target is to be sustained. The survey findings show that developers in both the commercial and residential sectors are clearly not averse to developing on contaminated sites. Practically all the survey respondents were prepared to undertake development on sites requiring remedial treatment and around three-quarters had actually developed on contaminated sites over the past year. Smaller developers are less likely to undertake schemes on contaminated sites; this is not unexpected given that they may not have the resources, the specialist knowledge or the financial reserves to carry the additional risks involved.

A majority of housebuilders (60%) were prepared to hold contaminated sites in their land banks (Figure 5.2). Attitudes towards contaminated land clearly appear to have changed as housebuilders have gained more experience of developing on brownfield sites. Appropriate insurance can reduce risk. However, although contractor warranties and insurance were commonplace, only 12% of developers used ('sometimes') environmental impairment liability insurance.

The readiness of the development industry to tackle contaminated sites could, however, be threatened by the impact of the EU Landfill Directive.[4]

Some 32% of developers and over two-fifths of housebuilders (37% of respondents overall) were likely to be discouraged from undertaking development on sites with contamination, following the implementation of the Directive (Figure 5.2). This was particularly true of smaller housebuilders and those without experience of commercial development. Commercial developers were less likely to be dissuaded from building on contaminated sites, but the Directive is clearly causing some uncertainty in the industry, because 'dig and dump' is still the most frequently used method of dealing with contamination, with some 57% using the technique, 'often' or 'always' (Figure 5.2). There is, however, evidence that some in situ treatments are being used, most commonly barrier methods and containment. Commercial developers typically had a greater awareness of alternative remediation techniques than housebuilders and were more likely to have experimented with them, particularly solidification/stabilisation and soil vapour extraction. Other techniques were generally used much less frequently.

The EU Directive does appear to have stimulated some interest in exploring alternatives to landfill; just over half of developers interviewed said they were doing this. Of the remainder, around half stated that they were also likely to continue developing on contaminated land, suggesting that they already have sufficient knowledge of alternatives to landfill. In terms of access to independent sources of information on remediation treatments, the majority of developers did not consider this to be a problem. Smaller housebuilders were less likely to share this view, and this could suggest that there is a greater role for government bodies such as the Environment Agency to publicise and disseminate information more widely.

Although the development industry is playing an influential role in the 'sustainability' agenda, there is a degree of scepticism over an agreed, industry-wide definition, and this may hinder its implementation. Indeed, only a third of respondents in the survey had a formal environmental policy (Figure 5.2). Developers seem to be adopting a proactive approach to defining sustainability on their own terms; developers' interpretations subsequently vary. Motivated by their efforts to comply with sustainability requirements for gaining planning applications, developers frequently concentrate on environmental and social objectives, although there is also a keen focus on the economic sustainability of the scheme, often limited to the end product itself rather than the economic vitality of the surrounding area.

The interviews also revealed varying perceptions regarding sustainability in brownfield regeneration. When developers were asked to define the term 'sustainability' in relation to their brownfield projects, the responses received were very varied, and a degree of scepticism was noted during the interviews on the sustainability concept promoted by government.

One housebuilder remarked 'the problem is that there is no one definition of sustainability' while another developer saw this concept as merely a 'political wish list'. However, in keeping with pressures from government policies and initiatives, developers have made attempts to formulate their own interpretation of the sustainability concept, which can be synthesised into the following main themes from the interviews, as follows (Dixon, 2006):

- **Environmental**
  - Sustainability related to the ecological impacts of the development project, which could mean either trying to impose as little environmental damage as possible during the course of development, or not leaving a long-term legacy of detrimental impact to the environment after completion.
  - Sustainability of the end products, translated as developing properties that are energy-efficient and made of sustainable materials; also promoting longevity to 'meet and exceed their planned life expectancy', as well as recognising the importance of the design that could actually address the real needs of the users.
- **Economic**
  - Sustainability of the development project. As one commercial developer stated, 'for our business to be sustainable, it has to operate within [the] financial framework'. This included looking at ways of developing in the most cost-effective way possible, and highlights the importance of the economic 'pillar' in the triple bottom line approach. This appeared to be more important than the wider economic benefits within a community.
- **Social**
  - Sustainability related to the site and its surroundings, such as location of development projects with high proximity to public transport links, schools and health services.
  - Sustainability related to the community in the development area, as implied by several housebuilders. As one suggested: 'keeping all the employment, keeping schools open, ensuring transports links are all in place'.

The results from this research also suggest that the development industry seems to be, in some instances, merely paying 'lip service' to the concept of integrating sustainability in residential and commercial developments, which remains 'contested' in nature. The relatively low level of uptake of environmental policies in the survey was symptomatic of this, for example. A similar picture emerged in relation to environmental standards, where the number of housebuilders incorporating Ecohome[5] standards was low. There are some important issues in relation to this, however (Dixon, 2006):

- There was a general lack of awareness, if not of BREEAM,[6] then of the exact standards required, especially among housebuilders.
- There was some evidence (although not statistically proven) that developers involved in commercial development (which includes a housing component) might be more likely to have had a higher proportion of houses meeting BREEAM or Ecohome standards than those focusing solely on housebuilding. This may perhaps reflect a greater commercial imperative, or simply a difference in approach to sustainability issues.
- Some developers claim they design to BREEAM standards but do not apply for accreditation owing to cost.
- Some housebuilders suggested that the cost of implementing higher environmental standards would not be met by house purchasers who are not prepared to pay extra for them.

Transport and accessibility were ranked highly as sustainability components, ahead of, for example, affordable housing in brownfield developments. More proactive developers have seized the opportunity to promote their own understanding of sustainability criteria, but there remains much uncertainty over the concept, and the research shows that a clearer definition of sustainability in relation to brownfield regeneration, and better guidance and standards for development projects, are needed to guide the development industry in this respect.

In relation to the survey work and national interviews, therefore, we see an industry coming to terms with brownfield regeneration, but with a dissipated approach towards full engagement with the sustainability agenda. But what is the reality on the ground at a case study level? The next section seeks to address this question in more detail.

5.5 Learning from practice: Thames Gateway and Greater Manchester

5.5.1 Conceptualising sustainable brownfield regeneration: a case study approach

A variety of models have been constructed to view the development of new buildings and spaces. Traditional models of the development process, from event sequence to agency, structure and institutional models, all offer different perspectives (see, for example, classifications by Healey, 1991; Guy and Henneberry, 2002). In contrast, conceptual models of brownfield regeneration have focused on particular aspects of redevelopment rather than seeking to examine the process as a whole. Some models have sought to explain the process in terms of 'stock and flow' (for example, CABERNET, 2003) or in terms of 'financial viability' (for example, English Partnerships, 2003).

An earlier model of brownfield regeneration developed by POST (1998) sought to adopt a more holistic stance, and is essentially a structure process-based model which identifies the drivers, barriers and risks involved in the policy/regulatory framework surrounding brownfield development. This model is described as a 'three-way dynamic', highlighting the tensions between the 'policy push' of regeneration and sustainability aims, the 'development frictions' in terms of the obstacles and uncertainties faced by developers, and the 'opportunity pull' of sustainable communities and investment returns (POST, 1998). The POST model does not attempt to define the interrelationships between different actors in the development process (see Chapter 4) or to identify resources flows between sectors but rather to provide a structure within which brownfield regeneration can be analysed either in a generic sense or in relation to a specific locality. It could be regarded as an alternative illustration of Healey's (1992) 'rules, resources and ideas', particularly in terms of how conflicting policies can cause development friction and how uncertainty caused by insufficient data and knowledge can hinder the progress of policy aspirations.

The research design for the case studies required a framework that combined the components of sustainable brownfield regeneration, and the drivers and barriers to that goal, with the focus of the research being on the development industry's responses to incorporating sustainability in brownfield regeneration projects. The framework in Figure 5.3 was therefore adopted because it combines sustainability with brownfield regeneration.

The overall aim of the research was to analyse the perceptions, attitudes and practices of the development industry in relation to sustainable development on brownfield sites. Therefore the research was not designed to measure sustainable development empirically, but rather to examine the extent to which developers are engaging with the sustainable development agenda, their attitudes to the agenda, and the drivers and barriers to it. The research was also designed to focus on the 'tensions' that can arise between these three facets of sustainability in the presence of 'push' factors (for example, policy, and contamination issues) and 'pull' factors (for example, profitability issues).[7]

The key questions addressed through the case studies (and which were triangulated from previous, related research) therefore were as follows:

- **Market impacts (within the 'economic' pillar).** For example:
  - How important are density and affordability of housing and supporting infrastructure for the viability of a brownfield scheme?

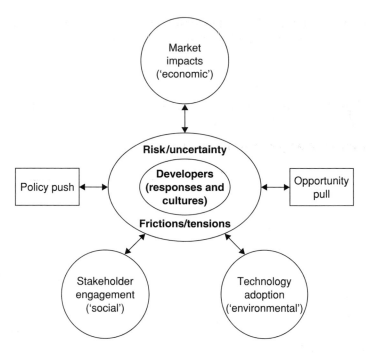

Figure 5.3 Conceptual framework.

- ○ What determines the economic viability of a brownfield scheme, and how best can a viable mix of housing be provided?
- ○ How important is the 'brand and image' of an area?
- **Stakeholder engagement (within the 'social' pillar).** For example:
  - ○ How do developers engage with other stakeholders (including national, local and regional government, agencies, remediation consultants, the public) during the brownfield regeneration process?
  - ○ What are developers' cultures and responses to brownfield regeneration?
  - ○ What are the most appropriate partnering/joint ventures schemes that are used?
  - ○ What are the drivers, tensions or frictions that may arise during the brownfield development process?
- **Clean-up and construction technology adoption (within the 'environmental pillar').** For example:
  - ○ What determines the type of remediation (or clean-up) technology used?
  - ○ How does the development industry view clean-up technology?
  - ○ How sustainable are the construction methods used?

Based on more than 50 interviews with key stakeholders, research at Oxford Brookes examined six sites in these two sub-regions and reveals some important implications for sustainable brownfield regeneration within the three pillars of sustainability (Dixon *et al.*, 2006).

5.5.2 The case study sub-regions

Building on the survey of developers in the UK, Thames Gateway and Greater Manchester provided the SUBR:IM research consortium with a rich laboratory for scientists and social scientists to study examples of best practice brownfield regeneration on a number of sites (New Islington, Higher Broughton and Hulme in Greater Manchester, and Barking Reach, Gascoigne Estate and South Dagenham (West) in the Thames Gateway), and to highlight those elements that work, and those that are not so successful.

Thames Gateway is perhaps the most ambitious regeneration programme undertaken in the UK. Set to deliver 120 000 new homes by 2016, with associated jobs and infrastructure, the development is a key part of the government's Sustainable Communities Plan. As one of three of our case study examples in Thames Gateway, Barking Reach, with its site conditions and related problems (for example, overhead pylons and layered peat) but with huge potential for growth, is the largest brownfield regeneration project in Europe (350 ha).

Within Greater Manchester, both Manchester and Salford have also received increased government and media attention as a result of the Northern Way initiative[8] and the Sustainable Communities agenda. Furthermore, the existence of a Housing Market Renewal Pathfinder[9] in Salford makes the locality a pertinent one to study. With three case studies located in close proximity to the city centre, these areas face many challenges. For example, the site for New Islington, part of English Partnership's Millennium Community portfolio, has suffered greatly from a lack of connectivity with the city centre and other growth areas, as well as issues of contamination related to Manchester's industrial past.

The characteristics of each sub-region also vary in relation to the nature and extent of brownfields (Figure 5.4). In 2004, there were some 3541 hectares of 'previously developed land' (brownfield) in Thames Gateway (TG), and 3354 hectares in Greater Manchester (GM). This represents more than 10% of all brownfield land in England. Analysis of the National Land Use Database (NLUD) for SUBR:IM revealed (see also Dixon *et al.*, 2005a; 2005b) the following:

- A significantly higher amount of brownfield land in GM is derelict/ vacant (73%, in 2004), compared with TG (44%, in the same year), which largely reflects the industrial legacy of the GM sub-region.
- In 2004, in relative land area terms, on average some 2% of the total land area in GM is derelict or vacant (2480 ha); in TG about 1% of the total

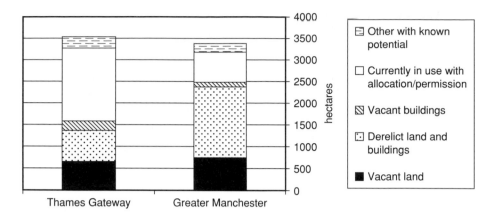

Figure 5.4 Brownfield land characteristics: Thames Gateway and Greater Manchester, 2004 (NLUD data).

land area is derelict or vacant (1580 ha in total). For England as a whole, the proportion is 0.3% (38 170 ha).

• In 2003, brownfield land was mainly in private ownership in both areas, although a substantial amount of ownership is unknown in GM, and dereliction is characterised by larger sites in TG (4.8 ha) than GM (3.0 ha).

Details of the case studies are provided in Table 5.2 and the locations are shown in Figures 5.5 and 5.6. Details of the interviews are provided in Appendix 2.

5.6 Towards best practice?

This section examines the main findings from the case studies interviews.

5.6.1 Environmental issues

Although contamination was still seen as an important challenge in both sub-regions, infrastructure, density and governance issues were considered more important. However, there was a view from the interviewees that contamination and waste legislation and guidance should be streamlined and rationalised, and that a single remediation permit system should be developed. Soil Guideline Values also need to be reviewed, to ensure that a sensible balance is created between safety and risk to public health (see Chapters 3 and 10). Not surprisingly, we also found that developers tend to

Table 5.2 Case study sites.

Thames Gateway	Type	Comments
Barking Riverside	First phase (residential) 1995. Latest phase is mixed use. 350 ha site	Largest brownfield site in UK. Previously contaminated. Joint Venture between Bellway Homes and English Partnerships.
South Dagenham West	Mixed use. 80 ha (whole site)	Key area for regeneration in Barking. Joint venture between London Development Agency and Axa Sun Life. Relatively 'clean' site (some manufacturing).
Gascoigne Estate	Residential (mixed tenure) 85 ha site	Part of Barking Town Centre Area. Major regeneration of existing housing estate. Site not contaminated.
Greater Manchester		
New Islington	Mixed use development 12.5 ha site	Designated as Millennium Community. Previously contaminated site. Partnership between New East Manchester (URC) and Urban Splash (with English Partnerships, Manchester City Council, and Manchester Methodist Housing Group).
Hulme	Residential (mixed tenure) 97 ha site	Government-backed Hulme City Challenge Initiative. 'Deck-access' flat demolition programme. Site not contaminated.
Higher Broughton	Residential (mixed tenure) 12 ha site	City Spirit and In Partnership collaboration. Forms part of market renewal pathfinder area. Previously contaminated site.

be cost-driven when it comes to remediation, although the case studies revealed several instances of innovative in situ techniques and a belief that 'soil hospitals' would become more common.

While there is a trend towards in situ methods driven by the EU Landfill Directive, stabilisation and solidification methods can still present regulatory problems because of their complex nature. As Chapter 10 demonstrates, the Environment Agency and the UK government have a key role to play in helping develop realistic risk guidelines for clean-up.

Our case studies also suggested that with limited gap funding now available, further public sector funding and improved grant regimes will be needed for 'hardcore' sites, if regeneration in these localities is to continue.

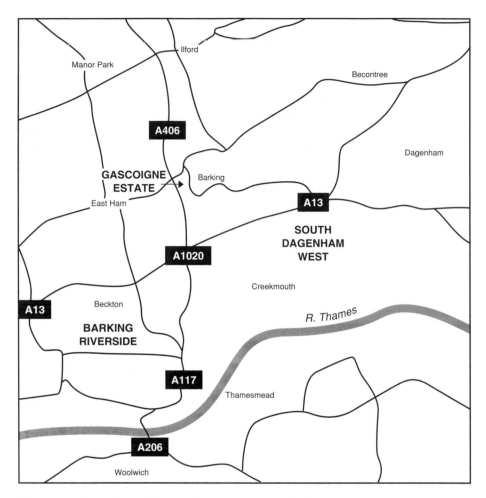

Figure 5.5   Location of Thames Gateway case study sites.

5.6.2 Economic issues

The research showed that there is a clear need for government and related agencies to ensure infrastructure is in place prior to development. In the absence of full government funding/support, this may mean that the introduction of a planning gain supplement (or equivalent) is inevitable. Already a number of local developer tax schemes exist, such as the Milton Keynes 'roof tax'. Further local schemes are likely, and English Partnerships (to be re-named Communities England in 2007) can play a key role here in providing local infrastructure and serviced sites. Local authorities may also have to 'sacrifice' land value on some sites to create the necessary education and health infrastructures required for communities.

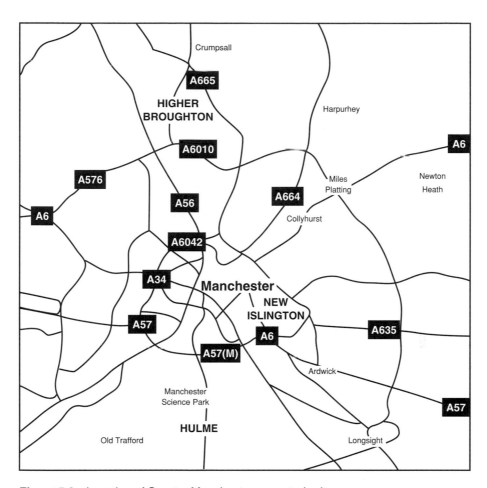

Figure 5.6 Location of Greater Manchester case study sites.

Creating a new image and brand was seen as a way of creating 'confidence' in an area to overcome perceived 'stigma'. However, this can create problems for local communities, as projects become victims of their own success and local people are priced out of the market, unless a sufficient amount of affordable housing is provided. In London there is currently a target set of 50% for affordability, although this may, conversely, create issues for developer confidence in the Thames Gateway, given the level of current residential values. It was also clear that there was an overemphasis on flats at the expense of housing in both sub-regions (in 2004–2005 some 46% of new dwellings in the UK were flats), and in some cases there was evidence of buy-to-let tipping the balance away from a diverse housing/tenure mix that would prove attractive to a range of age and income groups.

The research showed that area-based initiatives, based solely around property development, were more likely to fail in their aims, and so strong underpinning and support for people-based initiatives are needed to enable local people and businesses to thrive and flourish. This means regeneration must be based around jobs and re-skilling as well as housing provision.

5.6.3 Social issues

There is a need for a rationalisation of governance in the Thames Gateway. Clearer designation of responsibilities is required at national, regional and local levels, and although this is less of a problem in Greater Manchester, even here clearer designation of responsibilities is required, given the existence, for example, of two urban regeneration companies. At a national level, transport, environment and regeneration are currently undertaken by three separate departments (DfT, Defra and ODPM (now DCLG)) with fiscal arrangements being handled by two others (DTI and Treasury), which can lead to a lack of 'joining up' at national, regional and sub-regional level. Continuing planning delays and bureaucracy were also seen as key challenges by a number of stakeholders.

Our research also indicated that joint venture schemes are perceived as being generally successful. In both sub-regions, good examples of such schemes exist (for example, Barking Riverside and South Dagenham West, where the projects seem to be leading towards successful partnering), but there needs to be a balance between strong leadership and collaborative working to ensure success, and a fair risk/reward trade-off for those involved.

As far as community engagement and development are concerned, active dialogue with key elements in the community is needed. Several developers had used 'eco days' or 'green days' to highlight the benefits of sustainable communities which go beyond what might be considered 'conventional'. But education of consumers is key to highlighting the benefits of combined heat and power (CHP) (Figure 5.7), energy saving and the benefits of green construction. Community trusts may also become more common for community-based projects, founded on successful experiences in Thames Gateway.

There is also a major challenge for those involved in the sustainable development agenda to more closely define what 'sustainable communities' really comprise. Although ODPM produced a definition (ODPM, 2003), it was noticeable that stakeholders had developed an array of terminology (for example, 'liveability' and 'neighbourhoods of choice') to contextualise what they were trying to achieve. It is likely that those developers with a strong corporate social responsibility agenda are more likely to be fully committed to the sustainable communities agenda.

Figure 5.7   New Islington, Manchester, is one of seven English Partnership Millennium Villages. The £250 million development is on a 12.5-hectare site at Ancoats, East Manchester. The site includes the 1960s Cardroom Estate and the scheme is a partnership between New East Manchester (URC) and Urban Splash with English Partnerships, Manchester City Council and Manchester Methodist Housing Group. Innovative design, combined heat and power supplies, and a sustainable approach to remediation (using bioremediation) are features of the project. (www.newislington.com)

5.7 A checklist for developers

Clearly, valuable lessons have been learned from the experience on these sites. Table 5.3 provides a summary of key points that developers need to bear in mind when approaching brownfield development, and although this is not intended to be prescriptive, it can provide a useful tool for those seeking to develop in ways that really do provide for sustainable end products. In this sense the research is intended to help refine and complement existing 'sustainability checklists' such as the one produced by the South East England Development Agency (SEEDA). Indeed, SUBR:IM work has also developed a framework for assessing sustainability across the brownfield life cycle (see Chapter 12).

Table 5.3 Developer's checklist.

Stakeholder	Key roles/responsibilities	Comments
Environmental	• Use sustainable remediation techniques • Incorporate sustainable construction methods and high standards of design	• Engage with community and other stakeholders during and after clean-up • Driven by policy and guidance, design codes may be appropriate
Social	• Engage with community at an early stage of development	• A need to be proactive in design options • 'Eco Days' and 'Green Days' can help educate general public • Overseas best practice visits with community representatives • Promote risk transparency in clean-up (warranties on sale)
	• Focus on partnering and engaging with other stakeholders	• Joint Ventures and PPP-based schemes can offer advantages but require leadership and vision
Economic	• Promote a strong 'brand/image' for the project	• Sensitivity required because of the richness and diversity in the community
	• Incorporate a balance/mix of tenures and house types	• Affordable housing is key, and gated communities can create social exclusion • Mix of density, house and tenure type is vital
	• Focus on sustainable communities which provide 'liveability'	• Engagement with stakeholders to provide homes, where people want them, close to jobs and other services
All	• Measure sustainability proactively across the project life cycle	• Need to be consistent and to attempt to measure relevant sustainability components

Source: adapted from Dixon (2006)

5.8 Conclusions

There is no doubt that the development industry has a vital role to play in the brownfield regeneration process. It is difficult and oversimplifying to characterise developers as heroes or villains. But it is clear that, as a key risk-taker, the industry expects to receive a fair and equitable reward/return on development and investment, and the industry seems to have responded to opportunity 'pulls' and policy 'pushes' over the last few years. On the other hand, in relation to sustainable brownfield regeneration, 'reality' has not matched the 'rhetoric'. There needs to be greater awareness and understanding of alternative technologies to dig and dump, and better information and guidance for the development industry as a whole, for example.

Government must also bear responsibility. Although government policy seems to have been successful in shifting the pattern of development towards brownfield sites, conflicting policy aims may start to create difficulties and threaten the continued success of the regeneration agenda. For example, the attempt to reduce the amount of contaminated material going to landfill sites may slow down the development of brownfield sites, as alternative methods of remediation have to be sourced and implemented and costs of disposal rise. Higher costs for dealing with contamination may therefore threaten the viability of some brownfield redevelopments, thus increasing reliance on public sector intervention.

There also appears to be a greater need for the public sector to take the lead in disseminating and publicising the information that is available on alternative remediation treatments. It is clear therefore that the EU Landfill Directive, and the recent European Court of Justice case *Van der Walle and others* v. *Région de Bruxelles-Capitale* (C-1/03) can exacerbate tensions that already exist between brownfield and contaminated land policy 'layers'. Nonetheless, more sustainable methods of remediation may be promoted as a result, and it may be the case that the sustainable development agenda really does now become a main focus for debate within the property industry, as other environmental directives and legislation start to bite.

In policy terms, other issues highlighted by the research include the following:

- The contamination and waste legislation and guidance should be streamlined, and the remediation permit system needs to be simplified. Soil Guideline Values need closer review to ensure efficient operation.
- Sufficient amounts of affordable housing need to be provided. In London there is currently a target set of 50%, although this may create issues for developer 'confidence' on some sites. The case studies suggested there

is an overemphasis on flats at the expense of houses in both Thames Gateway and Greater Manchester, and 'buy-to-let' may create additional pressures.

It is also clear that the brownfield projects developers are engaging with today are often complex, have long life cycles, and involve peoples' homes, jobs and future lives. The starting point for improved success has to be a greater role for the public sector, especially on 'hardcore' sites. If policies are truly integrated at a national level this should help, and getting the right infrastructure will require a true partnership between public and private sectors. If funding and tax issues can be reviewed, perhaps in terms of a 'carrot and stick' approach, then increasing the rate of brownfield redevelopment becomes a reality. Finally, risk management, of course, should not be ignored. Costs and value are important considerations, but developers require better guidance, and better site characterisation on difficult sites. In short, there is a need to create a 'virtuous circle' for brownfield development (Figure 5.8).

The challenge will be to incorporate innovative and sustainable products and designs throughout the brownfield life cycle, from clean-up through to development and construction, in order to provide truly sustainable communities. But this will present major challenges over the next 20 years and the development industry has a key role to play in this working with other stakeholders. As one of our community representative interviewees put it:

> I worry really what we are creating – it's almost like scientists really: testing out design, testing out living materials and new products, but we're testing out on peoples' lives really, I think, and I just worry that are we creating a new Hulme, or a new area, that in 20 to 30 years we are going to knock . . . down again because it wasn't sustainable now. But I

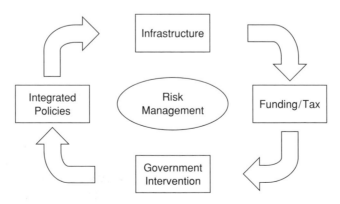

Figure 5.8   A virtuous circle for brownfields (adapted from Dixon *et al.*, 2006).

also think, on the other side, that it's important to test out new ideas and push the boundaries.

Acknowledgements

SUBR:IM is funded by EPSRC (Grant Number GR/S148809/01), with additional support from the Environment Agency. My thanks go to all those who were interviewed but who, for reasons of confidentiality, must remain anonymous. The project was completed at Oxford Brookes University, but a major role was played by Jude Shephard, Yasmin Pocock and Mike Waters of the College of Estate Management, Reading, in completing this research. The kind support of the RICS Foundation is also acknowledged.

5A.1 Appendix 1 National developer interviewees and questionnaire sample

5A.1.1 Developer interviewees

Details are provided in Table 5A1.1.

5A.1.2 Questionnaire sample

The response rate for the first phase of the questionnaire was 16% (158 useable responses – comprising 112 housebuilders and 46 commercial

Table 5A1.1 Summary of interviewees.

Development sector	Size of company*	% brownfield completions	Interviewee(s) (company position)	Location
Commercial	n/a	100	Director	London
Commercial	n/a	85	Development Manager	South East
Residential	Medium–large	90	Environmental Manager	South East
Residential	Medium–large	70	1) Finance Director	South East
			2) Land Manager	North West
Residential	Medium–large	70	Land Director	South East
Residential	Medium–large	95	Director of Innovation	South East
Residential	Medium–large	50	Planning Director	South West
Residential	Medium–large	70	Development Director	North
Residential	Medium–large	55	Managing Director	North
Residential	Medium–large	70	Managing Director	South East
Residential	Medium–large	95	Managing Director	North West

* Size of company based on data provided in *Private Housebuilding Annual 2004*. Size determined by number of units completed (small = <31 units pa; medium–large = 31 units or more pa).

developers[10] – from an overall sample size of 987). The second phase of the questionnaire was sent out to all those who responded to the first survey and supplied an appropriate contact; 65% of these were returned (94 useable responses representing 10% of the original sample) (Dixon, 2006).

A comparison of the survey respondents by number and size of output against NHBC data gives an indication of how representative the sample of those undertaking residential development is against the industry as a whole. Table 5A1.2 shows that the sample is more representative of medium- to large-sized housebuilders (31+ units p.a.) than smaller operators (<31 units p.a.). However, while housebuilders producing fewer than 11 units per annum account for 80% of registered and active housebuilders, NHBC data (2003) shows that their contribution to the industry's total output is small (Table 5A1.3).

Table 5A1.3 compares the sample in terms of the number of units produced annually by respondents against the industry structure. Again, the sample is clearly more representative of medium- to large-sized housebuilders.

In interpreting the results, therefore, it should be borne in mind that the survey captured a relatively small proportion of developers building up to 30 housing units per annum, although this group accounts for only 15% of the industry's annual output.[11] The survey captured over a fifth of housebuilders producing 31–100 units per annum rising to over two-fifths of volume housebuilders completing over 2000 units every year.

A further concern relating to non-response bias is that developers undertaking mainly greenfield development may not have felt the survey was

Table 5A1.2   Sample compared to industry structure: number of developers undertaking residential development.

Size band	NHBC data			Sample (phase 1)		
	Number registered (2003)	% by size band		Count of sample by size band*	% by size band	Sample count as % NHBC
0 units	10 188					
1–10 units	4 421	80%		11	9%	**0.2%**
11–30 units	712	13%		26	20%	**4%**
31–100 units	264	5%		48	38%	**18%**
101–500 units	112	2%		30	23%	**27%**
501–2000 units	20	0.4%		7	5%	**35%**
2000+ units	14	0.3%		6	5%	**43%**
Total active	5 543	100%		128	100%	**2.3%**
Total registered	*15 731*					***0.8%***

* Housebuilders and commercial developers undertaking residential development.
Source: NHBC New House-Building Statistics (2003 Q4)/CEM Survey

Table 5A1.3   Sample compared to industry structure: by size of output.

Size band	NHBC data		Sample				
	Starts by size band	% output by size	Output by size band*	Sample %	Ratio of sample output: industry output	Ratio of sample count: industry output	% sample output by NHBC reg
1–10 units	13 488	8%	66	0.1%	0.02	0.03	**0.5%**
11–30 units	11 802	7%	528	1%	0.15	0.5	**4.5%**
31–100 units	13 488	8%	2 900	6%	0.70	2.3	**22%**
101–500 units	23 604	14%	7 220	14%	1.00	1.9	**31%**
501–2000 units	23 604	14%	7 346	14%	1.01	2.5	**31%**
2000+ units	80 928	48%	33 662	65%	1.36	0.9	**42%**
	168 600	99%	51 722	100%			

Total UK completions 183 071 28%

Source: NHBC New House-Building Statistics (2003 Q4)/CEM Survey

relevant to them and failed to return a response. To investigate this issue a small-scale telephone survey was undertaken to gather information on non-response. An attempt was made to contact the original recipient of the survey and answers to the following questions were sought:

- Why were you unable to return the survey? (e.g. too busy, not relevant)
- Do you undertake brownfield development?
- How many units do you complete per annum?
- What proportion of these are on brownfield sites?

This information proved very difficult to gather with a hit rate of less than 1 in 10 calls. In cases where contact was made, the respondents had been too busy to return the survey or were in the process of winding down their companies because of retirement, but all undertook brownfield development. However, the sample was not large enough to allow any robust conclusions to be drawn. Further investigation of the questionnaires returned by type of developer shows that 16% of the sample derive at least 50% of their output from greenfield development, which does lessen concern about sample bias to some extent.[12] A review of the websites of non-respondents also indicates that many are clearly engaged in brownfield development.[13]

The proportionately lower response from smaller housebuilders discussed above is unsurprising given that smaller companies are perhaps less likely per se to respond to surveys owing to limited resources and no perceived benefits from participating (larger companies might be more concerned about their public image, for example). Data constraints and a lack of information regarding the underlying population also made it difficult to boost response among smaller housebuilders.

Although the survey did not manage to obtain a particularly high response rate from the housebuilding industry's smallest operators (those building up to 30 units per annum), it did achieve good penetration among medium- and larger-sized housebuilders who account for the vast majority of the industry's output. Therefore the sample does contain fewer small house-builders in percentage terms than the national proportion. However, our analysis also showed that non-response bias because of possible focus on greenfield sites was not an issue.

In total, the annual output of the survey respondents accounts for some 28% of annual housing completions in the UK; consequently the survey can be said to represent a valuable snapshot of the industry, bearing in mind the caveat relating to smaller operators.

5A.2 Appendix 2 Details of case study interviews

Some 54 face-to-face, structured interviews were conducted during 2004–2005 with key stakeholders on six sites in the Thames Gateway and Greater Manchester (Tables 5A2.1 and 5A2.2). In the Thames Gateway the sites comprised:

- Barking Riverside
- South Dagenham West
- Gascoigne Estate

In Greater Manchester the sites comprised:

- New Islington
- Hulme
- Higher Broughton

The key stakeholders included local authority planning and regeneration representatives; government agencies; developers; consultants; community representatives; housing associations; and others.

Table 5A2.1   Interviews with key stakeholders (details anonymised): Thames Gateway.

	Interviewee	Number
Barking Riverside	Local authority	2
	Government agency	1
	Development industry	3
	Consultant	1
	Community	1
South Dagenham West	Local authority	1
	Government agency	1
	Design	1
	Development industry	1
	Community	1
Gascoigne Estate	Local authority	2
	Developer*	1
General (all three case studies)	Local authority; agencies; development industry; housing associations; surveyor; estate agents; consultants.	13
Total		**29**

* As the scheme was in its very early stages, the interview covered other, generic issues in the Thames Gateway.

Table 5A2.2   Interviews with key stakeholders (details anonymised): Greater Manchester.

	Interviewee	Number
New Islington	Developer/partnership	5
	Housing association	1
	Design	1
	Community	1
Hulme	Developer/partnership	2
	Local authority	1
	Design	1
Higher Broughton	Development industry	1
	Surveyor/planner	2
	Design	1
	Local authority	3
	Remediation consultant	1
General (all three case studies)	Agencies and technical experts	5
Total		**25**

Notes

1. Any land that has been previously developed, including derelict and vacant land, which may or may not be contaminated (see ODPM, 2000).
2. In February 2005 EP and the ODPM launched the setting up of 12 pilot brownfield projects (which include hardcore sites) as part of the next stage of creating the National Brownfield Strategy.
3. The equivalent figure excluding conversions was 68%. For consistency, Figure 5.1 shows new dwellings, excluding conversions, and so figures are correspondingly lower.
4. From 16 July 2004 the Directive banned the co-disposal of hazardous and non-hazardous waste, resulting in a radically reduced number of sites permitted to accept hazardous waste. The aim of the Directive is to encourage waste reduction and wider adoption of more sustainable methods of dealing with contamination. A new Mobile Treatment Licence regime in 2006 may also help promote onsite treatment, and exemptions for contaminated waste still exist for Landfill Tax (under sections 43A and 43B of the Finance Act 1996 (as amended by the Landfill Tax (Contaminated Land) Order 1996 (SI 1996 No. 1529))).
5. BREEAM (BRE's Environmental Assessment Method) provides a broad-ranging assessment of the environmental impact of a building. Issues covered include those relating to the global, local and internal environments. BREEAM relates to design stage assessments (i.e. new build and refurbishment) and to the ongoing operation and management of the building. Assessors operating under licence from BRE carry out the assessments. Ecohomes is the homes version of BREEAM. It provides an authoritative rating for new, converted or renovated homes, and covers both houses and apartments.
6. See note 5 above.
7. This 'inductive' approach should be contrasted with the 'deductive' approach of Dair and Williams (2005; see also Williams and Dair, 2005), who developed a conceptual model to assess the sustainability of five English brownfield case studies (according to predetermined measures) and determine how different stakeholders influence the sustainability of completed brownfield developments.
8. The Northern Way is a collaboration between the three northern regional development agencies: Yorkshire Forward, Northwest Development Agency and One NorthEast. This is a 20-year strategy to transform the economy of the north of England.
9. The Sustainable Communities Plan, published on 5 February 2003, provides the government framework for a major programme of action that will, over the next 15 to 20 years, tackle the pressing problems of communities across England. One of the key areas forming the basis for the action programme is the tackling of low housing demand and housing abandonment: sustained action to turn round areas where housing markets have failed. From 2005 to 2008, £500 million is being made available for some of the worst affected areas, known as Pathfinder market renewal areas (of which nine are designated), with the intention of reversing low demand by 2010 (sourced from English Partnerships' website).
10. Some 26 of these also undertook residential development.
11. The results of the survey presented in this paper have therefore not been weighted to give more emphasis to smaller housebuilders, because of the importance of larger companies in volume terms.
12. The size structure of this sub-group was similar to that of the total sample.
13. A review of 80 websites selected randomly from non-respondents revealed that at least 90% of these housebuilders undertook brownfield development to a greater or lesser degree. Where this was not stated explicitly it was judged on factors including location of developments and land requirements. In comparison, the proportion of survey respondents undertaking *any* brownfield development was 96%.

References

Barker, K. (2003) *Review of Housing Supply. Delivering Stability: Securing Our Future Housing Needs (Interim Report – Analysis).* HM Treasury, London.

Barker, K. (2004) *Review of Housing Supply. Delivering Stability: Securing Our Future Housing Needs. Final Report: Recommendations.* HM Treasury, London.

CABERNET (2003) *Conceptual Model of Brownfields Regeneration.* Report No. BIS 11, Baseline Information Sheet. CABERNET, Nottingham.

Dair, C. and Williams, K. (2005) *Sustainable Land Re-use: The Influence of Different Stakeholders in Achieving Sustainable Brownfield Development in England.* Working Paper 4. Oxford Institute for Sustainable Development, Cities Unit, Oxford.

Davis Langdon Consultancy (2003) *Investing in Sustainable Developments Key Players Group.* Workshop 1 Report, Davis Langdon, London.

Department for Regional Development (2001) *Regional Development Strategy: Shaping Our Future.* Department for Regional Development, Belfast.

DETR (1998) *Planning for Communities of the Future.* The Stationery Office, London.

DETR (2000) *Our Towns and Cities: the Future – Delivering the Urban Renaissance.* HMSO, London.

Dixon, T. (2006) *The Role of the Development Industry in Brownfield Regeneration, Stage 3: Best Practice Checklists for Key Brownfield Stakeholders.* Oxford Brookes University, Oxford.

Dixon, T., Pocock, Y. and Waters, M. (2005a) *The Role of the Development Industry in Brownfield Regeneration, Stage 2 Report, Volume 1: Literature Review, National Developer Interviews, Planning Permission Analysis and NLUD Analysis.* College of Estate Management, Reading.

Dixon, T., Pocock, Y. and Waters, M. (2005b) *The Role of the Development Industry in Brownfield Regeneration, Stage 2 Report, Volume 2: Sub-regional Context (Thames Gateway and Greater Manchester).* College of Estate Management, Reading.

Dixon, T. (with Pocock, Y. and Waters, M.) (2006) *The Role of the Development Industry in Brownfield Regeneration, Stage 2 Report, Volume 3: Case Studies (Thames Gateway and Greater Manchester).* Oxford Brookes University, Oxford.

English Partnerships (2003) *Towards a National Brownfield Strategy: Research Findings for the Deputy Prime Minister.* English Partnerships, London.

Entec (2004) *Study into the Environmental Impacts of Increasing the Supply of Housing in the UK: Final Report.* Defra, London.

Environmental Audit Committee (2005) *Housing: Building a Sustainable Future,* Volume I. House of Commons, Stationery Office, London.

Guy, S. and Henneberry, J. (eds) (2002) *Development and Developers: Perspectives on Property.* Blackwell, Oxford.

Healey, P. (1991) Models of the development process: a review. *Journal of Property Research,* **8,** 219–38.

Healey, P. (1992) An institutional model of the development process. *Journal of Property Research,* **9,** 33–44.

HM Government (2005) *Securing the Future: Delivering UK Sustainable Development Strategy.* Defra, London.

National Round Table on Environment and the Economy (2003) *Cleaning Up the Past,*

Building the Future, A National Brownfield Redevelopment Strategy for Canada. National Round Table on Environment and the Economy, Ontario, Canada.

NHBC (2003) www.nhbc.co.uk

NLUD http://www.nlud.org.uk/

ODPM (2000) *Planning Policy Guidance 3: Housing.* Office of the Deputy Prime Minister, London.

ODPM (2001) *Land Use Planning Research Programme.* Office of the Deputy Prime Minister, London.

ODPM (2003) *Sustainable Communities: Building for the Future.* Office of the Deputy Prime Minister, London.

ODPM (2005) *Land Use Change in England: Residential Development to 2004 (LUCS-20).* Office of the Deputy Prime Minister, London.

Pahlen, G. and Franz, M. (2005) Sustainable regeneration of European brownfield sites: criteria for future funding decisions. In: *Proceedings of CABERNET 2005: the International Conference on Managing Urban Land.* Land Quality Press, Nottingham.

POST (1998) *A Brown and Pleasant Land.* London Parliamentary Office of Science and Technology (POST), London.

RESCUE (2003) *Analytical Sustainability Framework in the Context of Brownfield Regeneration in France, Germany, Poland and the UK.* Final Report of Work Package 1, www.rescue-europe.com

Shephard, J. and Dixon, T. (2004) *The Role of the UK Development Industry in Brownfield Regeneration, Stage 1 Report.* College of Estate Management, Reading.

Sustainable Construction Taskforce (2001) *Reputation, Risk and Reward: the Business Case for Sustainability in the UK Property Sector.* DTI, London.

Williams, K. and Dair, C. (2005) *A Conceptual Model of Sustainable Development.* Working Paper 3: Oxford Institute for Sustainable Development, Cities Unit, Oxford.

WWF (2003) *Building Sustainably: How to Plan and Construct New Housing for the 21st Century.* WWF, London.

6

Delivering Brownfield Regeneration: Sustainable Community-Building in London and Manchester

Mike Raco, Steven Henderson and Sophie Bowlby

6.1 Introduction

Since the late 1990s brownfield regeneration agendas in the UK have become increasingly bound up with broader questions concerning sustainable urban development and sustainable community-building. Brownfield redevelopment, it is argued, simultaneously solves environmental, technical, social and economic problems by enabling high-density redevelopment to take place on redundant and/or contaminated sites (see Chapter 5). In so doing, it enables the (re)formation of tight-knit urban communities and reduces the need for wasteful forms of transport and service provision. In short, the reuse of brownfields plays an important part in the creation of more sustainable and compact urban forms and delivers a range of socio-economic and environmental benefits. Some of the controversies that characterised urban regeneration policy in the 1980s, where brownfield sites were re-capitalised as part of a wider agenda for the regeneration of the inner cities, have been replaced by the positive, win–win–win discourses of urban sustainability and brownfield regeneration.

And yet, at the same time, there has been relatively little recognition that there is nothing inevitable about the outcomes derived from brownfield regeneration. Any form of regeneration requires policy makers and others to think about the types of problems that policy has to solve and the form and character of policy solutions. In some cases new developments will be implemented that are underpinned by strong economic rationalities even though they may be brownfield-led and 'sustainable' in name. Such developments may succeed in generating new forms of environmental exploitation and socio-economic diversification, rather than providing a bedrock for new forms of sustainability and sustainable community-building. In political terms development projects may also involve little direct input from local communities or others such as environmental groups, as economic objectives are prioritised. In different places, at different times, urban projects will often draw on contrasting rationalities that may be more or less conducive to the construction of sustainable communities and cities.

This chapter draws on research undertaken in London and Manchester to examine broader questions concerning brownfield regeneration processes and their wider sustainability. It argues that policy makers and others need to be more open to the possibility that brownfield development may represent only one part of a broader set of sustainable development agendas. It does not *necessarily* engender new forms of urban sustainability, and indeed, as we will show, particular forms of development may well exacerbate existing problems and difficulties.

The chapter is divided into three principal sections. The first examines the government's sustainable communities agendas and outlines some of the key questions that they raise for urban policy. This is followed by an exploration of the two case study sites, Paddington Basin and Salford Quays. The sites were selected as they were both seen as 'flagship' areas in which significant and long-term urban development programmes had been in operation since the mid 1980s; in both there had been a significant re-engineering of brownfield development sites; surrounding neighbourhoods contained deprived and marginalised communities; and, perhaps most importantly, both had been widely lauded as prime examples of sustainable urban regeneration. A substantial concluding section addresses some of the key findings from the case studies and highlights some of the core dilemmas to be resolved and thought through if brownfield development is to be made more cohesive and successful.

6.2 Building for the future: visions, practices and the delivery of sustainable urban regeneration

During the late 1990s and early 2000s, the discourses that have underpinned urban regeneration in the UK have been subject to significant change. Across western Europe governments have extended some of the principles of capacity-building established during the 1990s and focused more explicitly on the concepts of *sustainability* and its spatial policy manifestation, the *sustainable community* (SC). The wider discourse of sustainability emerged during the 1990s as part of a wider international debate over how economic development could be made more inclusive, democratic and environmentally sustainable (see Kenny and Meadowcroft, 1999; Whitehead, 2003; Henderson *et al.*, 2007). In the UK the Labour government has adopted elements of these wider ways of thinking and increasingly argued that development should be underpinned by new temporal and spatial imaginations of what places should be like now and in the future. In launching its SC strategy in 2003, for example, it defined SCs as 'places where people want to live and work . . . [and they] meet the diverse needs of existing and future residents, are sensitive to their environment, and contribute to a high quality of life' (ODPM, 2003, p. 1). The emphasis is on what places *can* become through the effective implementation of sustainability-oriented planning processes.

SC-building is underpinned by a particular politics of time and of future imaginations, as highlighted in Table 6.1. As Kenny and Meadowcroft note, 'nearly all the definitions [of sustainability] conceded that it involves the re-orientation of the meta objectives of a given society – by raising questions about different possible social trajectories through which the society may move' (Kenny and Meadowcroft, 1999, p. 4). However, they go on to argue that this 'does *not* in any way imply only one kind of social future as the embodiment of "the" sustainable society' (Kenny and Meadowcroft, 1999, p. 4; see also Maloutas, 2003). Within such discourses 'the future', therefore, becomes less of a vague alibi for short-term programmes and instead takes on the features of *a definable object* that should be integrated into planning frameworks at all stages of their development and design. It becomes something tangible to be controlled, ordered and worked towards, with active citizens expected to consider their own actions in the context of how they might impact on 'the future'.

The process of turning these policy aspirations into practical development agendas is a complex and challenging task. It forces development agencies, planners, communities and other stakeholders to engage with key questions such as the following:

- How do you define a 'problem' place and what would make it a 'better' place?
- Who are the target groups that regeneration is aimed at (e.g. developers, investors, house buyers, local communities)?
- How can infrastructure be created that supports the formation of sustainable communities?
- What types of urban environment should be developed, and for whom?
- What role does brownfield regeneration play in the successful implementation of development projects?

Table 6.1   The central features of sustainable and unsustainable communities.

Criteria	A sustainable community	An unsustainable community
Citizenship and governance	Active citizens and communities; long-term community stewardship; effective political engagement; representative, accountable governance systems; balance of strategic, top down visionary politics and bottom up emphasis on inclusion.	Passive and dependent citizens and communities; lack of community engagement or ownership; low levels of voluntary activity and/or social capital; closed, unaccountable systems of governance; over-reliance on passive, representative forms of democracy.
Economic change and development	Flourishing economic base; built on long-term commitments; stability; inclusive of a broad range of workers; a steady transition in economic activity away from declining sectors to growth sectors.	Domination by dependent forms of development; lack of employment opportunities; vulnerable; insecure, short-term, and divisive labour markets.
Environmental dimensions	Reuse of brownfield sites; minimisation of transport journeys; good quality public transport.	Expansion into greenfield sites; maximisation of transport journeys; car dependence and the absence of public transport.
Community change and identity	Broad range of skills within workforce; ethnically and socially diverse; mixture of socio-economic types of inhabitants; balanced community; well-populated neighbourhoods; sense of community identity and belonging.	Absence of skills within workforce; ill-balanced communities of place; high levels of (physical) separation between groups; lack of diversity; formal and informal segregation; lack of population, lack of local associational culture and ownership of public space.

Source: adapted from Raco, 2007, p. 173

The delivery of sustainable regeneration and its impacts on people and places are directly related to the ways in which such questions are addressed and the imaginations and visions that are established before policy is initiated. These visions, in turn, are created through a combination of theoretical, professional and academic knowledge, often gained through an understanding of practice from elsewhere and an assessment of the specific local contexts that exist in a development area. At the same time the institutional, political and economic circumstances within which development programmes are initiated are subject to frequent change and contestation so that answers that seem appropriate in one context at a particular time can quickly become outdated or be seen (ironically) as a 'problem' at a later date. The remainder of the chapter now turns to our case studies to address these issues and questions in greater detail. It begins by discussing the research and some of the key developments that have taken place in the two areas before assessing their broader impacts and the extent to which they contribute to sustainable urban development.

6.3 Flagship urban brownfield regeneration in the UK: the redevelopment of Salford Quays and Paddington Basin

The research was undertaken in two case study sites from the SUBR:IM portfolio – Salford Quays in Greater Manchester and Paddington Basin in west London. During 2003–2005 the authors undertook a detailed investigation of both developments. The research adopted a mixed-methods approach in order both to provide historical context to the case studies and to capture some of the ongoing and rapidly changing processes of social and economic change that were taking place. The first stage of the work involved the collection of archival and documentary sources in order to provide some background on the changes that had taken place in the development sites, and the dominant policy visions and strategies that had shaped the form and character of the regeneration. A second stage then involved the interview of key stakeholders, including planners, developers, investors, community representatives and business support groups. Eighty interviews were conducted in the two sites. The interviewees were uncovered using a 'snowballing' method, and most interviews lasted approximately one hour. In addition, focus groups were held with community representatives and a survey of businesses was conducted in Salford Quays. In employing these methods the intention was to obtain a broad range of opinions and perspectives from a variety of groups, many of whom were relatively difficult to track down and were wary of the formal research process (see May, 2003).

6.3.1 Paddington Basin

The Paddington redevelopment area lies a short distance from the commercial heart of London, within the Conservative party stronghold of Westminster City. It is an area of regeneration that incorporates Paddington train station and the Paddington Canal. In 1988, in response to wider development pressures in Westminster, the local authority created a new development area that it called the Paddington Special Policy Area (PSPA; see Figure 6.1). This 30.35-hectare PSPA was, and still is, fractured by roads, railways and a canal, and is characterised by complex forms of public and private sector landownership. By the mid 1980s the site was either derelict or occupied by low rental land uses, such as warehousing, car parking or waste disposal. The canal itself was not open to pedestrians, and the

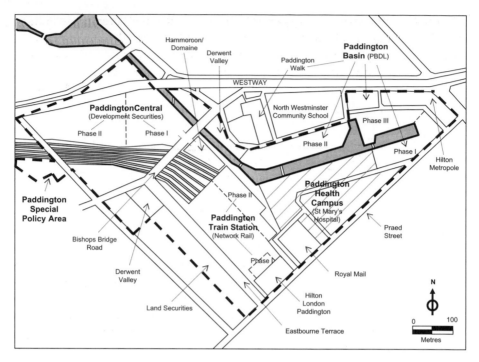

Figure 6.1 Landowners and development schemes in the PSPA 2005 (adapted from Paddington Waterside Partnership, 2001, p. 8).

buildings that remained turned their backs to the water. Other than providing a site for a small number of canal boat residents, it had little community use except for the occasional recreational canoeist. From the public's perspective it was an area to be avoided and was 'right up there as one of the infamous blighted sites of London' (Sadek, 2002, p. 233).

During the 1980s, despite these structural difficulties, the area had been characterised by significant and aesthetically spectacular regeneration. There have been two principal development phases of the PSPA. The first took place during the late 1980s when plans for the wholesale, property-led regeneration of the area were announced. This initial period of interest was subdued in the early 1990s by a national economic recession but a second, more substantial phase of development took place from the mid 1990s onwards. This renewed activity was fostered by a different set of developers, encouraged by the announcement and eventual completion of the Heathrow Express rail link in 1998. Since then the Paddington area has been physically transformed, with massive changes in its economic structure and use. By the middle of the following decade approximately one third of the planned development had been completed including 7.32 hectares of office space, 687 new homes and 1.5 hectares of retail/leisure use

(Paddington Waterside Partnership, 2001, p. 1). The development has been so significant that in February 2004 the Mayor of London labelled Paddington one of nine 'Opportunity Areas' in the capital, or spaces in which sustainable forms of brownfield regeneration had and continued to take place.

6.3.2 Salford Quays

The redevelopment of Salford Quays (SQ) has also been among the most significant developments of its type anywhere in the UK. The SQ project emerged in a context of decline in an area whose fortunes have always been closely tied to its docks. Until the late 1960s Salford Docks had a successful history as an inland port, following the opening of the Manchester Ship Canal in 1894. Local industry thrived and communities of workers migrated into the area, attracted by the availability of work and the unusual stability of local dock labour. However, by the early 1970s changing shipping technology and trade patterns saw activity in the docks decline. Their eventual closure in 1982 symbolised the wider process of de-industrialisation that was affecting Greater Manchester and other industrial cities, and the area became blighted by high levels of unemployment. In the neighbouring area of Ordsall, which had supplied many of the dock workers, unemployment was registered at 32% in 1985 compared to 15% in Greater Manchester. Social problems increased and Salford became a classic example of a deprived inner urban area with relatively high rates of crime, drug abuse, and (selective) out-migration (see Robson *et al.*, 1994). In the case of Ordsall these culminated in a series of riots in the summer of 1992.

The visible extent of decline in Salford made it a target for policy makers from the mid 1970s onwards. In 1978 parts of Manchester and Salford were designated under the Inner Areas Act and in 1981 a significant portion of what was to become Salford Quays (as well as land in Trafford Park) became an Enterprise Zone for a ten-year period (see Figure 6.2). In 1983 Salford City Council (SCC) purchased the dock site and associated land (a site of 37 ha) from the Manchester Ship Canal Company, and even at this early stage the idea of a water-based development, influenced by North American ideas, emerged onto the agenda. The subsequent redevelopment of SQ required a large input of up-front finance. The bulk of this funding was provided by central government with a £25 million rolling grant from 1985–1990/91 from the Derelict Land Grant. This was used for land and water clearance and remediation. Money also came from the Urban Programme for the provision of infrastructure, landscaping and roads. SCC were so successful in putting together development bids that in the period 1985–2002, £145 million in public funds had been invested into Salford Quays, including a £65 million National Lottery Grant for the Lowry – a performance arts theatre and gallery (Deas *et al.*, 2000, p. 10).

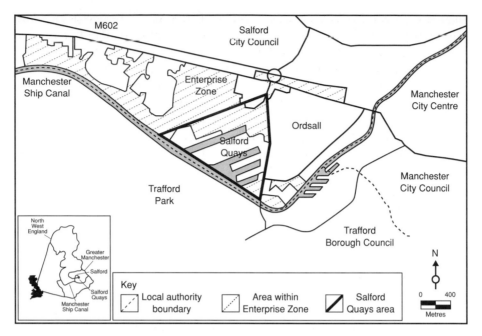

Figure 6.2   Salford Quays and its location in Greater Manchester (adapted from Shepheard *et al.*, 1985, p. 4).

The remainder of this chapter draws from our research findings to examine some of the key lessons for contemporary brownfield regeneration and sustainable community-building in the UK. The discussion is framed around the four core themes analysed in Table 6.1: citizenship and governance; changing economic landscapes; environmental dimensions; and community change and identity.

6.3.2.1 Citizenship and governance

The formation of sustainable communities and the wider 'success' of any brownfield development programme are in large part dependent upon the mobilisation and inculcation of new modes of active citizenship and new processes of inclusive and engaging local governance. It is increasingly argued by policy makers, academics and others that for any development to be effective it needs to be 'owned' by local communities who take enhanced responsibility for ensuring that development agendas are inclusive, wide-ranging, and targeted on local problems (see Urban Task Force, 1999). Encouraging new, more positive forms of community and individual engagement in the governance of sustainable communities is seen as one of the core building blocks of urban regeneration.

In the PB and SQ cases there was mixed evidence over the character of local decision-making processes and the creation of new forms of active citizenship. For instance, many local residents and businesses in both areas complained that there had been a lack of *consultation* throughout the redevelopment and that major decisions were consistently taken behind closed doors. The early phases of the regeneration, in particular, required the mobilisation of technical forms of expertise and development knowledge. During the late 1980s and the 1990s engineers, planners, investors and developers drew on their own skills and knowledge to plan and structure the development. The initial priority was to convert the areas' brownfield sites into new investment spaces that would be re-labelled and re-marketed as Paddington Basin and Salford Quays. This was seen as the primary objective of policy, with community inclusion and the structure and character of processes of governance seen as relatively less important.

This focus on getting the development 'up and running' had significant implications for broader processes of community involvement and sustainable governance. The research showed that existing, long-term residents often perceived the new developments as a threat to their communities and to their future residence in the area. There was a general feeling that developments were 'out of their control' and that technical, political and financial considerations dominated decision-making processes and frameworks. As with development projects elsewhere, many residents felt that their very presence in the development area was increasingly a 'problem' for development agencies who had become focused on a process of urban gentrification in the name of making local communities more 'sustainable'.

Ironically these views were also expressed by incoming residents, many of whom have been professional, reasonably affluent property buyers. In the same way that established residents in neighbouring communities have been complaining about the priority that has been given to the 'needs' of the new incomers, those moving into the area felt that their social and economic needs had been ignored and that the area contained too little social infrastructure to encourage them to stay in the longer term if, for example, they decided to have children. This was particularly true in SQ where new residents complained that they were seen as little more than investment pawns in a wider regeneration process. Generating new forms of sustainable and active citizenship in such a context has proved to be a significant challenge. None of the respondents in PB or SQ felt that they had played a 'stewardship' role over the developments thus far. There was little sense that new forms of governance were being introduced that would bring this about. Without this sense of ownership and control there is little prospect of encouraging new types of community-led engagement.

A further insight from the research has related to broader questions of *partnership-working* and the experiences of different agencies in PB and SQ. One of the challenges for development bodies in both locations has been the fragmented nature of landownership and institutional powers and resources. In many urban areas it has proved relatively difficult to bring together all those involved in the development process and encourage them to work together towards a coherent and agreed end. However, in PB and SQ partnership formation was developed under the close supervision of the local authorities. In SQ in particular, Salford City Council took a series of innovative and politically and financially risky steps during the 1980s and 1990s that have underpinned the subsequent redevelopment of the area (see Henderson *et al.*, 2007). Despite the massive investment of private sector resources into SQ, the regeneration owes much to public sector support and the setting out of development visions. Salford City Council also involved itself directly in the purchasing of land for the original developments and sought to build up a sense of confidence in the developer community. Similar processes have also been evident in PB where Westminster City Council has been at the heart of partnership-building processes and has managed to negotiate the complex and fragmented politics of landowner-ship and development unique to London. In broader sustainability policy frameworks there is relatively little recognition that local authorities could and should play such active roles; indeed, too much 'interference' with property markets through the planning process is criticised for limiting the extent to which sustainable urban redevelopment can take place.

Finally, the research has shown that there is a contested *politics of time* concerning the development, with different groups having very different understandings of exactly how and when the development should deliver on its objectives. One of the difficulties of the wider sustainability policy discourse is that it requires local communities and citizens in particular to develop longer-term perspectives, rather than concentrating on immediate and less sustainable concerns. And yet in both sites there has been conflict between developers, many local communities and investors on the one hand who are looking for shorter-term gains and the early construction of high-return projects, and planners and policy makers who ostensibly are looking to promote development that will bring benefits in the longer term. In addition, there have also been questions concerning the *liabilities* that different agencies will possess and the timing and phasing of specific benefits. In SQ and PB investors have been keen to promote high-return developments that may or may not be optimal for local communities and other stakeholders. Developing new forms of sustainable citizenship will require the resolution of such conflicts.

If the new development visions are to encourage a greater mixing of communities then this will have to be implemented through carefully crafted strategies that are not simply driven by the profiteering demands of developers. The SQ case, in particular, exemplifies what can happen when planning agencies prioritise investment and the bringing in of middle-class, mainly childless residents, at the expense of wider community development programmes. Indeed, the research encourages us to question the government's wider ambition of developing brownfield sites and the implementation of new development projects in the name of sustainability. The drive for such developments is likely to make planning agencies even more willing to attract development, whatever the time frames and needs of others.

6.3.2.2 Changing economic landscapes

In an era in which UK manufacturing jobs have been lost, new employment opportunities are emerging in service-based, high-tech and cultural industries and businesses experience greater freedom in terms of their geographical mobility, brownfield redevelopments are associated with the creation of new sustainable economic landscapes. The environmental credentials of emerging employment spaces is linked to their location on previously used land as opposed to greenfield business parks, and their accessibility by transport modes other than private automobiles. From an economic perspective brownfield redevelopments help to broaden and strengthen urban economies, and thus assist in reducing the medium- to long-term risk associated with dependence on individual industries. The national and international competitiveness of business occupiers ensures that brownfield spaces are stable, if not expanding, in terms of their employment base. They can also support the wider community through their employment practices and engagement in neighbourhood support. From a social sustainability perspective, emerging economic spaces are inclusive of a wide range of workers, including employees varying in age, ethnicity, gender and skill levels. The potential for local residents to find employment on regenerated brownfield sites scores highly in a sustainability sense, as it enables employees to walk or cycle to work.

As large-scale brownfield development projects, PB and SQ are promoted in terms of their contribution towards urban sustainability. Both regeneration schemes are located in inner city locations, and thus provide important site alternatives, especially for office-based businesses seeking large low-cost floor spaces in locations proximate to their respective city centres. SQ opened up new locations for office-based and high-tech businesses. It showed that they did not need to relocate to the urban edge to secure a spacious, high-amenity business park that offered good transport connections. PB also provided an alternative office base for London's corporate headquarters, and thus helped to support London's international position as a 'world city'. At the same time it deflected development pressures away from areas of heritage protection by providing development capital in a brownfield zone where development applications would be treated more favourably. The potential for larger office floor spaces and building sizes in turn provided the opportunity to increase the diversity of office spaces available in the national capital. In terms of sustainable transport, both

PB and SQ have significant advantages. PB has been developed around Paddington train station, which provides access both to London's underground Tube network and to regional commuter trains. SQ, although this was not the case initially, is connected to Manchester city centre and beyond by Manchester's Metro (Rapid Transit) tram network.

However, despite these visible changes, there was nothing inevitable about the types of economic spaces that have emerged in PB or SQ. Both locations experienced significant difficulty in convincing the development and commercial sectors of their viability as investment and business locations. Positive promotion, up-front investment and a willingness to support catalyst-style investments underpinned initial development activity on both sites (see Chapter 4). In the case of Salford, the legacy of de-industrialisation in terms of dereliction and lower socio-economic neighbourhoods discouraged developer and investor interest. Though the surrounding context differed in the case of PB, commercial interest was slow to emerge as Paddington was initially considered too far from the existing commercial and retail core of London. The sensitivity of both development sites to prevailing market attitudes was further reinforced in the early 1990s when a national recession halted development progress on both sites. Equally, it is reflected in 'trend-based' planning, with the desire to build office space outstripping that for other types of development in the late 1980s. In the case of SQ the diversity of land uses that is evident today would have had a different look had the national recession not brought a halt to speculative office-based developments. Market slow-down provided Salford City Council with the opportunity to re-examine the initial objective for a rough land-use split of a third residential, a third employment and a third leisure, and to recognise the need for additional leisure uses. PB differs as from an early stage Westminster City Council had insisted on a mixed residential–employment-based scheme, in reflection of the predominantly residential character of the surrounding area.

Within the broader context of market constraints, interviewees expressed reservations at the approach adopted by the local authorities and key development interests to regenerate the respective development areas. In the discussion that follows two key findings are highlighted which have resonance for brownfield regeneration projects more generally. A first conclusion is that the orientation of brownfield spaces towards high-quality amenity, positive promotional images, and the desire to attract successful, growth-oriented and high-profile businesses, is reflected in various biases that counteract calls for diversified economic activity and employment markets. The high commercial letting cost and corporate orientation of brownfield redevelopments, which partly reflects what is required for such projects to be financially viable, mean that emerging spaces provide few options for small-scale firms or new entrepreneurs. Interviews also highlighted how existing firms can feel excluded from what is going on around them, encouraged by a lack of communication on the part of key decision makers and the feeling that the objective of attracting new economic activity means that the continuing and long-term contribution of existing businesses is overlooked and undervalued (see also Turok, 1992). To the extent that this discourages existing businesses from investing in growth and technology, it potentially renders them uneconomic in the future. Equally it was felt that emerging brownfield development schemes, with their focus on middle- to high-income workers and consumers, and the prospect of commercial and residential gentrification, may undermine existing local establishments catering for low-income or minority customers.

A second finding relates to the difficulty of linking local residents to the emerging employment opportunities and thus the potential of sustainable communities based on tight-knit forms of living and working. The same industrial legacy that has created the opportunity to redevelop low-cost derelict land is also reflected in the socio-economic conditions in adjacent areas and mismatches between employees' low levels of skills and the requirements of newly emerging businesses. Where employment connections have been forged, they often occur at a relatively low skill level, including cleaning, labouring and security activities. Evidence of limited local employment connections was collected through a postal survey of SQ businesses in 2005. Respondents were asked to indicate what percentage of their employees originated locally as opposed to coming from more distant areas, and their total number of employees. The results indicated that 3.1% of total employees arrived from adjacent neighbourhoods, 2.9% from Salford Quays itself, 6.8% from Salford, 45.8% from Greater Manchester, and 41.3% from further afield. Thus 20 years on from the commencement of the SQ development, local employment linkages remain extremely limited. Key reasons included the absence of positive measures, including investment in training and education, to link local people to emerging employment opportunities.

In large part this reflected a faith in trickle-down policies, including the positive role of the physical regeneration in inspiring the local labour force. Yet when the jobs that were advertised to attract local support did not emerge, mental barriers were reinforced, including the attitude that while SQ is a spectacular development, it is not necessarily for local people. In the case of PB greater attention has been given to the softer, human aspects of successful urban regeneration schemes. Paddington First, for example, has been established as an employment agency to link local people to local jobs, and to formalise the word-of-mouth recruitment strategies that predominated in the local communities around SQ. This difference aside, the likelihood that business occupiers were relocating from sites elsewhere within the metropolitan area, and thus bringing existing employees, limited the range of initial jobs to low-skill activities. As in SQ, there remain doubts over the extent to which training and educational programmes are sufficient to enable local residents to move beyond the lower rungs of employment activity in the emerging economic landscapes. Without significant resources to support sustainable communities, the prospect remains one of spatially segmented workforces, with managers and higher-skill workers commuting over significant distances.

6.3.2.3 Environmental dimensions

One of the most striking features of the British government's prescriptions for building 'sustainable communities' has been the lack of emphasis on what are conventionally thought of as 'green' policies and practices. For example, in the government's Sustainable Communities Plan (ODPM, 2003) there is no discussion of energy-saving building designs, carbon-neutral methods of supplying energy or environmentally friendly waste disposal. The conventional mantra that sustainable development involves 'balancing and integrating the social, economic and environmental components of their community' and 'meeting the needs of existing and future generations' is repeated on the Communities and Local Government website (http://www.communities.gov.uk). However, as indicated in Table 6.1, the economic and social aspects of a sustainable community are discussed at length while environmental components are presented as concerning only the reduction of car travel and access to open space for its residents. (In fact, the need for, and justification of, reduction of car travel on environmental grounds is not discussed explicitly but is simply implied by the endorsement of good public transport access as a characteristic of a sustainable community.)

These two aims are to be achieved, first, through provision of good access to public transport and 'the opportunity to live and work within close proximity' (ODPM 2003, p. 51), and second, by ensuring that people have access to open space without irreversible damage to the ecology. For example, in relation to Ebbsfleet and Eastern Quarry in the Thames Gateway, measures to ensure the quality of life for existing and future residents and the sustainability of the communities will include the provision of 'extensive open space and parkland to support recreation, leisure and sport, while conserving and enhancing the ecology of the sites' (ODPM 2003, p. 53; see also Chapter 7).

So how far have the developments in our two case study sites met these objectives? Both sites are easily accessible by public transport. In this respect, both sites do meet one important criterion for an environmentally sustainable community. However, this public transport is being used to bring employees from far afield to work in the highly paid service occupations, which are the dominant type of employment on both sites. Thus on neither site has good access to public transport gone hand in hand with creating a place in which a high proportion of people both live and work. As we have already discussed in relation to the creation of a sustainable economic landscape, both SQ and PB have offered residents in surrounding areas only limited access to jobs – although more attempts have been made to do so in the case of PB than SQ. The desire to attract growth-oriented and successful businesses militated against the development of a diversified local economy with a high proportion of local employment for local residents.

This situation, in which there is a high proportion of fairly long-distance commuters among employees on the sites and relatively few workers who live nearby does call into question the use of the singular term 'sustainable community'. Rather than an inclusive, socially diverse but interactive set of residents and workers, we found in both areas different communities with little interaction between them. We found little evidence on either site that workers who live some distance away were engaged with one another or with residents as active citizens contributing to debates about the future of the areas. In PB it is early to draw conclusions about links between residents and workers and their mutual involvement in decisions relating to the area. In SQ, as we have discussed, local residents and businesses both on and near the site felt excluded from decision-making and ignored by the local authority. Thus, for brownfield redevelopment of this type more thought needs to be given to devising ways of engaging both those who live and those who work in the areas in decision-making and interaction.

Access to attractive open space has been a significant element in the design in both SQ and PB since a high priority was given to creating an aesthetically attractive environment in order to attract development. Neither place had officially listed ecologically or culturally important sites. Both have made much of their waterside location by ensuring that housing and office developments included views of the water and providing pedestrian access to the waterside. In the case of SQ, the canal basins were developed to provide leisure and to create more aesthetically and appealing investor landscapes. This required the mobilisation of considerable technical expertise and money to clean the canals, and enormous efforts were made in the early days of SQ to improve the quality of the local water (Henderson *et al.*, 2007). In PB the canalside has been made accessible and public art, planting and cobbled surfaces have been used to create an attractive environment. In SQ there have been similar developments with, in addition, the Lowry Centre making a visually stunning landmark visible from many parts of the site.

Although attracting development was a major motive in redeveloping the sites aesthetically, in both places there was also a strong commitment by local authority planners to ensuring that the sites were open to outsiders and provided space for recreation for both site residents and people living nearby. However, it is notable that little provision for children's informal leisure is evident in either place. This has generated conflicts over the use of private resources for public activities. In SQ some children use the site for illegal summertime swimming, a practice that has infuriated local developers and the local authority as it is perceived to threaten the environment's aesthetic value. Other tensions have also emerged, and these are most evident in SQ because of its longer history. Here, opening up the site has also facilitated a high incidence of petty crime, such as theft from cars and businesses. This eventually resulted in the setting up of a Quay Watch scheme and the newer buildings being made more secure. In SQ some local people from the surrounding estates use the area for walking and fishing, but further development of the site is, in the opinion of some residents, eroding its aesthetic appeal and reducing its wildlife.

The implications of our research for future development of major brownfield sites is that in order to create sustainable communities more attention needs to be given to social and economic links between those living on the site and residents outside; developing a means for workers on the site to interact with residents on issues relating to its future development; and ensuring that more employment on the site is potentially accessible to local and new residents. Finally, we suggest that in future policy should embrace a richer set of environmental concerns such as impacts and resource use relating to land, water, energy and air. These must be recognised as integral to developing sustainable communities.

6.3.2.4 Community change and identity

One of the justifications for new rounds of investment has been that derelict brownfield sites represent eyesores that undermine a sense of community pride and attachment to an area (see Turok, 1992). This can have wider negative effects as it may encourage skilled people to leave deprived areas and thus make them less and less attractive to potential external investors. The extent to which this is the case is, however, relatively unproven. Indeed, in some urban communities brownfield sites are, in fact, valued by local residents who see them as open spaces or important parts of an area's history, culture and traditions. Moreover, changing the physical environment is seen as an important mechanism for the creation of new community identities based on *aspirations*. It is argued that the presence of higher-quality infrastructure encourages individuals and communities to take greater responsibility for their own futures so that they can 'improve themselves' and aspire to a 'better' quality of life. In many ways, such programmes indicate the ways in which a form of aspirational economics has replaced trickle-down as the principal mechanism through which the benefits of development are to be enjoyed by a broader range of local people (see Miliband, 2005).

Our research found that despite their often limited feelings of 'ownership' over the contours of development, in many cases respondents felt a sense of pride in the regeneration projects that were taking place in their immediate neighbourhoods. In SQ there was much praise for the character of the new development spaces in relation to what existed on the site before. The Lowry Museum was singled out for particular praise as a symbol of the area's history that had been turned into a major attraction to be enjoyed. Similarly, in PB there was a general consensus among interviewees that the development had made the area 'better' and that, despite the lack of thought given to their inclusion, there was little doubt that the area had seen major improvements, and that once excluded physical spaces had now been opened up in new and potentially exciting ways. In both cases these positive views on the development in large part reflected the relative isolation of the sites from resident communities and their 'cut-off' status. Brownfield development is very different in different contexts and takes on a variety of meanings. In PB and SQ the perceived and physical disconnection of the existing post-industrial brownfield sites and their transformation into spaces that could be used for a variety of purposes had undoubtedly contributed to these positive perspectives.

However, despite these positive feelings, respondents in both SQ and PB also reported that investment on the sites had amplified their concerns that their own neighbourhoods had been neglected. In the presence of so much ostentatious investment, their growing problems of unemployment (particularly in the case of SQ), poor-quality housing, crime and disorder, and poor-quality urban environments had become more obvious and their feelings of exclusion increasingly apparent. Rather than encouraging new forms of aspiration-building in local communities, the existence of the development areas has created new forms of disconnection and the feeling that regeneration efforts are not 'for them'. Indeed, among some of the young people interviewed in SQ and PB, the new developments were seen as an expression of exclusion rather than inclusion; they had created a 'them-and-us' local culture, rather than acting as sites for new forms of positive identification and ownership.

In addition, the research also showed that *legacies* are critically important to the formation of sustainable communities. One of the limitations in urban development visions, in places like SQ and PB, is that the localities in which development takes place tend to be defined as 'blank slates', to be moulded and shaped to meet the wider ends of development policy. However, in both areas the legacies of earlier rounds of regeneration have had a significant bearing on the politics and practices of 'new' regeneration agendas. There is evidence that the fractured and contested histories of community development in the two areas has limited the willingness of established community groups to get involved in the latest rounds of development as there is an expectation that they are powerless and unable to influence the course of development. This was particularly true in SQ where perceptions of powerlessness came across strongly from research respondents. Building new forms of identity and engagement in such contexts is extremely challenging. At the same time expectations have been raised in both areas about the benefits that brownfield regeneration will bring, and in some senses the developments have become a victim of their own publicity and hype. The sustainable communities agenda makes powerful promises that can be translated into unrealistic, holistic agendas at the local level. However, in SQ and PB these agendas were not matched through practice.

Finally, one of the interesting dimensions to emerge from the research was the use and deployment of wider concepts of heritage in the brownfield development process. As English Heritage (2004) and others have argued, the sustainable communities agenda, and the wider push towards brownfield-led regeneration, has often led to the downgrading of the cultural value of development sites/areas. Once places become labelled as development spaces, their former uses are, by definition, re-classified as out of date and of little economic value. They may, however, be historically or culturally valuable to local communities and others in very different, non-market terms. In both PB and SQ the relationships between heritage, identity and regeneration were evident. The pre-industrial historical uses of the water-based infrastructure in both places has, for example, been reclassified and given a new prominent status in the marketing and (re)labelling of the areas. Paddington *Basin* and Salford *Quays* both derive their names from a particular vision of their heritage. Some of the rhetoric in the former has also been about recreating local communities in a context where motorway development and housing renewal projects in the 1950s and 1960s physically transformed the area and effectively removed community infrastructure.

However, in the main the new development discourses have been relatively uncontroversial and there has been relatively little protest over the re-branding of the areas as investment sites. Again, part of the reason for this is the relative lack of ownership over the brownfields as they were before the regeneration. Where there has been controversy it has been related to the commodification of the new developments and the ways in which Paddington and Salford are now being marketed as exclusive development areas, in a context where neighbouring communities are still suffering from acute disadvantage. While such places clearly do not represent 'blank slates' to be developed, the lack of community protest over their re-characterisation is indicative of very different place–community–site relationships found here compared to other development areas such as London Docklands and Cardiff Bay, where controversies over the history of development spaces have dominated local politics (see Imrie and Thomas, 1999).

6.4 Conclusions: lessons for urban development policy

This chapter has argued that the process of sustainable urban brownfield regeneration and the wider building of sustainable communities in Britain's town and cities is a far from straightforward process. New forms of development require a reassessment of the complex relationships between economic change, the construction and design of new urban environments, the reformation of structures and systems of governance, and the fostering of new forms of place identity. The extent to which these changes can be structured and coordinated in a coherent and inclusive manner will vary from context to context, given the legacies of earlier rounds of physical, social and economic regeneration. The chapter has highlighted some of the strengths and weaknesses of existing attempts to create sustainable communities in English cities and some of the threats and opportunities in future rounds of sustainability-driven development.

In conclusion, there are a number of lessons for policy and planning that can be drawn from the SUBR:IM research:

- Most significantly, there needs to be a *clear definition of institutional roles and responsibilities* so that local programmes are embedded in a wider set of development practices. *Duplication and overlap* between different development agencies needs to be minimised. All too often development practice becomes confused between different agencies each with their own targets, resources and priorities. These may or may not be mutually inclusive and at worst, as examples across Britain's urban areas have shown, can be flatly contradictory.
- Development also needs to be driven by clear *strategies* of action. Inevitably, as contexts change over time, new opportunities will emerge and some strategically managed adaptability is necessary. However, if it is not clear exactly what a development agency is responsible for and how its programmes interact with those of other agencies then the confusion is likely to impact on the effectiveness of policy practice.
- On a broader scale, the powers and responsibilities of different agencies need to be thought about and *coordinated*. In many cases development agencies possess limited powers and resources and are unable to effect change. For example, it is common for urban regeneration bodies to possess time-limited, highly focused funding with little responsibility for other factors that directly and indirectly impact on the local economy

(such as housing, transport, health and so on) but remain outside of their areas of responsibility. However, a balance needs to be struck between the prescription of clear roles and responsibilities from 'above' (i.e. DCLG) and allowing local agencies to be flexible in developing their own programmes, local networks and priorities of action from 'below'. Evidence suggests that where local agencies have a clear role and have been designated appropriate powers, resources and responsibilities they have been at their most effective.

- The form and character of *funding regimes* underpins all development practice. Parkinson and Robson (2001), for example, advocate a system of flexible funding for urban regeneration in which local agencies should be encouraged to generate their own income streams if they are successful at facilitating new investments. However, there can also be a danger for agencies in being seen as 'too successful', with support attracted to those areas that can demonstrate that they have the most severe socioeconomic problems. This has the danger of stifling local innovation for fear of withdrawal of funding.

- The significance of *place factors and boundaries* is critical in influencing the effectiveness of local practice. Development boundaries should be designated in such a way that they are 'coherent and logical'. There can be a tension between functional boundaries and those of local community imaginations and place 'association'. Defining problem places and where development boundaries should be drawn is essential to the implementation and effectiveness of development programmes. There is a danger of creating 'cliff edges' within cities between included and excluded places.

- Finally, d*evelopment visions* are critical to the character of development practice. Defining exactly where and when a development should take place, what the end product will 'look' like and how development policy should be implemented and governed will shape how development programmes emerge and what impacts they will have. Visions define success, failure and policy directions.

References

Deas, I., Robson, B. and Bradford, M. (2000) Re-thinking the urban development corporation 'experiment': the case of central Manchester, Leeds and Bristol. *Progress in Planning*, **54**, 1–72.

English Heritage (2004) *A Place Called Home? A Vision for the Thames Gateway*. English Heritage, London.

Henderson, S., Bowlby, S. and Raco, M. (2007) Re-fashioning local government and inner-city regeneration: the Salford experience. *Urban Studies*, in press.

Imrie, R. and Thomas, H. (eds) (1999) *British Urban Policy: An Evaluation of the Urban Development Corporations*, pp. 3–43. Sage, London.

Kenny, M. and Meadowcroft, J. (eds) (1999) *Planning for Sustainability*. Routledge, London.

Maloutas, T. (2003) Promoting social sustainability: the case of Athens. *City*, 7, 165–79.

May, T. (2003) *Social Research*, 2nd edn. Open University Press, Buckingham.

Miliband, D. (2005) 'Allies of Aspiration'. Speech to National Housing Federation Annual Conference, 16 September, found at http://www.odpm.gov.uk/index.asp?id=1122748

ODPM (2003) *Sustainable Communities: Building for the Future*. HMSO, London.

Paddington Waterside Partnership (2001) *Creating a Place*. Paddington Regeneration Partnership, London.

Parkinson, M. and Robson, B. (2001) *Urban Regeneration Companies: A Process Evaluation*. Department of the Environment, Transport and the Regions, London.

Raco, M. (2007) *Building Sustainable Communities: Spatial Policy, Place Imaginations and Labour Mobility in Post-War Britain*. Policy Press, Bristol.

Robson, B., Bradford, M., Deas, I., Hall, E., Harrison, E., Parkinson, M. *et al.* (1994) *Assessing the Impact of Urban Policy*. HMSO, London.

Sadek, J. (2002) *The Effectiveness of Government Regeneration Initiatives*. Select Committee on Office of the Deputy Prime Minister: Housing, Planning, Local Government and the Regions, Examination of Witnesses 2 December: 232–9. http://www.publications.parliament.uk/pa/cm200203/cmselect/cmodpm/76-ii/2120205.htm

Shepheard, Epstein and Hunter (1985) *Salford Quays: The Development Plan for Salford Docks*. Salford City Council.

Turok, I. (1992) Property-led regeneration: panacea or placebo? *Environment and Planning A*, **24**, 361–79.

Urban Task Force (1999) *Towards an Urban Renaissance*. The Urban Task Force chaired by Lord Rogers of Riverside. HMSO, London.

Whitehead, M. (2003) (Re)analysing the sustainable city: nature, urbanisation, and the regulation of socio-environmental relations in the UK. *Urban Studies*, **40**, 1183–206.

Part 3
Remediation

7

Greening Brownfield Land

Andy Moffat and Tony Hutchings

7.1 Introduction

In this chapter, we will discuss the challenges of establishing and maintaining valued greenspace on land previously used for industrial purposes, part of what is commonly known as 'brownfield land'. Other chapters have explained the derivation of this term and its various meanings today (see, for example, Chapters 1 and 5). It is our responsibility here to explore the various meanings around the term 'greening', and to provide our views, significantly reviewed and refined as a result of participation in the SUBR:IM consortium, on how greenspace can be created *sustainably*. Of necessity, our discussion will be dominated by our experience in the UK, but examples from other countries will be used where appropriate. Our appreciation of international scientific literature suggests that many of the issues discussed in the chapter will find resonance in other countries where the challenges of converting brownfield land into high-quality urban greenspace are being faced.

7.2 Background and context

7.2.1 Greening and greenspace

Despite indirect historical references going back several decades, the formal terms 'greening' and 'greenspace' have only comparatively recently been used explicitly to describe the purposeful creation of one or more forms of vegetation on land previously devoid of it, or where existing vegetation is fragmentary and/or failing to meet the 'new' needs of the site. Greening is used at various scales and intensities, from adding a significant greenspace component to regional urban regeneration (Defra/ODPM, 2004) to establishing a grass sward on a newly placed soil layer over a landfill site in order to reduce the risk of erosion. It is also used nowadays to describe a process towards environmental sustainability, with a view to the future, as in the development of an industry's environmental policy. Greening attracts very different interpretations in terms of the motives behind its importance, from the desire for public goods and services that properly executed greenspace might provide, persuading the public that industrial activities do not conflict with principles of sustainable development, to the need to prove compliance with planning conditions for revegetation after mineral extraction or landfill closure. 'Greening up' is often used in the latter context, but this can also convey an additional meaning of temporality, as a process to be accomplished with little or no concern for future permanence or success in meeting declared aims and objectives.

Greenspace is the outcome of the greening process, and is usually considered as a feature of the *urban* landscape. It can take many forms, and serve many purposes (Table 7.1). At its heart, there is the expectation that greenspace has been created and continues to be managed for the public good. In the UK there is renewed interest at government level in the quality of urban living, and greenspace has received considerable focus as a means for its improvement. In urban regeneration, almost inevitably involving the remediation and reclamation of brownfield land, greenspace is seen as an essential component of the new landscape.

The wide variety of greenspace types, coupled with the range of policy drivers and local interpretations, means that the process of conversion from brownfield land, itself infinitely variable in site properties and geographical context, is potentially a complex subject to discuss in a single chapter of a book. Nevertheless, we shall attempt to pull out the main issues and processes that apply to the generic conversion of one type of land to the other.

Table 7.1   Types of urban greenspace.

Public/local parks
Planned gardens
Domestic gardens
Churchyards and cemeteries
Urban forests/woodland parks
Allotments
Sports fields
City farms
Specialist parks, e.g. Victory parks, ecology parks
Riparian zones

7.2.2 History of urban greenspace

In the UK, parks and domestic gardens represent the most areally extensive forms of urban greenspace. In London, for example, major parks like Hyde Park and Regent's Park have evolved from royal hunting grounds adjacent or local to the seat of power. The Victorian era saw the greatest creation of formal open greenspace, with many areas purposefully secured from potential housing development in order to promote the physical and mental well-being of urban dwellers and encourage their moral and citizenship qualities (Reeder, 2006). Parks became increasingly open to the general public rather than merely to those living in the immediate surroundings. Even during these times, greenspace was regarded as important for its comparative lack of pollution and human pathogens, and its educational, recreational and sporting value was also used to promote greenspace 'causes' (Reeder, 2006).

In subsequent city expansion into the suburbs, and post-war reconstruction, there has been a need for more intervention under the planning system, requiring a formalisation of the purpose of greenspaces and desirable levels of provision. In suburban expansion, greenspace provision has, for example, mainly been based on the appropriate protection of existing parkland and farmland from the housing developer and road engineer. However, bombsite redevelopment stands out as an earlier example of where greenspace creation occurred on 'brownfield' land.

7.2.3 Modern greenspace

Today, advocates of greenspace point to a range of goods and services that it can provide. Development of moral fortitude is not on the 'A' list today, but to the many remaining benefits of greenspace espoused by the Victorians there are some positive environmental effects to add (Table 7.2). With the increase in concern over global warming, some authors have pointed to the potential benefits of greenspace for reducing the urban 'heat island effect' (Huang *et al.*, 1987), as well as providing shade against high temperature and potential diseases such as skin cancer caused by exposure to the sun (Tretheway and Manthe, 1999; Whitford *et al.*, 2001). Provision of wildlife habitat has been subsumed under the mantle of biodiversity, and artificial urban habitats are now recognised for their importance in protecting nationally as well as locally scarce species (Harrison and Davies, 2002). Biodiversity interests and ecological science are also the backbone behind so-called urban 'ecology parks' where certain habitats are deliberately created for educational purposes (Sellers *et al.*, 2006). Areas of greenspace within the town or city will reduce local rainwater run-off but they may also serve as zones where surface water can be allowed to flood, rather than doing so in

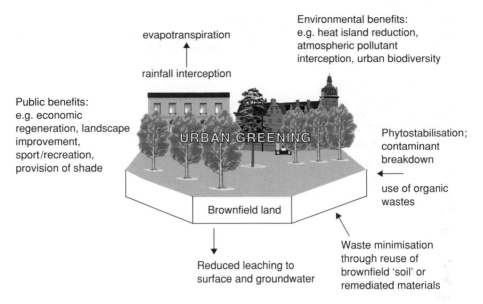

Figure 7.1   'Positive' greening.

residential or industrial sectors. Flood mitigation is increasingly important for urban areas where climate change scenarios predict higher winter rainfall, and where sea level rise will increase the risk of flooding in the future. Pollution modification is also seen as an important contribution that greenspace can make. Atmospheric pollutants can be intercepted by vegetation and concentrations in the air reduced, for the benefit of both human dwellers and the city infrastructure such as buildings (Broadmeadow and Freer-Smith, 1996). Certain forms of vegetation may also sequester soil contaminants such as heavy metals (Pulford and Watson, 2003), or help to break down organic ones (e.g. Lynch and Moffat, 2005). Vegetation, especially woodland, can also reduce noise 'pollution' and enhance quality of life (Martens, 1981; Bucur, 2006). Furthermore, urban greenspace may play a small part in carbon sequestration and in fossil fuel substitution if biomass is produced for heat and power generation (McPherson *et al.*, 1997). The concept of 'positive' greening is summarised in Figure 7.1.

7.2.4 Opportunities for greenspace in UK urban redevelopment

Modern estimates of the amounts of brownfield land vary considerably. On the conservative side, the National Land Use Database[1] suggests about 36 600 ha of the brownfield total is vacant and derelict land or buildings in England (in 2005), and Scottish sources[2] suggest 16 000 ha (in 2001) in

Table 7.2   Main benefits of urban greenspace.

Benefit	Useful references
Social	
Facilitates recreation and sport	
May encourage walking and cycling rather than using powered vehicles	
Enhances recovery from ill health	Ulrich (1984)
Encourages mental well-being	de Vries *et al.* (2003)
Provides community focus	
Opportunities for education, e.g. forest school	Westley (2003)
Economic	
Enhances property values of surrounding land	GLA Economics (2003); Laverne and Winson-Geideman (2003)
Encourages inward investment	
Encourages tourism	Villella *et al.* (2006)
SME spin-offs, e.g. cafes, crafts	
Environmental	
Atmospheric pollution abatement	Broadmeadow and Freer-Smith (1996); McPherson *et al.* (1997)
Soil pollution abatement	Hutchings (2002); Lynch and Moffat (2005)
Water pollution prevention	
Climate mitigation	Huang *et al.* (1987)
Wildlife habitat/biodiversity	Harrison and Davies (2002)
Flood abatement	Xiao *et al.* (1998)
Noise abatement	Kellomäki *et al.* (1976); Fang and Ling (2005)
Biomass production	Enviros Consulting Ltd and CL:AIRE (2006)
Carbon sequestration	Brack (2002)

equivalent terms. There are seemingly no equivalent data for Wales. However, these amounts contrast with the estimated 360 000 ha of contaminated land in Britain (ENDS, 2000), the 100 000 sites possibly affected by contamination in England and Wales (DTLR, 2003) and the estimated 200 000 ha of 'damaged and neglected' land in England and Wales (Perry and Handley, 2000). Perry and Handley considered that the latter 'clearly represents a potentially important opportunity for lowland afforestation' – urban greenspace, in more modern parlance.

Not all brownfield land is suitable for greenspace creation and, depending on geographical and economic circumstances, other land uses will have greater claims. For example, the chapters in Part 2 of this book have demonstrated the very different competition from hard-end development in the two SUBR:IM study areas, namely Greater Manchester and Thames Gateway. In addition, the balance between the budget for land remediation and greenspace infrastructure and that for maintenance and management may significantly affect decision-making about greenspace desirability in a particular area. Nevertheless, the above case studies suggest that brownfield land provides a considerable opportunity for greenspace creation, and thus enhancement of the urban environment where it occurs.

7.2.5   Greenspace policy drivers

In *Living Places. Cleaner, Safer, Greener* (ODPM, 2002a), the UK government set out its views on the importance of parks and greenspaces, recognising them as integral to the wider public space network, and 'as much a part of the urban fabric as its buildings'. Greenspace was seen to support a wide range of government priorities, including regeneration, renewal and housing programmes, supporting healthy living, and fostering neighbourhood pride and community cohesion. A more detailed expression of the government's view is found in *Greening the Gateway* (Defra/ODPM, 2004). This is specifically a greenspace strategy for Thames Gateway, but as an expansion of the higher-level policy in previous reports, it is more widely applicable. Although environmental benefits are specifically acknowledged in the Thames Gateway report, it is clear that the main drivers for greenspace are to enhance the 'liveability' of existing and newly created domestic and work places; environmental drivers are relevant inasmuch as they benefit mankind. Provision of 'user-friendly public and green spaces' is also one of the stipulations for a sustainable community, as defined in the government's Sustainable Development Strategy (Defra, 2005).

7.2.6   Other policy relevance

7.2.6.1   Contaminated land
The 'clean-up' of contaminated land has been an important environmental factor behind urban regeneration. With the advent of a modern definition of contaminated land in the UK (DETR, 2000a) and modern requirements for making such land safe once it is identified,[3] there has grown a greater appreciation of the possibility of contamination in previously used urban land, and the need for effective site and risk assessment before, during and after land remediation (see Chapter 3). The opportunity for sites or parts of sites to be restored to greenspace in order to reduce remediation costs has been advocated by some (e.g. Hutchings, 2002; Hutchings and Moffat, 2003). Although a renewed focus on contaminated land has occurred in recent years, this might be seen more as a catalyst than as a driver for greenspace creation.

7.2.6.2   Water management
Interest in sustainable urban drainage systems (SUDS) has also required an integrated approach to land use, which is theoretically at its most flexible in large redevelopment programmes. Amenity and biodiversity interests

are intrinsic to SUDS, and are likely to be maximised through spatial linkage with greenspace (Darlow *et al.*, 2003). As well as helping to improve amenity and ecological value, SUDS can help to achieve other requirements set for water quality and flooding under the Water Framework Directive (WFD) through impacting on the river basin management plans (RBMPs) for the region in which the urban area is located. It is still relatively early to detect the impact of the WFD on greenspace creation, location and type, but it is likely to become significant as RBMPs are developed from 2008, and subsequently evaluated. Greening brownfield land will inevitably be part of this process.

7.2.6.3 Biodiversity

Biodiversity is an issue which has enjoyed significant support in relation to the urban environment, and brownfield land has been viewed as a real and new prospect to enhance biodiversity through purposeful planning for wildlife (Box and Shirley, 1999). Government and regional guidance supports the need to assess impact on biodiversity when planning greenspace development on brownfield land (Mayor of London, 2002; ODPM, 2002b). In this sense, biodiversity considerations are more a constraint than a driver for greenspace development, though Local Agenda 21 interests may be interpreted through greenspace creation. However, the concept of urban biodiversity remains relatively ill-defined and misunderstood (Gyllin, 1999), and in practice is unlikely to strongly influence greenspace policy.

7.2.6.4 Waste minimisation

Since the EU Landfill Directive entered into force in 1999, the introduction of Landfill Regulations in 2002–2003[4] has increased the emphasis on waste minimisation and reuse of materials in land remediation, and limited the disposal of contaminated soils to hazardous waste sites only. There is thus increasing interest in in situ remediation technologies, to minimise costly 'dig and dump' operations. Research conducted under the SUBR:IM programme has shown that for some remediated materials using, for example, thermal or bioremediation, successful grass, wildflower and tree growth can be achieved (Sellers *et al.*, 2005). Such research may increasingly instil confidence in developers and regulators that greenspace is a cost-effective and safe option in land remediation. However, it is too early to see whether waste minimisation will become a genuine driver in this respect.

There is more evidence to suggest that the need to reuse or recycle other wastes, notably organic materials removed from the waste stream before landfilling or from the water industry, is supporting greenspace creation at a strategic level. In 2000, around 6% of sewage sludge was recycled in the reclamation of brownfield land nationally (DETR, 2000b), but in some regions such as north-west England where brownfield land occurs disproportionately, recycling is as high as 25%.[5] The opportunity for land recycling has been taken up by several companies who have taken on responsibility for incorporating waste organic material into the land reclamation strategy, notably to improve the fertility and water-holding capacity of the growing media. Such activities have, almost certainly, influenced the way local decisions have been taken on the desirability and practicability of brownfield conversion to greenspace in recent years.

7.3　A sustainable process for greenspace

Notwithstanding the policy relevance of greenspace and its avowed importance as a component in urban regeneration of brownfield land, the *sustainability* of new urban greenspace is rarely discussed (Moffat, 2004). This issue remains the poor relation compared to the larger-scaled issue of the sustainability of urban regeneration. Given the rhetoric from many quarters about the importance of greenspace, it is sad that investment in maintaining *existing* urban parks has been inadequate for many years, and although quality has improved nationally, there remain significant problems in one in six local authorities (Greenhalgh and Worpole, 1995; ILAM Services Ltd, 2000; House of Commons Committee of Public Accounts, 2006). It is vital that this issue is confronted; otherwise there is a large risk that the expectations of stakeholders with an interest in *new* urban greenspace will not be realised. This section deals with 'best practice principles' for greenspace creation on brownfield land, based on a combination of the benefit of the Forestry Commission's longstanding experience in creating urban forests and community woodlands on this land type, the work of the SUBR:IM consortium in looking holistically at the subject, and in the biophysical research undertaken by Forest Research and others to increase greenspace sustainability.

7.3.1　Where to site greenspace

There is significant benefit in considering strategic issues at the landscape scale when choosing the location of greenspace, rather than accepting piecemeal projects which result in an unplanned provision. In larger regeneration schemes, such as those in the two major SUBR:IM research locations (Thames Gateway and Greater Manchester), there is usually considerable scope to take this approach.

A major social sustainability criterion for new greenspace is that it should be required. This means that there should be an institutional champion for the greenspace project, a likelihood of successful passage through the planning and regulatory frameworks, a budget for the work and for management and maintenance thereafter, and public acceptance that the greenspace will be appreciated – and used. Community support for greenspace is probably strongest in areas where there is a current dearth and community needs are paramount. Nevertheless, achieving sustainable greenspace is most likely when it forms an integral part of a larger regeneration project or programme, rather than being a stand-alone idea. Opportunities must be taken to see the greenspace as increasing the overall value and delivery possibilities for regeneration, rather than as a token, or simple, contrast to the built environment. For example, location of new schools and hospitals should take the possible co-location of greenspace into account, in order to maximise the educational and recuperative properties that greenspace can offer. It is also important that new greenspace is easily accessible to those expected to use it (Villella *et al.*, 2006).

The desirability of the interconnectedness of urban greenspace has been promoted widely by several lobbies. 'Greenways' have been advocated as 'networks of land containing linear elements that are planned, designed and managed for multiple purposes including ecological, recreational, cultural, aesthetic, or other purposes compatible with the concept of sustainable land use' (Ahern, 1995). In the Thames Gateway, 'green grids' of interconnected greenspace have been promoted, to benefit from the presumed enhanced biodiversity, recreation and flood attenuation that this will give (Defra/ODPM, 2004). Depending on the spatial arrangement of these green 'routeways' or 'linear corridors', brownfield land may play its part in supporting the integral nature of the system through greenspace creation on it. However, the nature of the brownfield site and substrate (e.g. if seriously contaminated) may mean that the requirement for greenspace interconnectedness at a particular location demands inordinate resources, and necessitates a revision of the greenspace policy. In these circumstances, it is vital to evaluate options at a regional scale in as holistic a manner as possible. In addition, the scientific basis for the biodiversity benefits of urban greenways is equivocal (Angold *et al.*, 2006), though more research is warranted in this area. On contaminated brownfield sites, connectedness may actually increase the threat of toxicity to important fauna and flora (Lafortezza *et al.*, 2004).

The nature of brownfield land has often led to its dereliction, and in time, to vegetative colonisation. Many vegetated brownfield sites are now highly regarded for the greenspace they provide, whether for informal recreation such as dog-walking, or for unusual or rare plants and animals that the brownfield substrate supports. Several unreclaimed brownfield sites are designated as Sites of Special Scientific Interest (SSSI) for these reasons. In London, brownfield sites support the nationally scarce black redstart, burrowing bees and wasps as well as providing feeding habitat for the declining house sparrow (Mayor of London, 2002). Indeed, remediation and reclamation are regarded by many ecologists as counterproductive in their desire to maintain and enhance urban biodiversity (Harrison and Davies, 2002). Although it is important to balance ecological value against dangers presented by contamination or unstable structures (Mayor of London, 2002), current policy is to support the retention of some brownfield sites, or parts of them, in order to contribute to urban greenspace with little or no further engineering.

In large regeneration projects, it is certainly desirable to evaluate potential greenspace locations from a strategic viewpoint. For example, in the north-west England SUBR:IM study area, the Public Benefit Recording System (PBRS)[6] has been widely used to prioritise effort and expenditure in greenspace creation. PBRS is based on a simple additive scoring methodology, and Figure 7.2 is a summary of its methodology and criteria used to score sites being considered for greenspace. PBRS attempts to integrate potential 'social', 'access', 'economic' and 'environmental' benefits for each site under investigation, but it does not consider relative remediation and reclamation costs – these are determined later in the process. Hence, early decisions on the desirability of greenspace creation may need substantial modification as relative costs are factored in. Other decision support systems have been devised to support community consultation in the process of greenspace location (Tippett, 2004). However, there appears as yet no methodology that integrates the various determinants which ultimately affect greenspace location.

7.3.2 Site investigation for greenspace establishment on brownfield land

Achieving sustainable remediation of brownfield sites is highly challenging because of the heterogeneous nature of ground conditions and contamination types; the existing social, ecological and archaeological resources on site; the need for community engagement and involvement; and the need to ensure that the restoration will deliver the targeted functionality of the restored landscape.

Successful restoration requires that site commodities and liabilities are fully ascertained and understood prior to restoration or other forms of engineering. The site investigation process is an integral part of a greenspace development cycle as it provides foundation information which determines ecological and historical resources as well as hydrological, chemical and physical conditions. This will assist the developer in making informed choices when considering liabilities, remedial requirements and appropriate soil, habitat and species choices. In all cases, the restoration of a site should be considered in terms of the wider landscape and its context within locally, regionally and nationally defined regeneration strategies. Guidance

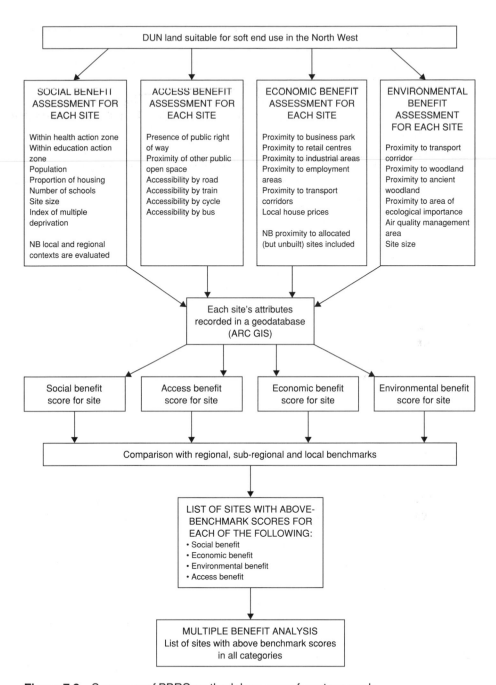

Figure 7.2 Summary of PBRS methodology. www.forestry.gov.uk

is available that defines the site investigation process in detail (Hutchings *et al.*, 2006; Doick and Hutchings, 2007).

7.3.3 The soil resource

Urban soils differ considerably from their rural counterparts because physical and chemical disturbance has often resulted in conditions which will constrain vegetation growth. It is vital that restored ground conditions allow adequate provision of a soil resource that is free of compaction and toxic contamination and has a water-holding capacity and nutrient levels to support the survival and growth of vegetation in an unhindered way. The site investigation process must involve a review of historical land use, which should guide the user in determining soil materials that are of concern or, alternatively, represent a significant resource. Earth movement during the site redevelopment process offers a unique opportunity to remediate physical and chemical constraints and add materials to 'construct' soils which will be sustainable in the long term.

The nature of soil, its potential to sustain vegetation and the consequences of these features are fundamental to the success and sustainability of a greenspace establishment project. Consequently, Doick and Hutchings (2007) advised that the soil environment should be considered the focal point of the site investigation and that four key questions should be *repeatedly* asked throughout the site selection and investigation processes:

(1) Will the site support trees/vegetation? Particular consideration should be given to drainage and water-holding capacity, presence of phytotoxic chemicals, fertility, physical soil characteristics, soil cover and rootable depth, and topography of the site.
(2) Will the establishment of the greenspace generate, amplify or negate risks and hazards?
(3) Will vegetation establishment adversely affect the site?
(4) How will the site be managed in the short and long term?

7.4 Contamination

Much of the 'vacant' land that is suitable for conversion to urban greenspace is likely to be contaminated. Previous industrial usage is the major cause of contamination and there is a strong link between the industrial history and the type of contamination found on sites. Certain industrial processes or mixed former land use can result in 'cocktails' of contaminants, which often compounds the challenge of clean-up.

7.4.1 Assessing contamination risk

The source–pathway–receptor model is used to evaluate risks of contamination on brownfield land destined for greenspace. As in any remediation project, fitness for purpose principles should be employed to ensure that users or components of the final greenspace are not receptors which will be exposed to further contamination and therefore risk of harm. In the greenspace context, potential receptors almost certainly include human visitors as well as ground and surface waters, ecology, livestock and buildings. Furthermore, components such as the created habitat constitute valid receptors which should not be ignored when assessing risks and devising suitable methods and targets for remediation.

7.4.1.1 Types of sampling

There are two main types of sampling:

(1) targeted: based on prior knowledge and professional judgement to investigate a given area
(2) non-targeted: sets out a defined sampling pattern and spacing to investigate an area

A combination of both targeted and non-targeted sampling should be used where there are obvious areas of potential contamination (targeted) *and* areas where contamination location is unknown (non-targeted).

Targeted sampling should only be employed where the conceptual site model has highlighted specific risks that the professional reviewer is confident can be resolved using a targeted sampling approach. For example, if the model had highlighted a medium to high risk of pollutant linkage from a point source to a specified receptor and the exact whereabouts of both source and receptor were known, then a targeted sampling approach could be implemented to test whether the pollutant pathway was significant.

Other examples where targeted sampling might be used include:

- areas of stressed vegetation
- areas where surface water has collected, which may indicate soil compaction
- very sensitive areas, e.g. planned picnic or children's play areas

Guidance on sampling design and the number of samples required to delineate hotspots of soil conditions which limit vegetation success, such as contamination, compaction and levels of plant nutrients, are given in Hutchings *et al.* (2006).

7.4.1.2 Analytical methods for metal contaminated soils

Many environmental benchmark standards (e.g. soil screening values and soil guideline values) are based on *aqua regia* extractable metal concentrations rather than total metal concentrations that would traditionally be determined by laboratory-based X-ray fluorescence (XRF) or following concentrated hydrofluoric acid digestion. *Aqua regia* methods do not constitute a complete digestion because the least acid-soluble components, for example metal silicates, are not wholly digested and are therefore not included within the analytical result. In contrast, data gathered by XRF are derived from all matrix material and thus represent a 'total' analysis.

Concentrated acid extractions and their laboratory analyses are time-consuming and expensive. In some cases, this expense may compromise the extent of sampling, analysis and quantification of contaminants, and variability across sites. Field-portable XRF systems (FPXRF) have been developed that offer a fast and efficient means of determining soil contaminant concentrations at a definitive or screening level. Several authors (Potts *et al.*, 1995; Kilbride *et al.*, 2006) have compared *aqua regia* extractable and FPXRF-analysed metal levels in soils and found that the detection limits of FPXRF were sufficient to 'screen' soils for heavy metals at levels which would be deemed to be contaminated under modern guidance (Table 7.3).

7.4.2 Assessing bioavailability and toxicity

Although useful for screening purposes, research and experience has shown that ascertaining total contamination concentrations alone does not give a true reflection of the exposure (leachability and bioavailability)

Table 7.3 Dual source FPXRF data showing the quality level for the entire concentration range of the sample set.

Analyte	*Aqua Regia* extractable concentration range (mg kg⁻¹)	FPXRF concentration range (mg kg⁻¹)	n*	r²	RSD	Slope	Y-intercept	Data quality level
Fe	482 – 91566	437 – 121958	81	0.97	1.7	0.99	n/a	Quantitative
Ni	13 – 84	67 – 258	16	0.10	6.7	0.34	1.64	Qualitative
Cu	29 – 785	55 – 10400	66	0.88	6.4	n/a	n/a	Definitive
Zn	8 – 25389	21 – 47180	80	0.94	5.7	0.88	0.33	Quantitative
Cd	3 – 447	21 – 287	11	0.61	12.7	0.46	0.98	Qualitative
Pb	9 – 40398	10 – 61286	79	0.97	4.0	1.01	n/a	Quantitative
As	8 – 4585	10 – 3160	47	0.92	8.2	0.87	n/a	Quantitative
Mn	84 – 38267	422 – 83354	19	0.88	5.3	0.80	0.74	Quantitative

* n < 81 signifies analyte concentration was below the level of detection of the FPXRF instrument.
Source: Kilbride *et al.* (2006)

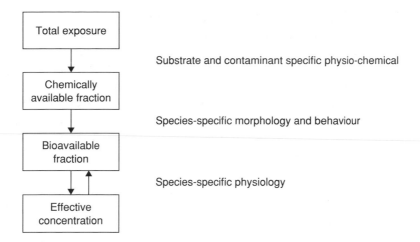

Figure 7.3 Influence variables on the biologically effective concentration of a contaminant in soil (after Herrchen *et al.*, 2000; Kördel and Hund-Rinke, 2001).

or toxicity of contaminants for vegetation. This represents a challenge to site redevelopers and environmental regulators who must demonstrate the degree of remediation required to reduce contamination to an acceptable end-point. Contaminant exposure is commonly divided into a chemically available fraction, a fraction available *to* an organism (bioavailable fraction) and the fraction taken up *by* the organism (effective concentration) (Figure 7.3). Toxicity is thereby most closely related to species-specific physiological behaviour of uptake and tolerance, and contaminant solubility and bioavailability. To achieve sustainable vegetation establishment on soils with elevated concentrations of contaminants, it is therefore important that the soluble and bioavailable fractions of the total soil concentration are quantified and that the physiological behaviour of the vegetation being established is fully appreciated.

As well as employing various chemical methods, another way of assessing contaminant availability is to employ toxicological testing. This gives an insight into how much contaminant will be taken up by the biota and how the organism of interest will react. Limited techniques are now available for such assessments to be made on humans (Wragg and Cave, 2002) and ecological receptors such as earthworms and other soil invertebrates (ISO, 1993; 1998; 1999) and plants (ASTM, 2002). There is broad consensus that these methodologies are very useful but research is required to generate techniques that have a greater ecological relevance.

The spatial variation of plant-available soil contaminants is also important to establish. Although there are numerous studies of the spatial variation of total metal concentrations in soil, there are few which describe the spatial variation of plant-available contaminants. Kilbride and Hutchings

(2005) tested the potential of poplar cuttings to act as biological indicators of the variability in plant availability of soil-borne metal contaminants across heavy-metal-contaminated sites. Geostatistical analysis of bioavailable foliar metals showed that there was spatial dependence at the sites for plant leaf concentrations of Cu, Cd, Zn and Pb of as little as 2.6 metres. The number of sampling positions which exceeded recommended toxicity threshold values for the plant tissues were far fewer than those which exceeded regulatory guideline values based on total concentrations. Although sampling by a site redeveloper at this scale will usually be impractical, the study shows the need for stringent site investigation which examines the risks that contaminants pose to vegetation as well as to other receptors. This suggests that greater resources given to targeted site investigation are likely to yield information that can help delineate significant risks, which may greatly limit the remediation required on sites and reduce the incidence and costs of vegetation failure.

7.4.3 Plant uptake, compartmentation and tolerance

Plant species have different contaminant tolerances, and uptake and compartmentation behaviours. Tolerance to metals varies between species and between populations of the same species (Dickinson *et al.*, 1992) and resistance to one particular metal does not imply resistance to others (Dickinson, 1996). For trees, tolerance, uptake and compartmentation may change with age (Turner, 1994). Although young stock must be used to establish trees on polluted land, changes in the tolerance of a tree should be considered. Successful establishment of seedlings probably indicates good tree survival in the long term; however, mature trees on a site cannot be used as an unconditional indicator that a seedling of the same species would also survive. As the root structure of a tree develops it may also encounter regions of greater contamination with an increased risk of root toxicity. This highlights the importance of good site survey prior to planting, and continued monitoring of tree growth and soil conditions after greenspace establishment.

Species vary considerably in the degree of metal uptake from the soil and the retention of the metal in the root system. It is important that consideration is given to metal compartmentation as the movement of metals into the upper biomass may pose a risk of food chain transfer and ecological harm.

7.4.3.1 Soil manufacture

Research in the SUBR:IM consortium has demonstrated that vegetation can be established on contaminated soil materials which have undergone a range of remediation technologies (Sellers *et al.*, 2005). Table 7.4 is an

Table 7.4 Effect of selected remediation technologies on some important soil properties.

Technique	pH	AWC	Nutrients	Organic matter
Thermal desorption	pH raised as base cations released from organic matter. This can lead to the binding up of any P that is left and non-availability of micronutrients such as Fe, Mn, Zn and Cu	Available water capacity will be reduced due to the absence of OM and reduced pore size.	Most major nutrients either mineralised or destroyed, in particular N	Loss of organic matter leading to poor soil structure, nutrient retention and reduced AWC
Bio-remediation	pH may not necessarily be affected but the correct pH is vital for microbial activity. Below about pH 6 the microbial activity is curtailed so low pH substrates need to be amended to raise the pH.	As microbial and faunal activity proceeds then OM will be created and soil fauna such as earthworms will increase. This should lead to improved available water capacity.	Bacterial activity, such as nitrogen fixation will improve N levels. However, bacteria degrading contaminants may require more nutrients than are available, especially N. Nutrient levels may have to be supplemented for bacterial activity to be optimised.	Organic matter will increase over time through microbial activity and possible faunal activity, but will depend on the time the soil spends being bioremediated.
Chemical extraction	If acidic solvents used then pH may drop considerably. Below 5.5, nutrient availability is restricted and bacterial activity curtailed. Low pH will destroy clay mineral structure and organic matter. Cation leaching will increase. At low pH, compounds are in a reduced state which can increase toxicity.	If soil structure is destroyed by strong REDOX acidic or organic reagents then AWC will be reduced through the collapse of pore space and size.	If the soil becomes acidified by the reagents used then nutrient availability will be reduced. There will also be increased loss through leaching as nutrient cations such as P, K and Ca become displaced and solubilised by H^+.	Organic matter can be reduced as it is destroyed by decreasing pH and by the use of organic solvents.

Source: Moffat and Hutchings (2005)

example of a review of the effect of various types of remediation on the capability of soils to sustain vegetation growth and thus support urban greenspace. The application of organic and inorganic amendments to remediated soil-forming materials can aid vegetation establishment and in some circumstances help to further remediate metal and organic contaminants. Remediated soils commonly lack organic matter, and have little or no topsoil and low levels of essential macro/micronutrients that plants need for sustained growth and health. In addition to probable contamination, they have characteristically poor physical structure and poor water-holding capacity. Incorporation of organic materials into these soil materials prior to vegetation establishment can help to restore soil structure by providing organic matter and sustenance to invertebrates, which can aerate and mix materials effectively over comparatively short time periods. Amendments also supply plant nutrients (Table 7.5) and act as slow release fertilisers.

Amendments can influence contaminant mobility and bioavailability. For example, the mobility of metal and organic contaminants may be reduced by the formation of insoluble complexes between the amendment and contaminant (Gadepalle *et al.*, 2007). In contrast, interactions of organic composts with some soils can cause the formation of soluble metal complexes which may increase the uptake of metals into vegetation. In contrast to metals, many organic compounds such as polyaromatic hydrocarbons (PAHs) can be degraded by microorganisms present in or whose populations are promoted by the added composted materials (Lynch and Moffat, 2005).

The use of composted organic soil amendments (e.g. municipal solid waste compost, biosolid compost, mature compost, cow manure) for restoring heavy metal- and arsenic-contaminated soils has been tested by several researchers. Most experiments demonstrated that the uptake of heavy metals and arsenic by plants was reduced by the addition of composted materials to the soil. However, experience gained during the SUBR:IM programme has shown that the response of soil contaminants to an amendment is both compost- and soil-specific, with some interactions causing an increase in metal solubility and bioavailability (see Chapter 8).

Natural zeolites have potential for metal immobilisation owing to their high ion exchange capacities and highly porous structures. Research has shown that clinoptilolite is most effective in immobilising lead and cadmium in soils (e.g. Coppola *et al.*, 2003). Care should be taken when using zeolites with high sodium content as they can increase soil salinity and exchangeable sodium concentrations which could cause plant toxicity (Stead, 2002). Soil amendment with 'red mud' and other iron-rich compounds has been shown to immobilise labile metals. For example, red mud is especially effective at reducing arsenic bioavailability and improving plant performance (Friesl *et al.*, 2003). Red gypsum, phosphogypsum and other phosphate-based amendments also have good abilities to absorb lead.

Table 7.5 Key properties of organic amendments (from Kilbride, 2006).

Type	Fertiliser value			C/N ratio	Organic matter content	Cost/availability	Other properties
	N	P	K				
Liquid sewage sludge	..	.	x	..	.	Free to users – transport costs to be paid	Generally unpleasant to use; biological pathogens likely although reduced in 'digested' sludges; heavy metals possible
Sewage sludge cake	..	..	x	..	..	Free. Most common biosolid currently available	Possible heavy metals, malodorous, pathogens likely if originated from undigested liquid sludge
Thermally dried sewage sludge	..	..	x	..	..	Comparatively expensive, limited availability	Liming properties
Alkali-conditioned sewage sludge	..	x	.	..	..	Cost charged, not commonly available	Increase soil pH, risk of nitrate leaching
Composted sewage sludge	.	.	.	..	..	Not commonly available, cost charged	Horticultural use
Greenwaste compost	..	.	.	..	..	Cost charged, available throughout UK	Mature composts provide slow release of N
Wood residues (various from forest industries)	xx	x	x	xx	..	Widely available in a range of types. High production costs.	Principally increases soil porosity and drainage, can cause N deficiency – addition of N fertiliser recommended
Animal manure (cow)	..	x	x	..	..	Readily available – mainly transportation costs	Different sources of origin e.g. cattle, pig. Properties vary dependent on source. Potential loss of N through volatilisation of NH_4-N
Industrial by-products (paper mill sludge)	xx	x	x	xx	..	De-watered papermill sludges are widely available. Generally provided free of charge	N immobilisation due to very high C/N ratio; possible presence of fungicides/bactericides
Spent mushroom compost	..	.	..	..	..	Readily available – cost charged	Used as a surface mulch, can be detrimental to young plants
Straw	xx	x	x	xx	..	Commonly available – low cost	Can cause N deficiency
Blood and guts	..	x	x	.	x		Unpleasant origin

.. Very good, . good or adequate, x no effect, xx may be detrimental
Source: Kilbride (2006)

Many potential amendments are based on materials derived as industrial by-products from the waste stream, and the use of specific materials in the greening of brownfield land can significantly help towards meeting waste utilisation targets of local authorities. For example, the Waste Resource Action Programme (WRAP) estimates that the use of source-segregated greenwaste composts (PAS 100) (British Standards Institution, 2005) on brownfield sites for greening purposes could utilise 163 000 tonnes of material in 2007 (WRAP, 2006). The potential use of an amendment at a site can be usefully examined based on a strengths, weaknesses, opportunities and threats (SWOT) analysis, as presented in Table 7.6.

7.4.3.2 Greenspace design

The conversion of greenspace from brownfield land must be a multidisciplinary team task if the greenspace is to succeed. It will entail the inputs of landscape architects; soil scientists, hydrologists and engineers; ecologists, silviculturalists and horticulturalists; and social scientists. And the design process probably represents where in the whole greening procedure this wide range of disciplines must work together. Greenspace design may be low key or statement-making, but it is essential that the various 'strands' that form the design process are all considered, even if on a particular project some are more important than others. Table 7.7 gives a summary of the most important, and the remainder of this section will put them in context.

In the past, greenspace design has probably been led by the landscape architect whose main concern was to create a pleasing and usable landscape primarily for the local community. Clearly, the main use or uses of the land must strongly influence its final appearance, and different forms of community engagement are usually used to inform the process. If the land is in public ownership, there will usually be a considered view of the range of possible land-use outcomes, and consultation with stakeholders will be used to refine and 'polish' these plans. If the land is in private ownership, the views of the landowner may be paramount. Depending on location, existing state and initial plans for the site, stakeholder consultation may be largely restricted to the local community, or may range more widely to include organisations interested in wildlife protection, sport and recreation, or history and archaeology, for example. Access to and use by the public are usually cited as the main reasons for community interest in a greenspace project, though wildlife habitat creation or 'ecology' can also be popular with the local community.[7]

There are many types of community engagement (Hislop *et al.*, 2004), and some are more appropriate than others, depending on the desired outcome of this process. Table 7.8 highlights those types considered most appropriate for greenspace consultation, based on our experience in the two SUBR:IM research areas.

Table 7.6 Potential strengths, weaknesses, opportunities and threats of amendment usage.

Strengths
- Can potentially immobilise metals and persistent organics
- Can potentially provide environment where biodegradation of organics is speeded up
- Can utilise sustainable sources of materials e.g. 'waste' derived materials
- Low cost of materials e.g. may only have to pay for transportation and incorporation
- Low site disturbance
- In situ processes which are simple to operationalise
- Can be tested at a laboratory scale prior to full-scale usage
- Can remediate contaminants whilst also adding nutrients to sustain vegetative growth
- Low input energy or water requirement
- Could act as a life belt for mitigation against potential future mobilisation of contaminants
- Could act as a final 'polishing' system after other remediation techniques
- Can be used to convert materials affected by other techniques e.g. thermal desorption back to soils

Weaknesses
- Amendments can potentially mobilise contaminants especially at the application stage
- Amendments may legally be defined as 'wastes' rather than products
- Amended soil material may legally be defined as 'waste'
- Liabilities remain i.e. metal and persistent organic contaminants remain on site
- Geotechnical strength of the material may be too low to allow use under the foundations of buildings or pathways
- Fluxes in soil conditions may cause a release in metals or gasses e.g. changes in redox (flooding) could cause release of methane and/or affect water quality where organic amendments have been applied inappropriately
- Potential remobilisation of contaminants as organic additives degrade
- Needs to be tested at the laboratory scale prior to site usage
- Difficulty mixing consistently within heavier soils

Opportunities
- Could revolutionise in situ contaminated land treatment
- Significant diversion of materials from the waste stream
- Environmental clean-up of 'low-value' sites may become cost-effective
- Could be used to treat catchment scale problems e.g. mining site remediation
- Potential use for soil remediation or soil formation at ex-situ treatment centres
- Joining of three areas of predicted high economic growth – remediation, waste reduction and urban greenspace sectors

Threats
- Over or improper specification of materials by material providers or contractors to create a 'waste' sink
- Statutory authorities seeing amended material as a waste
- Declassification of some amendments from wastes to products is difficult
- Risk of over-application
- Incineration of 'waste' derived amendments may provide cheaper option
- Materials could become too sought after, their value spirals upwards

At its most polarised, greenspace design can be either 'naturalistic' or 'artificial' (Kendle, 1997), the former approach deliberately working with nature to establish vegetation that most suits the site conditions, usually in the expectation of minimal ongoing management after vegetation has

Table 7.7   Greenspace design 'strands'.

Consideration of substrate(s) via SI; ability to support vegetation and of what types/species (edaphic issues); degree of contamination and need for remediation/engineering (e.g. capping)
Likely residual contamination issues; human exposure
Community consultation
Consideration of constraints (e.g. wayleaves, hotspots)
Site characteristics (topography, rainfall, soil erodibility, fertility, water holding capacity)
Predicted climatic conditions and likely water supply
Possibilities for organic waste importation, and types
Integration with existing greenspace, and access from surrounding housing
Planning conditions (?)
Discharge consent(s)
Roading and pathways; existing design lines
Range of vegetation type options and soft end uses
Need for car parking and other hard standing
Existing archaeological and ecological value
Existing surface water and drainage requirements; need for water features
Primary purpose(s) of greenspace; likely usage
Likely management structures and maintenance budgets

Table 7.8   Suitable methods for consultation about greenspace establishment.

Technique	Informing	Consulting	Involving
Advertisements	**		
Advisory committees			***
Displays	**		
Events	**	**	
Focus groups		**	
Forums		**	**
Interviews		**	
Leaflets	**	*	
Media	**		
Newsletters	***		
Openhouse	**	**	
Open space		**	**
Participatory appraisal		**	***
Public hearings	**	*	
Questionnaires		**	
Site visits	**	***	**
Staffed displays	**	**	
Task force		*	***
Websites	***	*	
Working groups		**	***
Workshops		**	***

* suitable, ** very suitable, *** highly suitable
Source: derived from Hislop *et al.* (2004)

been established. The 'community woodland' movement of the 1990s is a good example of this approach. In contrast, artificial design owes more to the tradition of park design championed by the Victorians (Reeder, 2006), with an acceptance of non-native species, geometrical layout of planting areas, and a landscape of novel or unique appearance. This approach usually depends on the drawdown of significant maintenance costs, for example in grass-mowing or irrigation, and Thames Barrier Park is a good example (Villella *et al.*, 2006). Of course, a brownfield site may be designed across this spectrum, depending on sub-site characteristics and the interests of the stakeholders.

The naturalistic approach to vegetation establishment is usually cheaper than the artificial one because it does not need to employ the use of comparatively expensive non-native plants. But the main reason for its economic advantage lies in the tailoring of species to site conditions as far as possible. This means that the information gained through site investigation is fully used to identify species which are edaphically[8] tolerant, or that will be once the substrate is modified to address issues of *extreme* acidity, infertility or toxicity. Essentially it is the approach of modifying the species list to suit the site, rather than attempting to modify the site to permit the growth of chosen species. Matching species to site will also involve an assessment of the water needs of the plants and the ability of the site to supply them, usually through plant-available water stored in the soil. Irrigation should be avoided if at all possible, as such costs can be large and a significant annual overhead. Water provision will, of course, be affected by climate change (Hulme *et al.*, 2002), notably in the south and east of England. Hence, design should increasingly take future climate change scenarios into account when planning vegetation choice, notably in the context of soil materials available to support summer water demand (Moffat, 1995). Other aspects of the future climate, such as temperature and windiness, may also affect the choice of vegetation for new greenspace on urban brownfield sites.

Design for wildlife habitat or biodiversity is likely to be very different from that for formal recreation, such as the provision of sports pitches, and many brownfield sites may require compromise in order to attempt to achieve both aims. Obviously, the naturalistic approach to design will favour habitat creation. Nevertheless, biodiversity objectives, although often strongly advocated by some, will pose real challenges to their successful execution in an urban environment, with obvious competing pressures.

Many have also suggested that the provision of water features can be advantageous. However, the creation of ponds and lakes on sites capped to restrict ingress of water and plant roots into the contaminated material below may be problematic, with some regulatory bodies ill-disposed to them on these substrate types. Even after remediation, surface waters on contaminated brownfield land may contain elevated metal concentrations which could pose a danger to visiting humans and their pets. In these cases, constructed wetlands may be important to 'polish' the water quality. Water features which depend on pumping have had a chequered history, requiring significant maintenance and energy resources to sustain them, and there are several examples of failure in the SUBR:IM study areas (Sellers *et al.*, 2006). Health and safety issues must also be taken fully into account during water feature design.

Purposeful design to emulate particular ecosystems is at the heart of the 'ecology park' concept. Originating in the Netherlands, this approach has been adopted when greening several brownfield sites, for example Russia Dock Ecology Park in London (Sellers *et al.*, 2006). While ecology parks may have intrinsic educational value, it is very doubtful whether a series of fully functional ecosystems can be created on most brownfield sites, given the probably limited size available and other constraints and expectations placed upon them. Such an approach is also at odds with modern understanding of landscape ecology and the need for such ecosystems to be supported by other similar ones nearby. Nevertheless, popular demands to support wildlife may encourage greenspace planners to include ecology-park thinking into the design, even if this is unsustainable in the medium to long term.

7.5 Sustainable greenspace

Previous sections of this chapter have discussed factors that must be taken into account when planning and executing greenspace development on brownfield land in order to achieve a sustainable solution. This section will look at evidence to assess the current state of progress towards the sustainability 'goal' and future prospects for urban greenspace on such sites, given likely or possible environmental, social and economic changes to urban centres and the people who live in them.

7.5.1 Greenspace sustainability – myth or reality?

Sustainability has become engrained in UK planning legislation (ODPM, 2005), although the central concept remains somewhat elusive and open to personal interpretation. Because academics have had most time to consider it, there perhaps appears a greater understanding and commitment to the principle than is apparent in regional and local authorities where the complexity of the real world can severely test its relevance and applicability (Selman, 2000). In the context of urban greenspace, there remains considerable confusion over what sustainability means. For engineers, the main

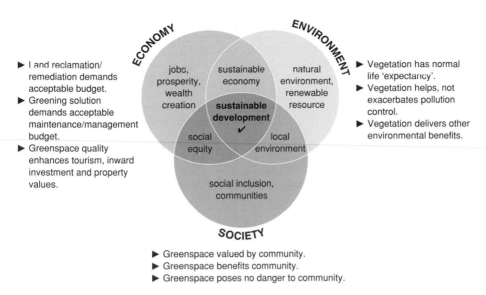

► Land reclamation/ remediation demands acceptable budget.
► Greening solution demands acceptable maintenance/management budget.
► Greenspace quality enhances tourism, inward investment and property values.

► Vegetation has normal life 'expectancy'.
► Vegetation helps, not exacerbates pollution control.
► Vegetation delivers other environmental benefits.

► Greenspace valued by community.
► Greenspace benefits community.
► Greenspace poses no danger to community.

Figure 7.4 Sustainable greenspace.

concerns will be the effectiveness and longevity of pollution-control measures, for example after in situ remediation or capping. For silviculturalists and horticulturists, the most important issue might be the longevity of the vegetation planted. For community leaders, it might be focused on the utilisation of the greenspace by the public and the degree of social cohesion engendered by it. For the local authority, it might be the balance between the enhancement of the value of the properties adjoining the greenspace and the cost of running it. And developers will inevitably be most interested in supporting site sustainability design criteria so long as budgets and profits aren't threatened. A simple model is shown in Figure 7.4. There remain no easily usable tools to integrate these aspects of sustainability so that trade-offs between conflicting demands can be analysed and holistic decisions made, although work is currently ongoing to develop these.[9] In their absence, greenspace projects will continue to suffer from subjective interpretation and fail to reach the potential that could otherwise be achieved.

Many examples of urban greenspace established on brownfield land have been visited during the SUBR:IM programme, and in other similar research campaigns (e.g. Dobson and Moffat, 1993, 1999; Bending *et al.*, 1999; Moffat and Laing, 2003). Examples of long-term sites have been identified where comparatively low levels of maintenance are required to support vegetation that appears to be serving the needs of the local community well. The Russia Dock Ecology Park in Southwark, London, and Eastbrookend Country Park in the London Borough of Barking and Dagenham are good examples of this type. However, our research has revealed many more sites where

Table 7.9   Examples of greenspace created on brownfield sites.

Queen Elizabeth Park, Vancouver, Canada
Parc des Buttes-Chaumont, Paris, France
Parc André-Citroën, Paris, France
Landschaftspark, Duisburg-Nord, Germany
Thames Barrier Park, London, UK
Gas Works Park, Seattle, USA
Russia Dock Ecology Park, London, UK

serious failings have resulted in essentially *un*sustainable greenspace (e.g. Moffat and Laing, 2003). Iconic sites that exemplify the problem of retaining sustainability include the 1984 International Garden Festival site in Liverpool.[10] Nevertheless, there are some notable city parks around the globe sited on brownfield land, which demonstrate that sustainable goals can be met (Table 7.9).

Guidance on greenspace creation on brownfield land is increasingly tied into the sustainability model. From a very technical position in the 1980s and 1990s (e.g. Bradshaw and Chadwick, 1980), modern guidance now stresses the socio-economic and ecological benefits to be gained and how to achieve them. Nevertheless, there remain disciplinary divides which prevent uptake of best practice guidance, and thus its implementation. Engineers are largely ignorant of the benefits that some forms of vegetation can bring to the remediation and reclamation of land, despite occasional attempts to inform them (e.g. Coppin and Richards, 1990). And the SUBR:IM research experience has emphasised the very different approaches in the language, understanding and problem-solving of biophysical and social scientists which must be bridged in order to exploit guidance effectively (see Chapter 2).

In addition to the challenges of assimilating information across a wide range of disciplines, it is clear that in some cases, supposedly authoritative guidance, built upon sound science, is pushing in different directions. A major problem concerns the application of ecological theory in the urban centre – and how wildlife conservation can be balanced with the social demands of urban greenspace. Conservationists are naturally interested in the concept of greenspace creation for the opportunity that it might give them to deliver wildlife protection, given the sometimes limited opportunities in other land-use sectors. And they have imported landscape ecological thinking from rural environments into the city to defend their stance that urban planners should play a significant part in designing greenspace to that end (Angold *et al.*, 2006; Gyllin, 1999). However, there has been little research and less evidence that greenspace connectedness makes a significant difference to the ability of urban fauna and flora to survive, reproduce and extend their habitat boundaries.

7.5.2 Greenspace in the context of future societal and environmental change

Environmental and societal change does not appear to have been taken account of in current greenspace design on brownfield land, and may be at odds with community consultation and dated ecological thinking which pushes towards use of native species (Moffat and Laing, 2003) and stock of local provenance for naturalistic areas when climate change predictions suggest some of these may be ill-suited under different climatic conditions in future years. This issue is especially important when considering the planting of perennial crops such as trees that may live for many decades (Broadmeadow and Ray, 2005). The science behind local provenance selection has also been challenged recently in the context of a changing climate (Hubert and Cottrell, in press).

Within the SUBR:IM portfolio of work, Al-Tabbaa *et al.* (Chapter 11) have described the likely effects of climate change on engineered solutions to land contamination. They argue that vegetated greenspace may have a distinct place in helping to break the source–receptor linkage on some contaminated sites. However, it is clear that confidence in using vegetation will grow only if the longevity of vegetation in a changing climate can be assured. The main issue to address here is provision of plant-available soil water during the growing season, but this may require inordinate amounts of soil or soil-forming materials which may or may not be available. Hence, the desirability of creating greenspace must always be tempered by a realistic expectation of future growth. There will also remain the risk or perception of risk that some forms of vegetation may challenge engineered pollution-control measures such as caps or barriers in containment systems, especially in periods of drought associated with climate change (e.g. Albrecht and Benson, 2001). Further research is required to explore UKCIP climate change scenarios in the context of these types of system.

7.5.3 Monitoring and evaluation

Despite modern encouragement by central government for urban expansion and regeneration to focus on brownfield land, including stated objectives for greenspace to enhance liveability, there has been little conspicuous attempt at monitoring or evaluating greenspace sustainability at a site, regional or national level. Monitoring and evaluation across the many impacts that greenspace can make in urban regeneration should be seen as an essential part of the overall process. Results allow sustainability to be evaluated and lessons learned for future projects to utilise. Doick *et al.* (2006) have proposed a methodology for integrated consideration of social, environmental and economic impacts which incorporates measurable objectives of greenspace establishment at a range of spatial scales. The methodology is based, in part, on the Redevelopment Assessment Framework (RAF) developed by SUBR:IM research (Pediaditi *et al.*, 2005; see also Chapter 12). Country-wide application of this monitoring strategy could provide a powerful evidence-base of greenspace sustainability and allow a comprehensive analysis of policy drivers while identifying knowledge gaps, research needs and best practice requirements.

7.5.4 An economic case for greenspace establishment

The economic case for greenspace establishment in urban areas is commonly misunderstood. Some developers consider greenspace establishment as a reduction in their financial margins, using it solely as a means of appeasing planning authorities when proposing hard-end development of brownfield sites. In reality, greenspace has been shown to increase property values within its surrounding area. For example, Tyrväinen and Miettinen (2000) examined the relationship between greenspaces and property values in the district of Salo, Finland. They concluded that a one kilometre increase to the nearest urban forest led to an average 5.9% decrease in the market value of properties, with properties with a view onto urban forests having an average premium of 4.9% over dwellings with otherwise similar characteristics. A study in London (GLA Economics, 2003) found that a 1% increase in greening in a typical ward is associated with a 0.3–0.5% increase in an average house price. In addition, greenspace was ranked as being the fifth most significant indicator in explaining the variation in average house prices in the same study.

7.6 The future of greenspace on brownfield land

Climate and societal change will inevitably occur in the next decades, and have a significant effect on the type and extent of greenspace sited on brownfield land. That brownfield land continues to be generated at a similar rate to that utilised during urban regeneration (Perry and Handley, 2000) suggests that there will remain significant opportunities for conversion to greenspace. Urban populations are predicted to increase, just as they have done over the last millennium, with London reaching over 7.7 million by 2021 (Virdee and Williams, 2003). Urban dwellers will need greenspace, perhaps more than they do today. For example, urban temperatures are set to rise, and shade provided by greenspace may offer much-needed respite from the heat and reduce the risk of skin cancers. Trends in obesity may be reversed, though this currently seems unlikely, and so greenspace provision will be increasingly necessary to enable authorities to encourage more physical activity. Increasing fuel charges, plus a likely future tax on use of fossil fuels or the road network, will restrict the movement of urban dwellers so that access to the countryside may become more difficult.

Engineering and ecological research will continue to explore how vegetation may serve to support land remediation and quantify its effects so that regulators become increasingly confident about adopting greenspace as an after-use for contaminated land. Harmonised monitoring and evaluation procedures, built on quality-management systems, will also increase confidence in current remediation and reclamation projects and help promote their successors.

Funding will remain an inevitable obstacle to increased conversion of brownfield land to greenspace and its long-term management. Local authority budgets will always experience stiff competition, and greenspace projects will only succeed if they are exemplary. More attention needs to be given to mechanisms for financial support for the maintenance and management of such sites. Certainly the Land Restoration Trust[11] has demonstrated the value of funding specifically for this purpose. Other forms of funding also need to be explored and the design of new urban greenspace reconsidered in order to achieve this more effectively.

7.7 Conclusions

This chapter has demonstrated real opportunities, and need, for continued greenspace creation during urban regeneration, inevitably involving brownfield land. There is good evidence that greenspace creation can serve a range of useful purposes, though some claims, notably those related to urban biodiversity, require more science behind them before they can be factored into greenspace spatial planning. Research into the ability to grow vegetation on remediated soil materials has begun, and although there is more to do, results during the SUBR:IM programme suggest significant possibilities for using these materials. In contrast, land reclamation technologies for sustainable greenspace creation are well understood and tested, though there is considerable evidence of malpractice. Site investigation procedures are demanding in order to reduce risk of unsustainable vegetation on (remediated) contaminated land and the risk to human visitors. New techniques have been developed to enable more effective survey (e.g. FPXRF) but more guidance needs to be developed, for example on likely bio-uptake.

There have been some attempts at maximising the value of greenspace through regional analysis, and some tools have been developed. Yet there appears no widespread coherent understanding of the totality of the challenge and opportunity. This chapter identifies the need for agencies to work together more strategically in future. The problem remains that greenspace does and should involve many strategic partners and many academic disciplines in order to achieve sustainable status. Without an appropriate holistic conceptual model, and an effective 'chain of command', it remains all too easy for failure to occur during greenspace implementation. The importance of greenspace is such that it probably demands integration of some of the most important 'players' so that sustainable best practice can be more often achieved.

Notes

1. http://www.nlud.org.uk/
2. http://www.scotland.gov.uk/Publications/2002/03/14412/1531#4
3. Statutory Instrument 2000 No. 227, The Contaminated Land (England) Regulations 2000.
4. Statutory Instrument 2002, No. 1559, The Landfill (England and Wales) Regulations 2002; Scottish Statutory Instrument 2003, No. 235, The Landfill (Scotland) Regulations 2003.
5. http://uucsr.lfdev.net/resource-use.asp
6. http://www.pbrs.org.uk
7. http://www.crystal.dircon.co.uk/report21_intro.htm
8. That is, the effect of soil characteristics, especially chemical or physical properties, on plants and animals.
9. http://www.zalf.de/home_ip-sensor/products/SENSOR_del_5.2.2_complete_appr_format_small.pdf
10. http://www.bbc.co.uk/liverpool/capital_culture/2004/04/garden_festival/index.shtml
11. http://www.landrestorationtrust.org.uk

References

Ahern, J. (1995) Greenways as a planning strategy. *Landscape and Urban Planning*, **33**, 131–55.

Albrecht, B. and Benson, C. (2001) Effect of desiccation on compacted natural clays. *Journal of Geotechnical and Geoenvironmental Engineering-ASCE*, **128**, 67–75.

Angold, P.G., Sadler, J.P., Hill, M.O., Pullin, A., Rushton, S., Austin, K. *et al.* (2006) Biodiversity in urban habitat patches. *Science of the Total Environment*, **360**, 196–204.

ASTM (2002) E 1963–98 standard guide for conducting terrestrial plant toxicity tests. In: *Annual Book of Standards*. Vol. 11.05 Biological Effects and Environmental Fate; Biotechnology; Pesticides. ASTM International, West Conshohocken, PA.

Bending, N.A.D., McRae, S.G. and Moffat, A.J. (1999) *Soil-Forming Materials: Their Use in Land Reclamation*. The Stationery Office, London.

Box, J. and Shirley, P. (1999) Biodiversity, brownfield sites and housing. *Town and Country Planning*, **68**, 306–309.

Brack, C.L. (2002) Pollution mitigation and carbon sequestration by an urban forest. *Env Poll*, **116**, 195–200.

Bradshaw, A.D. and Chadwick, M.J. (1980) *The Restoration of Land*. Blackwell Scientific Publications, Oxford.

British Standards Institution (2005) *Specification for Composted Materials*. British Standards Institution Publicly Available Specification 100:2005, London.

Broadmeadow, M.S.J. and Freer-Smith, P.H. (1996) *Urban Woodland and the Benefits for Local Air Quality*. Research for Amenity Trees 5. The Stationery Office, London.

Broadmeadow, M. and Ray, D. (2005) *Climate Change and British Woodland*. Forestry Commission Information Note 69. Forestry Commission, Edinburgh.

Bucur, V. (2006) *Urban Forest Acoustics*, p. 191. Springer-Verlag, Berlin.

Coppin, N.J. and Richards, I.G. (eds) (1990) *Use of Vegetation in Civil Engineering*. Construction Industry Research and Information Association (CIRIA)/Butterworths, London.

Coppola, E.I., Battaglia, G., Bucci, M., Ceglie, D., Colella, A., Langella, A. *et al.* (2003) Remediation of Cd- and Pb-polluted soil by treatment with organozeolite conditioner. *Clays and Clay Minerals*, **51**, 609–15.

Darlow, T., Garden, M., Wild, T. and Walker, K. (2003) Maximising the benefits of SUDS by taking an integrated approach to planning. In: *Diffuse Pollution and Basin Management* (ed. M. Bruen), pp. 4-32–4-37. Proceedings of the 7th International Specialised IWA Conference, Dublin, Ireland.

Defra (2005) *The UK Government Sustainable Development Strategy*. The Stationery Office, London.

Defra/ODPM (2004) *Creating Sustainable Communities: Greening the Gateway*. Office of the Deputy Prime Minister, London.

DETR (2000a) *Contaminated Land: Environmental Protection Act 1990 Part IIA*. DETR Circular 2/2000, The Stationery Office, London.

DETR (2000b) *Waste Strategy 2000 for England and Wales*. The Stationery Office, London.

De Vries, S., Verheij, R.A., Groenewegen, P.P. and Spreeuwenberg, P. (2003) Natural environments – healthy environments? An exploratory analysis of the relationship between greenspace and health. *Environment and Planning* A, **35**, 1717–31.

Dickinson, N.M. (1996) Metal resistance in trees. In: *Heavy Metals and Trees* (ed. I. Glimmerveen), pp. 85–92. Proceedings of a discussion meeting. ISF Press, Edinburgh.

Dickinson, N.M., Turner, A.P. and Lepp, N.W. (1992) How do trees and other long-lived plants survive in polluted environments? *Functional Ecology*, **5**, 5–11.

Dobson, M.C. and Moffat, A.J. (1993) *The Potential for Woodland Establishment on Landfill Sites*. HMSO, London.

Dobson, M.C. and Moffat, A.J. (1999) Examination of tree and root performance on closed landfills in Merseyside. *Arboricultural Journal*, **23**, 261–72.

Doick, K.J. and Hutchings, T.R. (2007) *Greenspace Establishment on Brownfield Land: The Site Selection and Investigation Process*. Information Note. Forestry Commission, Edinburgh.

Doick, K., Sellers, G., Hutchings, T.R. and Moffat, A.J. (2006) Brownfield sites turned green: realising sustainability in urban revival. In: *Brownfields III. Prevention, Assessment, Rehabilitation and Development of Brownfield Sites* (eds C.A. Brebbia and U. Mander), pp. 131–40. WIT Press, Southampton.

DTLR (2003) *Development on Land Affected by Contamination*. Consultation paper on draft planning technical advice. Department for Transport, Local Government and the Regions, London.

ENDS (2000) *Britain's contaminated land liabilities put at £15 billion*. ENDS Report 305 (June 2000), 4–5.

Enviros Consulting Ltd and CL:AIRE (2006) *Uses of Compost in Regeneration and Remediation of Brownfield Sites in the UK*. WRAP, Banbury.

Fang, C.F. and Ling, D.L. (2005) Guidance for noise reduction provided by tree belts. *Landscape and Urban Planning*, **71**, 29–34.

Friesl, W., Lombi, E., Horak, O. and Wenzel, W.W. (2003) Immobilisation of heavy metals in soils using inorganic amendments in a greenhouse study. *Journal of Plant Nutrition and Soil Science*, **166**, 191–6.

Gadepalle, V.P., Ouki, S.K., van Herwijnen, R. and Hutchings, T.R. (2007) Immobilisation of heavy metals in soil using natural and waste materials for vegetation establishment on contaminated sites. *Soil and Sediment Contamination*, in press.

GLA Economics (2003) *Valuing Greenness: Green Spaces, House Prices and Londoners' Priorities*. GLA, London.

Greenhalgh, L. and Worpole, K. (1995) *Urban Parks and Social Renewal*. Comedia and Demos, Stroud and London.

Gyllin, M. (1999) Integrating biodiversity in urban planning. In: *Communication in Urban Planning*, Proceedings of the Conference of the Urban Density and Green Structure, European Research Network held in Gothenburg, Sweden 2–5 October 1999. http://www.arbeer/demon.co.uk/MAPweb/Goteb/got-mats.htm

Harrison, C. and Davies, G. (2002) Conserving biodiversity that matters: practitioners' perspectives on brownfield development and urban nature conservation in London. *Journal of Environmental Management*, **65**, 95–108.

Herrchen, M., Kratz, W., Marschner, A., Necker, U., Pieper, S., Römbke, J., Riepert, F., Rück, F., Terytze, K., Throl, C. and Wilke, B-M. (2000) Eckpunkte zur Gefahren-beurteilung des Wirkungspfades Bodenverunreinigungen–Bodenorganismen. Fachausschuss. *Biologische Bewertung von Böden* der Fachgruppe 4 *Bodenfunktionen und – belastungen* des Bundesverband Boden (BVB), 83 S.

Hislop, M., Twery, M. and Vihemäki, H. (2004) *Involving People in Forestry. A Toolbox for Public Involvement in Forest and Woodland Planning*. Forestry Commission, Edinburgh. http://www.forestresearch.gov.uk/forestry/infd-5xmf8l

House of Commons Committee of Public Accounts (2006) *Enhancing Urban Green Space*. Fifty-eighth Report of Session 2005/06. Stationery Office, London.

Huang, Y.J., Akbari, H., Taha, H. and Rosenfeld, A.H. (1987) The potential of vegetation in reducing summer cooling loads in residential buildings. *Journal of Applied Meteorology*, **26**, 1103–16.

Hubert, J. and Cottrell, J. (in press) *The Role of Forest Genetic Resources in Helping British Forests Respond to the Effects of Climate Change*. Forestry Commission Information Note. Forestry Commission, Edinburgh.

Hulme, M., Jenkins, G., Lu, X. *et al.* (2002) *Climate Change Scenarios for the United Kingdom: The UKCIP02 Scientific Report*. Tyndall Centre, University of East Anglia, Norwich.

Hutchings, T.R. (2002) *The Opportunities for Woodland on Contaminated Land*. Forestry Commission Information Note 44, Forestry Commission, Edinburgh.

Hutchings, T.R. and Moffat, A.J. (2003) *Greening Brownfield and Contaminated Sites*. Environmental Data Interactive Exchange. http://www.edie.net/library/view_article.asp?id=319&channel=0

Hutchings, T.R., Sinnett, D. and Doick, K.J. (2006) Best Guidance Note for Land Regeneration No. 1: *Soil sampling derelict, underused and neglected land prior to greenspace establishment*. Forest Research, Farnham.

ILAM Services Ltd (2000) *Local Authority Owned Parks. Needs Assessment. Phase 1*. ILAM Services, Reading. http://www.ilam.co.uk/downloads/parksrep.pdf

ISO (1993) ISO11268 *Soil Quality: Effects of Pollutants on Earthworms (Eisenia fetida) Part 1: Determination of Acute Toxicity Using Artificial Soil Substrate*. ISO, Geneva.

ISO (1998) ISO11268-2 *Soil Quality: Effects of Pollutants on Earthworms (Eisenia fetida) Part 2: Determination of Effects on Reproduction*. ISO, Geneva.

ISO (1999) ISO11268-3 *Soil Quality: Effects of Pollutants on Earthworms Part 3: Guidance on the Determination of Effects in Field Situations*. ISO, Geneva.

Kellomäki, S., Haapanen, A. and Salonen, H. (1976) Tree stands in urban noise abatement. *Silva-Fennica*, **10**, 237–56.

Kendle, A.D. (1997) Natural versus artificial methods of woodland establishment. In: *Recycling Land for Forestry* (ed. A.J. Moffat). Forestry Commission Technical Publication No. 22, Forestry Commission, Edinburgh, 26–35.

Kilbride, C. (2006) Best Guidance Note for Land Regeneration No. 6: *Application of sewage sludges and composts*. Forest Research, Farnham.

Kilbride, C. and Hutchings, T.R. (2005) The variability in plant available potentially toxic elements across a metal contaminated site using a biological indicator. *SEESOIL*, **16**, 37–46.

Kilbride, C., Poole, J. and Hutchings, T.R. (2006) A comparison of Cu, Pb, As, Cd, Zn, Fe, Ni and Mn determined by acid extraction/ICP–OES and ex situ field portable X-ray fluorescence analyses. *Environmental Pollution*, **143**, 16–23.

Kördel, W. and Hund-Rinke, K. (2001) Ecotoxicological assessment of soils – bioavailability from an ecotoxicological point of view. In: *Treatment of Contaminated Soil* (eds R. Stegmann, G. Brunner, W. Calmano and G. Matz). Springer-Verlag, Berlin Heidelberg, Germany.

Lafortezza, R., Sanesi, G., Pace, B. and Corry, R.C. (2004) Planning for the rehabilitation of brownfield sites: a landscape ecological perspective. In: *Brownfield Sites II. Assessment, Rehabilitation and Development* (eds A. Donati, C. Rossi and C.A. Brebbia), pp. 21–30. WIT Press, Southampton.

Laverne, R.J. and Winson-Geideman, K. (2003) The influence of trees and landscaping on rental rates at office buildings. *Journal of Arboriculture*, **29**, 281–90.

Lynch, J.M. and Moffat, A.J. (2005) Bioremediation: prospects for the future application of innovative applied biological research. *Annals of Applied Biology*, **146**, 217–21.

McPherson, E.G., Nowak, D., Heisler, G., Grimmond, S., Souch, C., Grant, R. *et al.* (1997) Quantifying urban forest structure, function, and value: the Chicago Urban Forest Climate Project. *Urban Ecosystems*, **1**, 19–61.

Martens, M.J.M. (1981) Noise abatement in plant monocultures and plant communities. *Applied Acoustics*, **14**, 167–89.

Mayor of London (2002) *Connecting with London's Nature. The Mayor's Biodiversity Strategy*. Greater London Authority, London.

Moffat, A.J. (1995) Minimum soil depths for the establishment of woodland on disturbed ground. *Arboricultural Journal*, **19**, 19–27.

Moffat, A.J. (2004) Captain SUBR:IM gets seasick. http://www.subrim.org.uk/Articles/discussion/CaptainSUBRIMsseasickam.pdf

Moffat, A.J. and Hutchings, T.R. (2005) *Greening of Brownfield Land*. Proceedings of the SUBR:IM – Sustainable Urban Brownfield Regeneration: Integrated Management 1st Public Conference 1 March 2005, Natural History Museum, London. CL:AIRE, London.

Moffat, A.J. and Laing, J. (2003) An audit of woodland performance on reclaimed land in England. *Arboricultural Journal*, **27**, 11–25.

ODPM (2002a) *Living Places. Cleaner, Safer, Greener*. Office of the Deputy Prime Minister, London.

ODPM (2002b) Policy Planning Guidance Note 17: *Planning for Open Space, Sport and Recreation*. The Stationery Office, London.

ODPM (2005) Planning Policy Statement 1: *Delivering Sustainable Development*. Office of the Deputy Prime Minister, London.

Pediaditi, K., Wehrmeyer, W. and Chenoweth, J. (2005) Monitoring the sustainability of brownfield redevelopment projects: the redevelopment assessment framework (RAF). *Contaminated Land & Reclamation*, **13**, 173–83.

Perry, D. and Handley, J. (2000) *The Potential for Woodland on Urban and Industrial Wasteland.* Forestry Commission Technical Paper 29. Forestry Commission, Edinburgh.

Potts, P.J., Webb, P.C. and Williams-Thorpe, O. (1995) Analysis of silicate rocks using field-portable X-ray fluorescence instrumentation incorporating a mercury (II) iodide detector: a preliminary assessment of analytical performance. *Analyst,* **120**, 1273–8.

Pulford, I.D. and Watson, C. (2003) Phytoremediation of heavy metal-contaminated land by trees – a review. *Environment International,* **29**, 529–40.

Reeder, D.A. (2006) The social construction of green space in London prior to the Second World War. In: *The European City and Green Space* (ed. P. Clark), pp. 41–67. Ashgate Publishing Ltd, Aldershot.

Sellers, G., Hutchings, T.R. and Moffat, A.J. (2005) Remediated materials: their potential use in urban greening. *SEESOIL,* **16**, 47–59.

Sellers, G., Hutchings, T.R. and Moffat, A.J. (2006) Learning from experience: creating sustainable urban greenspaces from brownfield sites. In: *Brownfields III. Prevention, Assessment, Rehabilitation and Development of Brownfield Sites* (eds C.A. Brebbia and U. Mander), pp. 163–72. WIT Press, Southampton.

Selman, P. (2000) *Environmental Planning: The Conservation and Development of Biophysical Resources.* Sage Publications, London.

Stead, K. (2002) *Environmental implications of using the natural zeolite clinoptilolite for the remediation of sludge amended soils.* PhD thesis, University of Surrey, UK.

Tippett, J. (2004) *A participatory protocol for ecologically informed design within river catchments.* PhD thesis, School of Planning and Landscape, University of Manchester. www.holocene.net/research/phd.htm

Tretheway, R. and Manthe, A. (1999) Skin cancer prevention: another good reason to plant trees. In: *Proceedings of the Best of the West Summit, San Francisco, California September 15–17, 1998* (eds E.G. McPherson and S. Mathis), pp. 72–5.

Turner, A.P. (1994) The responses of plants to heavy metals. In: *Toxic Metals in Soil-Plant Systems* (ed. S.M. Ross), pp. 153–88. Wiley, Chichester.

Tyrväinen, L. and Miettinen, A. (2000) Property prices and urban forest amenities. *Journal of Environmental Economics and Management,* **39**, 205–23.

Ulrich, R.S. (1984) View through a window may influence recovery from surgery. *Science,* **224**, 420–21.

Villella, J., Sellers, G., Moffat, A.J. and Hutchings, T.R. (2006) From contaminated site to premier urban greenspace: investigating the success of Thames Barrier Park, London. In: *Brownfields III. Prevention, Assessment, Rehabilitation and Development of Brownfield Sites* (eds C.A. Brebbia and U. Mander), pp. 153–62. WIT Press, Southampton.

Virdee, D. and Williams, T. (2003) *National Statistics: Focus on London 2003.* The Stationery Office, London.

Westley, M. (2003) Sensory-rich education. *Landscape Design,* **317**, 31–5.

Whitford, V., Ennos, R. and Handley, J. (2001) City form and natural process: indicators for the ecological performance of urban areas and their application to Merseyside, UK. *Landscape and Urban Planning,* **57**, 91–103.

Wragg, J. and Cave, M.R. (2002) *In-vitro Methods for the Measurement of the Oral Bioaccessibility of Selected Metals and Metalloids in Soils: A Critical Review.* R&D Technical Report P5-062/TR/01. Environment Agency, Bristol.

WRAP (2006) *Uses of compost in regeneration and remediation of brownfield sites in the UK.* The Waste & Resources Action Programme, Banbury, Oxon.

Xiao, Q., McPherson, E.G., Simpson, J.R. and Ustin, S.L. (1998) Rainfall interception by Sacramento's urban forest. *Journal of Arboriculture,* **24**, 235–44.

8

Novel Special-purpose Composts for Sustainable Remediation

Sabeha Ouki, René van Herwijnen, Michael Harbottle, Tony Hutchings, Abir Al-Tabbaa, Mike Johns and Andy Moffat

8.1 Introduction

Soil remediation using compost is an emerging technology that is gaining worldwide acceptance because of its potential for the treatment of a wide variety of contaminants and its environmentally friendly benefits. The success of this technique strongly suggests that specific characteristics of composts can be enhanced to increase their palliative effect. Special-purpose compost can be developed to enhance specific attributes, produced from a particular feed to increase activity and sorption. Most of the research published in this field is based on the use of a fairly unsystematic selection of unamended composts. In addition, substantial literature indicates that naturally occurring minerals can play a major role in controlling the environmental fate and reduce the availability of both organic and inorganic contaminants (Gadepalle *et al.*, 2007). Very little of this work, however, has been applied to improve the remediation capability of organic compost.

One particular area of interest is the improvement of both the metal- and organic-binding capacity of composts by the addition of inorganic materials. Published research indicates that some naturally occurring minerals such as clays and zeolites interact with metals to form a matrix in which the bioavailability of the metals is remarkably decreased (Ouki *et al.*, 1994; Ouki and Kavannagh, 1999; Ponizovsky and Tsadilas, 2003; Castaldi *et al.*, 2005) and their pathways to potential receptors of the contaminants reduced or blocked. This attribute, coupled with the sorption capability of the compost, could provide a unique and novel remediation technique. The ultimate goal of this technique will be to return the soil to a condition in which it can sustainably support vegetation and present no significant risk to the surrounding environment. In addition to the immobilisation of contaminants, the compost will advance this goal by facilitating plant growth and providing soil conditioning and nutrients.

This chapter provides an overview of the major results obtained during the experimental investigations to test the potential of composts combined with naturally occurring minerals (clays, zeolites) for their ability to reduce plant availability and leachability of heavy metals when applied to contaminated soils. A variety of commercially available composts with or without inorganic additions were examined and their potential assessed via a series of laboratory, ecotoxicity and nursery trials. The following experiments were undertaken in order to determine the viability of the technique:

(1) biomass growth
(2) bioavailibity
(3) batch leaching

In addition to the applied research, we have also conducted more fundamental research. Magnetic resonance imaging (MRI) and nuclear magnetic resonance (NMR) spectroscopy have been used here in a novel way to track both the uptake of soil contaminants by higher plants with and without added zeolite, and the effect of a zeolite barrier on contaminant transport within soil.

8.2 Materials characterisation

8.2.1 Soils

Contaminated soils were sampled from various parts in England. Soil was collected from Avonmouth (AML, AMH1, AMH2), which originated from a former zinc smelter near Bristol, UK (Figure 8.1) and was highly contaminated with copper, lead and zinc. The AMH1 and AMH2 soils greatly exceeded soil guideline values (SGV) (Table 8.1). Other soils were sampled from a spoil heap on the Devon Great Consols (DGC) former tin mine in south Devon, UK (Figure 8.2). The spoil material was highly contaminated with arsenic and copper. Arsenic was especially high compared to its soil guideline value (Table 8.1). A third material used was foundry sand of a former arsenal in Thamesmead (TM) London, UK (based in one of the two main SUBR:IM case study areas). This contained elevated levels of copper, lead and zinc but only copper was slightly exceeding its SGV. The final material was sourced from a site historically used for sewage sludge disposal

Table 8.1   Chemical and physical properties of the used soils.

Characteristic	Avonmouth			Devon Great Consols (DGC)	Thames-mead (TM)	Kelham Bridge (KB)	Soil guideline value[c]
	Low (AML)	High 1 (AMH1)	High 2 (AMH2)				
Available phosphorus[a]	11.2	9.0	6.9	16.6	6.4	208	
Available potassium[a]	240.5	147	224.5	29	112	225	
Available magnesium[a]	266.5	91	157.5	12	53	118	
Organic matter (%)	8.7	6.7	3.8	0.9	0.8	16.5	
C:N ratio	10.6	10.7	8.9	17.4	7.3	15.0	
Sand (%)	6	16	2.5	94	96	68	
Silt (%)	56	50	57	3	3	23	
Clay (%)	38	34	40.5	3	1	9	
Textural class	Silty/ clay	Silty/ clay	Silty/ clay	Sand	Sand	Sandy/ loam	
pH	6.8	6.6	7.0	4.1	7.3	4.8	
Total copper[b]	46.7	1 274	218	1 641	137	167	75[d]
Total zinc[b]	1 075	9 414	9 581	47.2	339	255	800[d]
Total lead[b]	315	16 300	1 688	189	323	172	450
Total arsenic[b]	19.7	191	30.6	34 470	12.3	3.3	20
Total cadmium[b]	10.9	239	128	813	2.4	0.5	30

[a] mg l^{-1}, analysed as described in Defra (2000)

[b] mg kg^{-1}

[c] Values given according to Environment Agency (2002); where no UK values present ([d]) Dutch intervention levels are given (VROM, 2000).

Figure 8.1   Avonmouth site near Bristol: dead trees and absence of vegetation at various locations indicate the high level of contamination. (Photo: Frans de Leij)

Figure 8.2   Bare heaps of mine spoil at the Devon Great Consols site in south Devon. (Photo: Frans de Leij)

for a period of over ten years (Kelham Bridge) and had elevated levels of copper, zinc and lead. Again, only copper exceeded its SGV (Table 8.1).

8.2.2 Composts

The composts used in this study were a commercially available green-waste-based compost (GWC), spent mushroom compost (SMC), a commercially available sewage sludge compost (SC) consisting of a mixture of composted sewage sludge and wood chips, a commercially available coir compost (CC) made from composted coconut husks, and a commercially available mixture of GWC and SC (GWSC) that was matured for at least two months after mixing. Detailed physical and chemical characteristics of the composts used in these experiments are given in Table 8.2. Only the SC had metal concentrations that exceeded the PAS100 criteria for heavy metals (BSI, 2005).

8.2.3 Minerals

The minerals used in this study were two natural zeolitic tuffs, a clinoptilolite of Australian origin and another of Turkish origin, both supplied by

Table 8.2   Chemical properties of the composts as determined in one sample.

Characteristics	Green-waste compost (GWC)	Sewage sludge compost (SC)	Spent mushroom compost (SMC)	Coir compost (CC)	Greenwaste/sewage sludge compost (GWSC)
Available phosphorus*	11.9	65.9	74.2	1618	<0.01
Available potassium*	6 745	593	18 989	13 533	974
Available magnesium*	65.7	785	1 701	1 677	347
Total cyanide**	5.03	20.4	9.52	31.5	7.7
Total nitrogen (% w/w)	2.01	2.93	2.65	1.01	2.01
Total phosphorus**	2 628	38 787	5 054	1 770	20 058
Total sulphate**	1 171	17 200	81 329	6 481	9 533
C:N ratio	10.9	7.0	12.2	44.9	10.9
pH	8.3	5.0	6.9	6.0	6.8
Total arsenic**	7.4	7.2	2.6	0.73	7.0
Total cadmium**	0.78	2.1	0.46	0.35	1.3
Total copper**	83.8	613	54.7	41.0	349
Total lead**	110	176	<2	<2	81.3
Total zinc**	253	871	205	44.9	527

* mg l^{-1}, analysed as described in Defra (2000)
** mg kg^{-1}

Euremica Environmental Ltd, Cleveland, UK, and a clay, mainly bentonite, obtained from Colin Stewart Minchem Ltd, Winsford, UK. The corresponding chemical characteristics of these minerals can be found in Table 8.3. The most important characteristic here is the cation exchange capacity (CEC), indicating the ability of the material to immobilise heavy metals. Both zeolites had a relatively high CEC for a natural zeolite and from Table 8.3 it can be seen that the Turkish clinoptilolite had the highest CEC. Since

Table 8.3   Chemical properties of the minerals as determined in one sample.

Characteristic	Australian clinoptilolite	Turkish clinoptilolite	Bentonite
Cation Exchange Capacity	150	160–220	70
Available phosphorus*	1.5	4	1.4
Available potassium*	168.5	9 686	530
Available magnesium*	454	517	6 385
Total cyanide**	<1	<1	<1
Total nitrogen (% w/w)	0.02	0.07	0.01
Total phosphorus**	190	166	42
Total sulphate**	88.5	996	690.5
pH	9.2	7.5	9.1
Total arsenic**	4.4	53.9	10.4
Total cadmium**	0.07	3.12	0.3
Total copper**	2.6	14.9	0.9
Total lead**	27.9	192	38.6
Total zinc**	28	122	34.4

* mg l^{-1}, analysed as described in Defra (2000)
** mg kg^{-1}

the zeolites used in our experiments are natural products and are not of pure zeolite, it is also notable that this zeolite had a very high lead concentration, almost half its soil guideline value of 450 mg kg^{-1}.

8.3 Experimental design

The potential of the composts was assessed on the basis of their capability to contain and reduce the toxicity of the metal-contaminated soils. For that purpose a series of nursery trials with Greek cress (*Lepidium sativum* L.), Perennial ryegrass (*Lolium perenne* L.) and a poplar species (*Populus trichocarpa* Torr. & Gray) were carried out. Ryegrass was selected because it is a good accumulator of metals and, as such, reduced bioavailability should show clearly as reduced concentrations of the heavy metals in this plant. Poplar was selected as a completely different plant species and because this tree is also widely tested on contaminated soils, enabling cross-referencing with similar experiments. The nursery trials were followed by a systematic sequence of batch leaching tests carried out on soils amended and unamended with composts only, or composts in combination with bentonite or zeolite.

The nursery trials were performed in a greenhouse (Figure 8.3) for a minimum period of eight weeks at a minimum night temperature of 10°C

Figure 8.3   Nursery trial with ryegrass (front) and poplar (back) growing on Avonmouth soil amended with compost. (Photo: René van Herwijnen)

and day temperature varying between 20 and 45°C. During the growing period, the water content of the soil in all pots was maintained between 0.35 and 0.5 m^3 m^{-3} by manual watering. At the end of the trial the plants were harvested for chemical analysis and the above-ground biomass was determined.

The batch leaching tests were carried out according to a standard European test (BSI, 2002) using slightly acidic water (pH 5.6) that was mixed with the soil mixtures at a solid:liquid ratio of 1:10 and agitated for 24 hours. Analysis of copper, lead and zinc levels in leachates of soil matrices containing both amended and unamended composts enabled the evaluation of the capability of this technique for metal immobilisation. Appropriate analytical measures were taken in order to ensure the quality assurance of the data.

8.4 Heavy metals containment in soils

8.4.1 Leaching and extractability of metals with compost amendment

Composts' abilities to improve fertility and water- and nutrient-holding capacity, stabilise pH, improve aeration and enhance revegetation of soils and spoils are well documented (Bradshaw and Chadwick, 1980; Bending *et al.*, 1999; Roman *et al.*, 2003). In addition, composts are also quite often referred to as good metal binders and as having properties that reduce the bioavailability and leachability of toxic metals (Geebelen *et al.*, 2002; Walker *et al.*, 2004; Castaldi *et al.*, 2005; Simon, 2005). The metal-binding capacity is usually investigated by soil extraction, leaching tests and concentrations in the soil solution as well as by analysis of leaf concentrations of plants grown on amended soil. For example, Kiikkila *et al.* (2002b) tested the effects of composted sewage sludge and composted organic household waste in combination with wood chips and bark chips on a copper-contaminated (1600 mg kg^{-1}) soil. The amended soils were tested for their exchangeable copper concentration in the soil solution. All compost combinations reduced the concentration of the exchangeable copper by at least 35%. A recent evaluation of the use of extractants for such an assessment (Menzies *et al.*, 2007) suggested that generally total acid extractants (e.g. 0.1 M HCl) and chelating agents (such as the widely used DTPA and EDTA extractants) were only poorly correlated with the availability of trace metals to plants. In contrast, the results of the study suggested that neutral salt extractants (e.g. 0.01 M CaCl$_2$ or 0.1 M NaNO$_3$) provided the most useful indication of plant availability.

In comparison to the large number of studies that report the immobilisation of metals with composts, fewer note the possibility that compost

amendments increase the leaching or extractability of metals. For example, it has been shown that certain composts in combination with specific
soils increase arsenic leaching (Cao *et al.*, 2003; Mench *et al.*, 2003). There
are, however, limited references on the mobilisation effects of composts on
other heavy metals of concern. Of the few papers published, Clemente *et al.*
(2006) reported increased bioavailability of copper after amendment of a
soil from a lead-zinc mine area with a compost made from olive leaves and
the solid fraction of olive-mill waste water. Interestingly, both zinc and
lead were effectively immobilised by the compost. Similarly Kiikkila *et al.*
(2002a) reported increased leaching of copper after mulching of a polluted
forest soil with an undefined compost.

For the purpose of this study, four different composts were tested on three
different soils to examine their effect on the leachability and extractability
of copper, lead and zinc (van Herwijnen *et al.*, 2006). The concentrations
of the three metals in the leachates are presented in Figure 8.4 and the
leachate pH is given in Table 8.4. From Figure 8.4, it can be observed that
SC and SMC both increased the leaching of zinc that was originally bound
to the AMH1 soil by a factor of 1.8 (from 17.4 mg l^{-1} to 31.4 mg l^{-1}) and 2.0
(from 17.4 mg l^{-1} to 35.1 mg l^{-1}) respectively while GWC and CC reduced
the leaching of zinc by 85% (from 17.4 mg l^{-1} to 2.6 mg l^{-1}) and 40% (from
17.4 mg l^{-1} to 10.2 mg l^{-1}) respectively. GWC-amended AMH1 soils reduced
the leachate concentration of zinc almost to its WHO drinking-water standard (WHO, 2004) of 3 mg l^{-1}, which means a reduced health risk, while SMC
increased the leachate concentration of copper from 0.09 mg l^{-1} to 2.2 mg l^{-1}
and as such exceeding its WHO drinking-water standard of 2 mg l^{-1}.

Mobilisation of zinc was also observed for DGC and TM soils. However,
in these cases the zinc concentration did not exceed that of the leachates

Table 8.4 Mean pH the leachates of soils amended with four different kinds of
compost and of the amendments on their own.

	Avonmouth (AMH1)		Devon Great Consols (DGC)		Thamesmead (TM)		Amendment only	
	mean	se.	mean	se.	mean	se.	mean	se.
Control	6.6	0.09	4.1	0.02	7.3	0.02		
CC	6.2[b]	0.10	5.7[a]	0.01	7.1[b]	0.04	7.0	0.07
GWC	7.0[a]	0.08	7.2[a]	0.08	7.6[a]	0.12	8.4	0.07
SMC	6.7	0.02	6.5[a]	0.06	6.9[b]	0.06	7.5	0.07
SC	5.8[b]	0.07	5.3[a]	0.04	6.9[b]	0.06	6.3	0.07

[a] Significantly higher than control (p < 0.05)
[b] Significantly lower than control (p < 0.05)
CC = coir compost, GWC = green waste compost, SMC = spent mushroom compost, SC = sewage sludge
compost, se. = Standard error of means for each specific soil, n = 4.

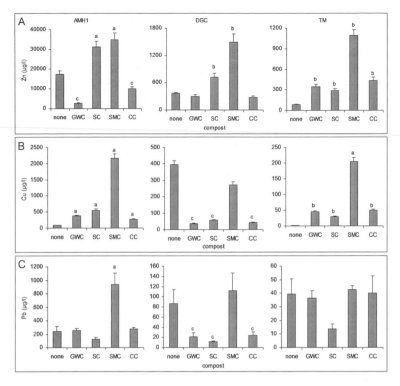

Figure 8.4 Mean concentrations (μg l^{-1}) of zinc (A), copper (B), and lead (C) in leachates from compost amended soils: (a) significantly higher than the control ($p < 0.05$) and significantly higher than the amendment only ($p < 0.05$); (b) significantly higher compared to the control ($p < 0.05$) but not significantly different from or lower than the leachate of the amendment only ($p < 0.05$); (c) significantly lower than the control ($p < 0.05$). GWC = green waste compost, SC = composted sewage sludge, SMC = spent mushroom compost, CC = coir compost. Error bars are standard error of the means, $n = 4$.

of the compost alone and therefore it is most probably originating from the composts rather than the soils. Although in most cases leachate concentrations did not exceed World Health Organization (WHO) drinking-water standards, the potential of increased leaching still means a potential increased risk, and the extent of that risk is dependent on the soil type and the compost type.

We saw that the magnitude of the effects of composts differed between the soils. On AM soil, for example, the effect of SC was almost ten times higher in magnitude than that of GWC while on TM soil the effect of SC and GWC were comparable. When examining AM soil it was noticed that all compost amendments increased the leaching of the soil-bound copper. GWC, which reduced the leaching of zinc from this soil, increased the leaching of copper at relatively the same order of magnitude as SC. In contrast, all amendments apart from SMC reduced the leaching of copper on DGC soil significantly by 90%. For lead, only SMC showed a concentration increase from the AM soil, but the other three composts appeared to have no significant effect. Reduced levels of lead leaching were observed with all compost amendments in the DGC soil.

Overall, it can be stated that the leaching behaviour of copper, lead and zinc is dependent on the type of soil and compost and their interaction. The difference can, in some cases, be explained by reduced immobilisation capacity of a particular compost. For example, GWC reduced the leaching of lead from DGC soil from 87 mg l^{-1} to 22 mg l^{-1} while it had no significant effect on leaching from the AM soil. It is a possibility that when mixed with the soil a large amount of zinc was sorbed onto the GWC, saturating its sorption sites and reduced its immobilisation capacity for lead. This would indicate a competitive interaction between the different metals influencing the immobilising effect of the composts for individual metals.

A reduction in pH is often given as a factor causing increased mobility of metals in soils (Alloway, 1994), but examples from our results show that changes in pH cannot be the sole cause of the observed increase in leaching. For example, SC caused increased leaching of zinc from the AM soil, similar to the effect observed for SMC. SC amendment also caused a lower leachate pH than the control, but leachates of SMC-amended AM soil had a similar pH to the controls.

Another explanation for the observed differences in metal-leaching behaviour could be associated with the different compositions of organic compounds found in the composts. Organic acids such as formic, succinic, oxalic and especially citric acid have been shown to increase zinc leaching from mine tailings under neutral conditions (Burckhard *et al.*, 1995; Chen *et al.*, 2003). In addition, one metal could be more susceptible to sorption by organic matter from the compost than another (Chien *et al.*, 2006). It has also been reported that humic acid from organic amendments can reduce the sorption capacity for copper and zinc of the humic acid originally present in the soil itself (Hernandez *et al.*, 2006). Other effects for organic matter have been reported for copper where a mixture of organic household waste compost and wood chips reduced the concentration of free copper ions in the soil water but increased the concentration of copper complexed with DOC (Kiikkila *et al.*, 2001). However, these changes did not affect the bacterial population in the soil (Kiikkila *et al.*, 2002b). Gray and McLaren (2006) used semi-empirical models to predict water soluble copper, chromium, cadmium, lead, nickel and zinc concentrations in a range of naturally contaminated soils based on the competitive adsorption of H^+ and metal ions and their dependence on solution pH, total metal content, total carbon content and soil oxide content. Soluble metal concentration was significantly influenced by total metal concentration for all of the metals tested, total soil carbon influenced nickel, lead and cadmium solubility, soil pH influenced chromium, cadmium and zinc availability, while oxalate-extractable aluminium and iron influenced chromium and zinc solubility respectively. These reports provide a good explanation for the rather non-generic nature of some of the results observed with some amendments.

More research into the effect of organic compounds from the compost on the leaching of metals is recommended. If it were known which compounds are most responsible for the increase in leaching, composts could be analysed for these compounds and the effects could be predicted without extensive testing. Nonetheless, it is possible to find a compost that is capable of immobilising a variety of metals in a soil. As an example, Figure 8.5 demonstrates that GWSC compost can effectively reduce the leaching of cadmium, copper, lead and zinc from AMH1 soil. Overall, these results show that experimental trials will be necessary if compost is to be applied for any particular remedial application.

8.4.2 Plant uptake

The uptake of metals into the shoots of ryegrass and Greek cress and poplar leaves was examined on a variety of soils amended with a selection of the compost (van Herwijnen et al., 2006). The uptake of zinc on highly contaminated Avonmouth soil amended with SC or SMC is presented in Figure 8.6 (AMH1 with SMC for Greek cress; AMH2 with SC for ryegrass and poplar). Concentration of zinc in the shoots of ryegrass was reduced by SC amendment in both high- and low-level contaminated soils with 60% and 30% (from 3331 mg kg^{-1} to 1377 mg kg^{-1} and from 477 mg kg^{-1} to 318 mg kg^{-1}) respectively, but for the poplar the leaf concentrations were doubled (from 709 mg kg^{-1} to 1488 mg kg^{-1}) on the highly contaminated soil while they were reduced by 25% (from 932 mg kg^{-1} to 701 mg kg^{-1}) on the low level contaminated soil. In contrast, a reduction by 80% (from 3708 mg kg^{-1} to 869 mg kg^{-1}) in zinc levels in the shoots of Greek cress was observed in the highly contaminated soil, and an increase by a factor of 1.5 (from 191 mg kg^{-1} to 282 mg kg^{-1}) on the low-level contaminated soil after SC amendment.

These differences in metal uptake are comparable with others in the literature where both reduced and increased plant uptake has been reported after compost was applied to polluted soils (Castaldi et al., 2005; Clemente et al., 2005; Neagoe et al., 2005). Nothing has been reported in the literature that demonstrates the variability in metal uptake by plant species growing on different composts and soils. One explanation for differences in plant uptake could be root exudates which are known to form complexes with metals and as such increase metal uptake (Tao et al., 2004; Tudoreanu and Phillips, 2004). These exudates may vary between plant species but do

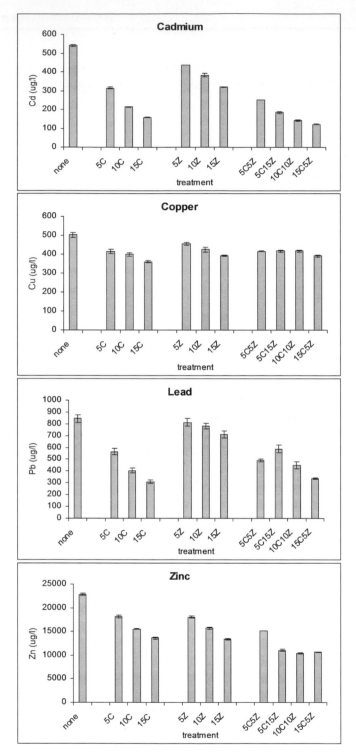

Figure 8.5 Concentrations of cadmium, copper, lead and zinc in leachates of AMH1 soil amended with GWSC (C) or zeolite (Z) at 5, 10 or 15% of the soil mixture (w/w). Error bars are the standard error of the means, n = 3.

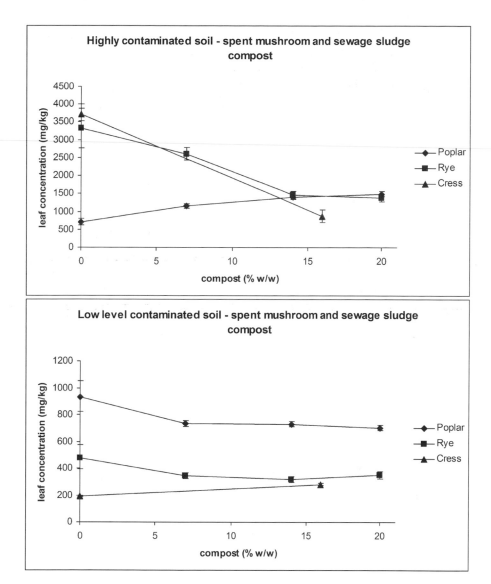

Figure 8.6 Concentrations of zinc in shoots of rye grass and Greek cress and leaves of poplar. The highly contaminated soil was AMH2 for poplar and ryegrass and AMH1 for Greek cress. The low level contaminated soil was AML for poplar and ryegrass and TM for Greek cress. Soils for poplar and ryegrass were amended with SC and soils for Greek cress were amended with SMC. Both composts increased the leaching of zinc from AMH1 and AMH2. Error bars are the standard error on the means, for the poplar and ryegrass n = 21, for the cress n = 4.

also depend on the health of the plant. Since different plants can react differently on specific amendments, the production of root exudates may therefore vary in a large number of ways, and as a consequence so will the metal uptake.

The results also clearly showed that the leaching tests performed in section 8.4.1 are not necessarily representative of the actual plant uptake and therefore of the bioavailablility. Quite often, extractions with EDTA (ethylenediaminetetraacetic acid) or DTPA (diethylenetriaminepentaacetate) are taken to be more representative (Madyiwa *et al.*, 2003; Fuentes *et al.*, 2006) but are also reported to have a poor correlation with metal concentrations in plants of komatsuna (*Brassica rapa* L.) (Takeda *et al.*, 2006) and overestimate heavy metal availability for fat hen (*Chenopodium album* L.) (Walker *et al.*, 2004). Contrary to the EDTA and DTPA extractions, other researchers reported that extracts with weak salt solutions give a better correlation with metal uptake by ryegrass (*Lolium perenne* L.) and lettuce (*Lactuca sativa* L.) (Aten and Gupta, 1996). Our results indicate that contradictory responses between different plant species can be reached for the same amendment on the same soil and therefore it is advisable that nursery trials using standard plant indicator species should be performed to assess metal availability to plants rather than relying on extraction and/or leaching tests.

8.5 Biomass

Composts are often added to brownfield sites in order to support plant growth to stabilise contaminated soils with plants and trees and to create amenity greenspace. They improve soil fertility and water- and nutrient-holding capacity, stabilise soil pH, improve soil aeration and therefore enhance the revegetation of brownfield sites (Bradshaw and Chadwick, 1980; Roman *et al.*, 2003). Biosolid compost added to a copper-, chromium- and arsenic-contaminated soil has been shown to increase the biomass of carrot (*Daucus carota* L.) and lettuce (*Lactuca sativa* L.) by three to five times compared to the control (Cao and Ma, 2004). Cala *et al.* (2005) reported improved growth of rosemary (*Rosmarinus officinalis* L.) when biosolid waste compost and municipal solid waste compost were added to a degraded agricultural soil for restoration. Compost made of olive husks (50%), sewage sludge (25%) and vegetable waste (25%) (v/v) increased the above-ground plant biomass of white lupin (*Lupinus albus* L.) by a factor of 3.6 (Castaldi *et al.*, 2005). However, compost does not always improve growth on contaminated soils. Chiu *et al.* (2006) showed that a manure compost increased the yield of vetiver (*Vetiveria zizanioides* L.) and common reed (*Phragmities australis* (Cav.) Trin. ex. Steud.) on a copper mine tailing but not on the tailing of a lead and zinc mine while both were severely contaminated with copper or lead and zinc respectively.

Results from experiments have shown that a compost can either improve or hinder plant growth. Figure 8.7 illustrates the biomass of ryegrass, Greek

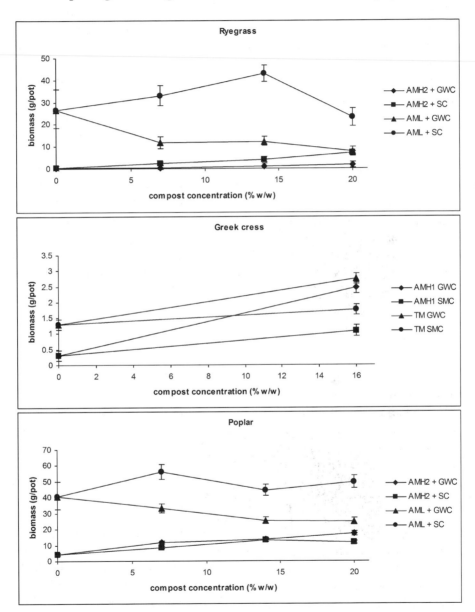

Figure 8.7 Biomass of ryegrass, poplar and Greek cress growing on highly contaminated soils (AMH1 and AMH2) and low level contaminated soils (AML and TM) amended with GWC and SC or SMC. Error bars are the standard error of the means, for the poplar and ryegrass n = 21, for the cress n = 4.

cress and poplar growing on low-level (AML or TM) or high-level (AMH1 or AMH2) contaminated soil amended with compost (GWC, SC or SMC). Both ryegrass and poplar growing on AML soil amended with GWC had a reduced biomass production. Amendment of 20% SC to AML soil seems to significantly reduce the biomass production of ryegrass compared to the 7% and 14% SC additions (Figure 8.8). However, on the AMH2 soil compost amendment at all levels improved the growth of both ryegrass and poplar. Similarly, improved growth was observed for Greek cress on AMH1 and TM soil for all compost amendments. High C:N ratios are known to cause reduced availability of nutrients to plants but this accounts for C:N ratios above 25 (Bending *et al.*, 1999). The C:N ratios of the soils and composts presented in Figure 8.7 are all below 15 (see Table 8.2) and therefore are not likely to pose a problem.

Figure 8.8   Ryegrass growing on AML soil amended with increasing concentrations of compost (% w/w).

8.6 Enhanced compost

The results obtained so far have shown that some composts have the potential to immobilise metals and reduce leaching and plant uptake. Naturally occurring minerals such as clays and zeolites also interact with metals to form a matrix in which the bioavailability of the metals is remarkably decreased (Ouki and Kavannagh, 1999; Simon, 2001; Alvarez-Ayuso and Garcia-Sanchez, 2003). Enrichment of compost with these minerals could provide long-term immobilisation of the metals because the minerals are only rarely and slowly degraded and as such will increase the long-term sustainability of the remediation method. So far, most research on combined use of composts and minerals has focused on the immobilisation of metals originating from compost (Nissen *et al.*, 2000; Zorpas *et al.*, 2000) or the stabilisation of nutrients during the composting of manure (Leggo and Ledesert, 2001). Coppola *et al.* (2003) successfully used a zeolite (Neapolitan yellow tuff) in combination with pellet manure to reduce the concentration of DPTA-extractable cadmium and lead from a polluted soil. Addition of 25% of this mixture to the soil also improved biomass production and reduced plant concentrations of lead in wheat (*Triticum aestivum* L.). An organo-zeolitic fertiliser has also been tested in combination with untreated zeolite to improve growth of ryegrass on mine waste contaminated with lead and zinc at levels of 915 mg kg^{-1} and 670 mg kg^{-1} respectively (Leggo *et al.*, 2006). Although the original fertiliser improved growth of wheat on this soil and reduced leachate concentrations of lead and zinc (Leggo and Ledesert, 2001), combinations of this fertiliser with untreated zeolite did not improve growth of ryegrass any further and leachate concentrations were not reported.

In the current study, two series of tests with combinations of compost and minerals have been performed. In the first series GWC and SC were tested in combination with an Australian clinoptilolite or bentonite added to AML and AMH2 soils. The maximum concentration of the minerals in the amended soils was 4% (w/w) in this series. In the second series, a Turkish clinoptilolite was tested at a maximum soil amendment rate of 15% (w/w), tested both on its own and in combination with GWSC. The compositions of the tested soil mixtures are given in Table 8.5. The maximum amendment rate of compost plus zeolite to the soil was 20% (w/w).

Our results show that the low-level mineral amendments had a very limited effect on plant performance and metal leachate concentrations. The only observation on plant performance was that the growth of ryegrass on AML soil amended with SC was slightly improved (21%) when the SC was amended in combination with zeolite. For the leaching results, the amendment of SC with bentonite and to a lesser extent with zeolite reduced the increasing effect that SC had on the leaching of zinc (as described in

Table 8.5   Composition of soil mixtures to test the effect of combinations of zeolite and GWSC amended to AMH1 and KB soils.

		Amount of compost in the soil mixture (% w/w)			
		0	5	10	15
Amount of zeolite in the soil mixture	0	X	X	X	X
	5	X	X		X
	10	X		X	
	15	X	X		

subsection 8.4.1) by an average of 24%. Both observations show an inter-action between the SC and the minerals rather than a direct effect of the minerals on the leaching and toxicity of the heavy metals.

Results of the higher level of zeolite addition to the soil showed that biomass production of ryegrass on AMH1 or KB was not significantly improved by combining the compost with the zeolite. On AMH1 soil mixtures not enough biomass was produced to examine the metal uptake, but on KB soil no significant effects on metal concentrations in the ryegrass tissue were observed when combining compost with the Turkish zeolite. Leachate results, however, provided a good reduction in leachate metal concentrations for the combined use of compost and zeolite. KB soil amended with GWSC and/or Turkish zeolite depicted reduced zinc concentrations in the leachate for all combinations by a maximum of 63% (from 322 μg l^{-1} to 118 μg l^{-1}) (Figure 8.9). Metal concentrations in leachates from AMH2 soil were effectively reduced for cadmium, copper, lead and zinc for the

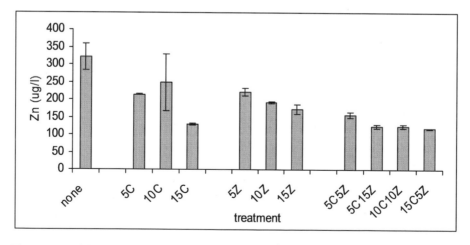

Figure 8.9   Concentrations of zinc in leachates of KB soil amended with GWSC (C) and/or zeolite (Z) at 5, 10 or 15% of the soil mixture (w/w). Error bars are the standard error of the means, n = 3.

individual amendments by maxima of 70% (from 541 µg l^{-1} to 158 µg l^{-1}), 30% (from 503 µg l^{-1} to 361 µg l^{-1}), 60% (from 844 µg l^{-1} to 310 µg l^{-1}) and 40% (from 22 805 µg l^{-1} to 13 374 µg l^{-1}) respectively (Figure 8.5). Combining GWSC with the zeolite as soil amendment worked very well for cadmium and zinc. For both metals, the reduction of the leachate concentration was significantly lower for the combined amendments (compost and zeolite) compared to the individual amendments. For cadmium, the maximum reduction using the combined mixture was 80% (from 541 µg l^{-1} to 121 µg l^{-1}) and for zinc 50% (from 22 805 µg l^{-1} to 10 424 µg l^{-1}). Addition of GWSC and the zeolite individually was found to reduce the leachate concentrations of lead but in general the combination of compost with zeolite was in most cases as effective as the compost on its own. The leachate concentrations of copper showed similar results as for lead but in this case the combination of compost and zeolite appeared to be less effective than either the compost or the zeolite as sole amendment.

Although the leachate concentrations of the metals did not go below WHO drinking-water standards (WHO, 2004), the efficient reduction obtained would provide successful remediation if such reductions in concentration were observed for soils where the levels of water pollution were closer to the WHO standard. The reduced leaching means a reduced transport of the metals towards receptors like groundwater and plants and therefore a reduced risk of food-chain transfer of the contaminants. Overall, it can be concluded that the combination of zeolite and composts demonstrated very good potential for metal containment of zinc and cadmium while the combination shows no ability to immobilise lead or cadmium over its individual constituents (i.e. individual amendments with compost or zeolite).

8.7 Magnetic resonance imaging

Traditionally, investigation into sub-surface environmental problems such as contaminated land requires relatively laborious analytical techniques along with intrusive sampling of the medium under investigation, such as leaching and plant uptake tests. Although well understood and accurate, such investigations are necessarily limited in time and space owing to the efforts required per sample, and can also lead to disturbance of the system under study. A non-destructive, non-intrusive method of investigating contaminant behaviour would be expected to complement these 'traditional' methods and provide information on processes occurring that would otherwise not be available.

Magnetic resonance imaging (MRI) and nuclear magnetic resonance (NMR) techniques have previously been used to study environmental problems (e.g. Nestle *et al.*, 2002). Such studies include tracking metallic contaminants in artificial soils (Nestle *et al.*, 2003), while several studies on plants have considered water stresses (Macfall *et al.*, 1990) as well as metal uptake (Nagata *et al.*, 1993) and the use of tracers to determine transport processes within the plant (Gussoni *et al.*, 2001). However, the use of these methods to investigate combined soil/plant/contaminant systems, as is the case in the work presented here, has not been studied previously.

Resonance imaging (MRI) and nuclear magnetic resonance (NMR) spectroscopy have been used here in a novel way to track:

(1) the uptake of soil contaminants by higher plants with and without added zeolite; lithium was chosen as a model contaminant as it can be directly imaged, unlike many pollutants
(2) the effect of a zeolite barrier on contaminant transport in soil; another metal, gadolinium, was used here

The initial results presented here are based on simplified systems that were used to overcome the complexity of real soils.

8.7.1 Studies of lithium uptake by poplar

Poplar (*Populus trichocarpa* Torr. & Gray; Fritzi Pauley clone) were grown for 14 days in either sand or a sand zeolite mix with quarter-strength Hoagland's solution as the pore fluid and nutrient source. Lithium chloride in solution was then added to the soil pore fluid (92 to 460 mM lithium) and the soil/plant/contaminant systems left for a further five days. The poplar stems were then removed from the soil, washed and imaged in a Bruker DMX 300 spectrometer with a dedicated ^{7}Li probe. Cross-sectional 2D images were obtained at several points on the stem of the plant. Further information on these investigations is available in Harbottle *et al.* (in press), from which these images were taken.

Figure 8.10 illustrates the effect of zeolite on the uptake of lithium by poplar at a higher level of lithium contamination. The two images are at identical locations in the two plants but the presence of zeolite had significantly reduced lithium uptake showing potential of the zeolite to break contaminant–receptor pathways. In addition, the location of lithium in the plant grown with zeolite was restricted to several clearly visible regions, possibly identifying plant structures involved in the uptake and transport of material from soil.

8.7.2 Studies of gadolinium transport in soil

Horizontal diffusion of a model metal contaminant (gadolinium) through water-saturated soil beds with and without zeolite was studied, to identify

(a)   (b)

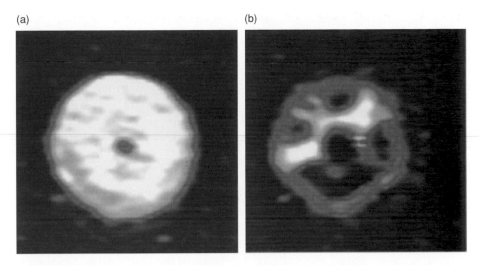

Figure 8.10   Lithium images across poplar stems at approximately 20 mm above the base of the stem after five days' exposure to lithium-contaminated soil (approximately 1000 mg/kg Li): (a) poplar grown in sand only; (b) poplar grown in 4:1 sand/zeolite mixture (both images have a 17.5 mm field of view).

the effects of zeolite on this model contaminant's movement. One bed consisted of sand only while another was primarily sand but with a 20 mm barrier of pure zeolite at 4 cm from the contaminant injection point. Gadolinium was injected into the soil and its movement tracked for a number of days.

Images of gadolinium movement through the soil are presented in Figure 8.11 (sand soil only) and Figure 8.12 (sand soil containing pure zeolite barrier). In both cases the imaged portion of the soil, including the zeolite barrier in the latter test, is clearly visible at the start of the test (0 hours). After 20 hours, the region containing gadolinium is visible as a darkened region in the soil, as this metal strongly affects the NMR signal. With sand soil only

(a)   (b)

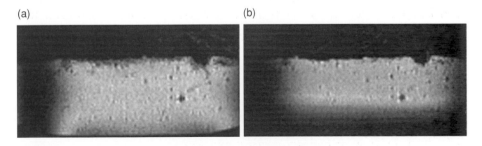

Figure 8.11   Movement of gadolinium tracer through soil pore water (injection point at left-hand side of each image. Dark regions in soil represent location of gadolinium) – soil consists of saturated sand only mix: (a) 0 hours; (b) 20 hours.

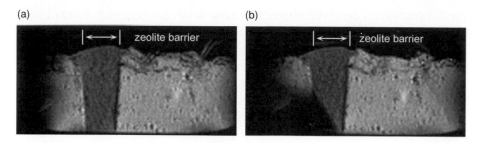

Figure 8.12   Movement of gadolinium tracer through soil pore water (injection point at left-hand side of each image. Dark regions in soil represent location of gadolinium) – soil consists of saturated sand/zeolite mix (4:1): (a) 0 hours; (b) 20 hours.

(Figure 8.11), the metal was found to have accumulated at the base of the bed, uniformly across the cross section. The presence of the zeolite barrier (Figure 8.12), however, was sufficient to restrain the contaminant as shown by the dark region to the left of the barrier.

8.8   Conclusions

The aim of the research presented here was to develop a compost amended with zeolites that has the potential to remediate various types of metal-contaminated sites. This compost should be able to improve plant growth and immobilise heavy metals. The results demonstrated that there are a range of parameters and mechanisms involved that can affect the generic performance of the composts. For example, we have observed that one particular compost can increase the leaching of a metal on one soil but reduce the leaching of this same element on another soil. Moreover, it has been shown that a leaching test may be representative for one plant species but not necessarily for another. The study has demonstrated that testing is imperative in order to predict the viability of a particular compost combination and its effects on metal mobility in soil and plant performance. Consequently, additional research is necessary in order to gain a better understanding of the sorption and leaching mechanisms involved and the bioavailability of the metals in the soils. A particular focus should be placed on the effect of organic compounds and organic matter originating from the compost. If we can understand which compounds are responsible for increased leaching, and under which circumstances this process takes place, composts could be analysed for these compounds and better predictions of their effects could be made.

Magnetic resonance imaging provided clear evidence of processes occurring in a range of environmental systems, specifically looking at the effect of zeolite on contaminant retardation in soils and plant uptake. As well as presenting evidence of the efficacy of zeolite in reducing the effect of contaminants in soils, the research has shown that particular contaminants can be tracked in real time and in a non-destructive fashion while providing detailed information that would otherwise not be accessible.

It is evident that there is a need for standard procedures to be developed and performed prior to the application of (amended) composts for remediation purposes. These procedures should consist of leaching tests that will predict the effect on the groundwater and toxicity tests involving more than one plant species. These will ensure that the novel remediation technique will not present any necessary risks that could possibly have adverse effects.

Overall we have shown that the combination of composts with minerals like zeolites proved to have a very good potential for the remediation of metals-contaminated sites. The minerals have the dual capability to reduce metal-leaching and stabilise the long-term release of nutrients from the compost into soil. In addition, in cases where there is a limit on the amount of compost that can be added to the soil in order not to inhibit the plant growth, more metal immobilisation can be achieved by combining the compost with a mineral like zeolite.

References

Alloway, B.J. (1994) *Heavy Metals in Soil.* Kluwer Academic Publishers, Dordrecht.

Alvarez-Ayuso, E. and Garcia-Sanchez, A. (2003) Palygorskite as a feasible amendment to stabilize heavy metal polluted soils. *Environmental Pollution,* **125**, 337–44.

Aten, C.F. and Gupta, S.K. (1996) On heavy metals in soil; rationalization of extractions by dilute salt solutions, comparison of the extracted concentrations with uptake by ryegrass and lettuce, and the possible influence of pyrophosphate on plant uptake. *Science of the Total Environment,* **178**, 45–53.

Bending, N.A.D., McRae, S.G. and Moffat, A.J. (1999) *Soil-forming Materials: Their Use in Land Reclamation.* The Stationery Office, London.

Bradshaw, A.D. and Chadwick, M.J. (1980) *The Restoration of Land: The Ecology and Reclamation of Derelict and Degraded Land.* University of California Press, Berkeley.

BSI (2002) *Characterisation of waste – Leaching – Compliance test for leaching of granular waste materials and sludges. BS En 12457: Part 2: One stage batch test at a liquid to solid ratio of 10 l/kg for materials with particle size below 4 mm (without or with size reduction).* British Standards Institution, London.

BSI (2005) *PAS 100:2005, Specification for composted materials.* British Standard Institution, London.

Burckhard, S.R., Schwab, A.P. and Banks, M.K. (1995) The effects of organic-acids on the leaching of heavy-metals from mine tailings. *Journal of Hazardous Materials,* **41**, 135–45.

Cala, V., Cases, M.A. and Walter, I. (2005) Biomass production and heavy metal content of *Rosmarinus officinalis* grown on organic waste-amended soil. *Journal of Arid Environments,* **62**, 401–12.

Cao, X.D. and Ma, L.Q. (2004) Effects of compost and phosphate on plant arsenic accumulation from soils near pressure-treated wood. *Environmental Pollution*, **132**, 435–42.

Cao, X.D., Ma, L.Q. and Shiralipour, A. (2003) Effects of compost and phosphate amendments on arsenic mobility in soils and arsenic uptake by the hyperaccumulator, *Pteris vittata* L. *Environmental Pollution*, **126**, 157–67.

Castaldi, P., Santona, L. and Melis, P. (2005) Heavy metal immobilization by chemical amendments in a polluted soil and influence on white lupin growth. *Chemosphere*, **60**, 365–71.

Chen, Y.X., Lin, Q., Luo, Y.M., He, Y.F., Zhen, S.J., Yu, Y.L. *et al.* (2003) The role of citric acid on the phytoremediation of heavy metal contaminated soil. *Chemosphere*, **50**, 807–11.

Chien, S.W.C., Wang, M.C. and Huang, C.C. (2006) Reactions of compost-derived humic substances with lead, copper, cadmium, and zinc. *Chemosphere*, **64**, 1353–61.

Chiu, K.K., Ye, Z.H. and Wong, M.H. (2006) Growth of *Vetiveria zizanioides* and *Phragmities australis* on Pb/Zinc and Cu mine tailings amended with manure compost and sewage sludge: A greenhouse study. *Bioresource Technology*, **97**, 158–70.

Clemente, R., Escolar, A. and Bernal, M.P. (2006) Heavy metals fractionation and organic matter mineralisation in contaminated calcareous soil amended with organic materials. *Bioresource Technology*, **97**, 1894–901.

Clemente, R., Walker, D.J. and Bernal, M.P. (2005) Uptake of heavy metals and As by *Brassica juncea* grown in a contaminated soil in Azincalcollar (Spain): The effect of soil amendments. *Environmental Pollution*, **138**, 46–58.

Coppola, E.I., Battaglia, G., Bucci, M., Ceglie, D., Colella, A., Langella, A. *et al.* (2003) Remediation of Cd- and Pb-polluted soil by treatment with organozeolite conditioner. *Clays and Clay Minerals*, **51**, 609–15.

Defra (2000) *The Analysis of Agricultural Materials*. Reference Book 427. The Stationary Office, London.

Environment Agency (2002) *Soil Guideline Values for Arsenic, Cadmium and Lead Contamination. SGV 1, 3 and 10.* Environment Agency, Bristol.

Fuentes, A., Llorens, M., Saez, M.J., Aguilar, I., Perez-Marin, A.B., Ortuno, J.F. *et al.* (2006) Ecotoxicity, phytotoxicity and extractability of heavy metals from different stabilised sewage sludges. *Environmental Pollution*, **143**, 355–60.

Gadepalle, V.P., Ouki, S.K., van Herwijnen, R. and Hutchings, T. (2007) Immobilisation of heavy metals in soil using natural and waste materials for vegetation establishment on contaminated sites. *Soil and Sediment Contamination*, **16**.

Geebelen, W., Vangronsveld, J., Adriano, D.C., Carleer, R. and Clijsters, H. (2002) Amendment-induced immobilization of lead in a lead-spiked soil: Evidence from phytotoxicity studies. *Water Air and Soil Pollution*, **140**, 261–77.

Gray, C.W. and McLaren, R.G. (2006) Soil factors affecting heavy metal solubility in some New Zealand soils. *Water, Air and Soil Pollution*, **175**, 3–14.

Gussoni, M., Greco, F., Vezzoli, A., Osuga, T. and Zetta, L. (2001) Magnetic resonance imaging of molecular transport in living morning glory stems. *Magnetic Resonance Imaging*, **19**, 1311–22.

Harbottle, M.J., Mantle, M.D., Johns, M.L., van Herwijnen, R., Al-Tabbaa, A., Hutchings, T.R. *et al.* (in press) Magnetic resonance imaging of the effect of zeolite on lithium uptake in poplar. *Environmental Science and Technology*.

Hernandez, D., Plaza, C., Senesi, N. and Polo, A. (2006) Detection of copper(II) and zinc(II) binding to humic acids from pig slurry and amended soils by fluorescence spectroscopy. *Environmental Pollution*, **143**, 212–20.

van Herwijnen, R., Laverye, T., Ouki, S.K., Al-Tabbaa, A., Hodson, M.E. and Hutchings, T.R. (2006) The effect of composts on the leaching of metals from contaminated soils. In: *Brownfields III: Proceedings of the Third International Conference on Prevention, Assessment, Rehabilitation and Development of Brownfield Sites* (eds C.A. Brebbia and U. Mander), pp. 99–108.Tallinn.

Kiikkila, O., Derome, J., Brugger, T., Uhlig, C. and Fritze, H. (2002a) Copper mobility and toxicity of soil percolation water to bacteria in a metal polluted forest soil. *Plant and Soil*, **238**, 273–80.

Kiikkila, O., Pennanen, T., Perkiomaki, J., Derome, J. and Fritze, H. (2002b) Organic material as a copper immobilising agent: a microcosm study on remediation. *Basic and Applied Ecology*, **3**, 245–53.

Kiikkila, O., Perkiomaki, J., Barnette, M., Derome, J., Pennanen, T., Tulisalo, E. *et al.* (2001) In situ bioremediation through mulching of soil polluted by a copper-nickel smelter. *Journal of Environmental Quality*, **30**, 1134–43.

Leggo, P.J. and Ledesert, B. (2001) Use of organo-zeolitic fertilizer to sustain plant growth and stabilize metallurgical and mine-waste sites. *Mineralogical Magazine*, **65**, 563–70.

Leggo, P.J., Ledesert, B. and Christie, G. (2006) The role of clinoptilolite in organo-zeolitic-soil systems used for phytoremediation. *Science of the Total Environment*, **363**, 1–10.

Macfall, J.S., Johnson, G.A. and Kramer, P.J. (1990) Observation of a water depletion region surrounding loblolly pine roots by magnetic resonance imaging. *Proceedings of the National Academy of Sciences of the United States of America*, **83**, 1203–7.

Madyiwa, S., Chimbari, M.J., Schutte, C.F. and Nyamangara, J. (2003) Greenhouse studies on the phyto-extraction capacity of *Cynodon nlemfuensis* for lead and cadmium under irrigation with treated wastewater. *Physics and Chemistry of the Earth*, **28**, 859–67.

Mench, M., Bussiere, S., Boisson, J., Castaing, E., Vangronsveld, J., Ruttens, A. *et al.* (2003) Progress in remediation and revegetation of the barren Jales gold mine spoil after in situ treatments. *Plant and Soil*, **249**, 187–202.

Menzies, N.W., Donn, M.J. and Kopittke, P.M. (2007) Evaluation of extractants for estimation of the phytoavailable trace metals in soils. *Environmental Pollution*, **145**, 121–30.

Nagata, T., Hayatsu, M. and Kosuge, N. (1993) Aluminium kinetics in the tea plant using 27Al and 19F NMR. *Phytochemistry*, **32**, 771–5.

Neagoe, A., Ebena, G. and Carlsson, E. (2005) The effect of soil amendments on plant performance in an area affected by acid mine drainage. *Chemie der Erde-Geochemistry*, **65**, 115–29.

Nestle, N., Baumann, T. and Niessner, R. (2002) Magnetic resonance imaging in environmental science. *Environmental Science and Technology*, **36**, 154A–160A.

Nestle, N., Wunderlich, A., Niessner, R., Baumann, T. (2003) Spatial and temporal observations of adsorption and remobilization of heavy metal ions in a sandy aquifer matrix using magnetic resonance imaging. *Environmental Science and Technology*, **37**, 3972–7.

Nissen, L.R., Lepp, N.W. and Edwards, R. (2000) Synthetic zeolites as amendments for sewage sludge-based compost. *Chemosphere*, **41**, 265–9.

Ouki, S.K. and Kavannagh, M. (1999) Treatment of metals-contaminated wastewaters by use of natural zeolites. *Water Science and Technology*, **39**, 115–22.

Ouki, S.K., Cheeseman, C.R. and Perry, R. (1994) Natural zeolite utilization in pollution-control – a review of applications to metals effluents. *Journal of Chemical Technology and Biotechnology*, **59**, 121–6.

Ponizovsky, A.A. and Tsadilas, C.D. (2003) Lead (II) retention by Alfisol and clinoptilolite: cation balance and pH effect. *Geoderma*, **115**, 303–12.

Roman, R., Fortun, C., De Sa, M. and Almendros, G. (2003) Successful soil remediation and reforestation of a calcic regosol amended with composted urban waste. *Arid Land Research and Management*, **17**, 297–311.

Simon, L. (2001) Effects of natural zeolite and bentonite on the phytoavailability of heavy metals in chicory. In: *Environmental Restoration of Metals-Contaminated Soils*, Vol. 1 (ed. I.K. Iskandar), pp. 261–71. Lewis Publishers, Boca Raton.

Simon, L. (2005) Stabilization of metals in acidic mine spoil with amendments and red fescue (*Festuca rubra* L.) growth. *Environmental Geochemistry and Health*, **27**, 289–300.

Takeda, A., Tsukada, H., Takaku, Y., Hisamatsu, S., Inaba, J. and Nanzyo, M. (2006) Extractability of major and trace elements from agricultural soils using chemical extraction methods: Application for phytoavailability assessment. *Soil Science and Plant Nutrition*, **52**, 406–17.

Tao, S., Liu, W.X., Chen, Y.J., Xu, F.L., Dawson, R.W., Li, B.G. *et al.* (2004) Evaluation of factors influencing root-induced changes of copper fractionation in rhizosphere of a calcareous soil. *Environmental Pollution*, **129**, 5–12.

Tudoreanu, L. and Phillips, C.J.C. (2004) Modeling cadmium uptake and accumulation in plants. *Advances in Agronomy*, **84**, 121–57.

VROM (2000) *Circular on target values and intervention values for soil remediation.* Dutch Ministry of Housing, Spatial Planning and Environment (VROM), The Hague.

Walker, D.J., Clemente, R. and Bernal, M.P. (2004) Contrasting effects of manure and compost on soil pH, heavy metal availability and growth of *Chenopodium album* L. in a soil contaminated by pyritic mine waste. *Chemosphere*, **57**, 215–24.

WHO (2004) *Guidelines for drinking-water quality. Recommendations.* World Health Organisation, Geneva.

Zorpas, A.A., Kapetanios, E., Zorpas, G.A., Karlis, P., Vlyssides, A., Haralambous, I. and Loizidou, M. (2000) Compost produced from organic fraction of municipal solid waste, primary stabilized sewage sludge and natural zeolite. *Journal of Hazardous Materials*, **77**, 149–59.

9

Robust Sustainable Technical Solutions

Abir Al-Tabbaa, Michael Harbottle and Chris Evans

9.1 Introduction

The remediation of contaminated land can have many impacts, both positive and negative. It can enable urban regeneration, allow sustainable land use and reduce human (or other) health risks to acceptable levels but it can also create waste and noise, and consume natural resources, and also may not remove the problem but simply postpone it. Implementing a truly sustainable remediation solution would mean reducing, removing or minimising these negative impacts in many areas of concern, including those that are not considered to be part of the 'core' objectives of remediation (i.e. cost and feasibility). Remediation of contaminated land in the UK is usually driven by a risk-based approach to the contamination alone, and is often very much site-specific. 'Non-core' objectives are therefore often neglected and possibilities for improving these methods, especially in terms of their wider environmental, social and economic impacts, may have been overlooked, by concentrating primarily on core objectives. In particular, durability over periods of more than a few years is commonly disregarded – a successful solution may often only be required to be effective for a few years. Inter- as well as intra-generational aspects are important in sustainable solutions and so durability has to be taken into account at the implementation stage.

This chapter focuses on:

- the assessment of sustainability of available containment and clean-up remediation methods
- the assessment of relative sustainability of a number of those remediation methods
- improvements to specific technical solutions based on sustainability assessment and experimental investigations

Sustainable technical solutions to brownfield land need to be 'robust', which as a working definition means that they must not be sensitive to small changes, must have a low risk of failure and need to be durable. Remediation solutions are sustainable if:

- future benefits outweigh cost of remediation
- environmental impact of the implementation of the remediation process is less than the impact of leaving the land untreated
- environmental impact of bringing about the remediation process is minimal and measurable
- the timescale over which the environmental consequences occur, and hence the inter-generational risk, is part of the decision-making process
- the decision-making process includes an appropriate level of engagement of all stakeholders

9.2 Sustainability assessment of currently available remediation technologies in the UK

9.2.1 Sustainability and the remediation industry

The sustainability of remediation is not just related to how quickly, cheaply and easily the risk of contamination can be ameliorated. There are many potential impacts that might arise from performing remediation that are not usually considered at all, unless they lead to delay or further cost. For example, emissions, waste production and the effect on stakeholders not directly involved with the process may often not be considered. Until recently, civil-engineering-based remediation methods, particularly 'dig and dump', were the most popular, owing to their low cost and ease of use. However, the implementation of the landfill tax and the European Union (EU) Landfill Directive has led to an increase in cost and a reduction in available landfill space for hazardous waste such as contaminated soil. As a result, other remediation methods are becoming established (see Chapter 5). Process-based techniques such as bioremediation, soil-washing and stabilisation/solidification are now increasing in popularity in the UK. Such techniques are often considered 'sustainable' as they have reduced impacts over excavation and disposal to landfill, but their full impacts may still be neglected.

A number of organisations have studied the potential of, and barriers to, the implementation of sustainable remediation, such as the EU-funded CLARINET (Contaminated Land Rehabilitation Network for Environmental Technologies) project. This included several studies on sustainable brownfield land management (Vegter *et al.*, 2002) and led to the development of frameworks that included sustainability as a major factor to be considered in the selection of remediation technologies (Bardos *et al.*, 2001; 2002b). However, the inclusion of sustainability is still at an early stage and as yet there has been little progress towards this at a practical level. Historically the major concerns in selecting a remediation technique have been cost and feasibility, although of course there are a range of wider environmental, social and economic impacts that would not therefore be considered. The area of sustainability that has received most attention to date is that of environmental impact. The acceptance of the need for assessment of these wider environmental effects is becoming more widespread in the regulatory and research arenas, and is becoming an important issue in the remainder of the industry (Bardos *et al.*, 2000a; 2000b).

A number of studies have attempted to determine the wider effects of remediation, particularly environmental, using some form of life cycle analysis (LCA), as summarised by Suèr *et al.* (2004). For example, Diamond *et al.* (1999) and Page *et al.* (1999) presented a method that assessed the true impacts of excavation and disposal to landfill, in terms of factors such as emissions, energy use and waste as well as human and ecological toxicity, on one case study. They also assessed generically and qualitatively, but less intensively, the effects of several other techniques. In a similar study, Bayer and Finkel (2006) compared groundwater remediation, using pump and treat and funnel and gate techniques, on a single site using LCA and were able to take into account long-term impacts to conclude that the former method, which had lower initial inputs, also had considerably greater impact over time.

Both Blanc *et al.* (2004) and Volkwein *et al.* (1999) also presented comparisons of the individual impacts of a number of techniques. However, both of these were designed as selection tools on one particular site, and did not look at past remediation projects. As such, projected rather than actual data were used. In addition, Blanc *et al.* (2004) applied a limited multi-criteria analysis to their data. Both Suèr *et al.* (2004) and Andersson (2003) assessed and compared a range of applications of similar techniques and found that different assumptions made in the methodologies meant that two assessment methods applied to the same project would come up with different suggested technologies.

Despite the work described above there is still no general consensus on a suitable technique for sustainability assessment in real projects; indeed, many of these techniques are complex and require considerable input of data and effort, and as such are unlikely to be used on any but the largest of remediation projects. They also relate to only a small part of the assessment of sustainability, and while they might be extended to take into account social and economic aspects, this would further increase their complexity.

In a survey of industry practitioners including local authorities, consultants, contractors and other interested parties, the awareness of sustainability and the extent to which it is considered in remediation projects was investigated based on 60 responses from a sample of 208 individuals involved in remediation in the UK in 2004. Figure 9.1 shows how the industry feels the tenets of sustainability are being considered in the selection of remediation technologies, incorporating wider environmental, social and economic impacts as well as what happens in the long term. All areas are apparently regularly taken into consideration, but to what extent? Commonly, the effects of contamination on these areas both immediately and in the long term are taken into account, but the effects of actually performing the remediation (transportation, waste, etc.) are often not, so progress towards a truly sustainable solution is hindered.

An important part of determining whether a project is to be sustainable is ensuring that *all* potential impacts are considered when the project is designed. The tools used to select a suitable remediation technology are therefore important, and considerable effort has gone into investigating the efficacy of different tools (Bardos *et al.*, 2002a). Figure 9.2, based on results taken from the same survey, shows the extent to which certain tools or approaches have been utilised (see also Chapter 5). As would be expected, professional judgement is exercised most commonly; concern might be raised over those who say they only use it 'often' or 'sometimes', but this is

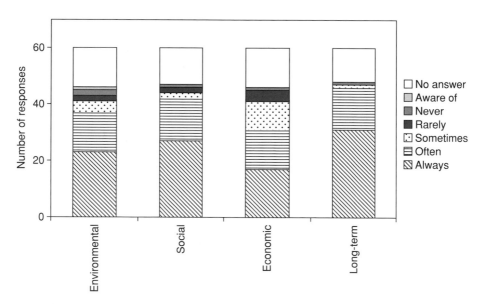

Figure 9.1   Summary of questionnaire responses to the question: 'Do you consider the sustainability of any aspects of a project in the selection of a remediation technology?'

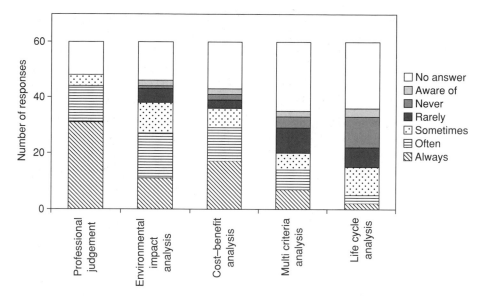

Figure 9.2   Summary of questionnaire responses to the question: 'What methodologies do you or have you used in helping you to determine the best remediation technology for a particular project?'

largely because other techniques are used to provide the answers. Environmental impact and cost–benefit analyses are frequently used, but more complex methods for which there is less guidance or legislation are less common. Life cycle analysis in particular is rarely used at present. However, a full life cycle-based approach is likely to be necessary in order to fully assess the true impacts of remediation, and therefore it is likely that the full impacts are to a certain extent neglected.

According to this survey, sustainability is clearly beginning to be appreciated by remediation practitioners, although the extent to which this is expressed in terms of building it into remediation projects is unclear from these results. It is likely that certain areas are incorporated, particularly in larger projects where the cost implications of assessment and monitoring are not so onerous, or where regulations require it. However, it is equally likely that many areas are not considered, perhaps because important data are not available or there is little appreciation of the potential effects.

9.2.2 *Technical and environmental sustainability methodology and application*

An appreciation of the true impacts of different remediation technologies would be extremely valuable in the design of sustainable remediation projects. The methodology presented here was developed and used to identify where technical and environmental impacts arise, using past remediation projects as case studies, in order to inform decision-making in such ventures, and is based around four broad criteria which address the physical impacts of remediation, and the effect on contamination, pathways and receptors:

- Criterion 1: future benefits outweigh cost of remediation.
  This requires any benefits of the remediation to outweigh any costs over the lifetime of the project and beyond. Benefits and costs measured in non-financial terms include risks to site users and public, quality and quantity of surface water, groundwater, air and soil, use of non-renewable resources, non-recyclable waste and potential range of future uses of the land. Financial benefits include economic value of the land, impact on surrounding areas and incentive/tax break. Costs include capital, operation and maintenance, labour, site investigation, monitoring/post-closure maintenance, professional fees, insurance/legal and offsite disposal.

- Criterion 2: environmental impact of the implementation of the remediation process is less than the impact of leaving the land untreated.
  The environmental impacts of the 'remediation' and 'no action' options, in terms of reducing or removing the risks of contamination to receptors should be measured and compared using factors such as future risk to human health, impact on ground conditions, impact on water flow, air pollution, flora and fauna, restriction on future use of the land, impact on other sites, landscape and fate of the contaminants.

- Criterion 3: environmental impact of bringing about the remediation process is minimal and measurable.
  This deals with the implementation of the remediation process itself, rather than the effect of contamination, and requires such impacts to be minimal. These includes impacts of all the processes involved such as transport, emissions to air, energy use, use of secondary materials, waste, direct use of natural resources and impact of the materials used in the remediation process. All impacts need to be measurable.

- Criterion 4: the timescale over which the environmental consequences occur, and hence inter-generational risk, is part of the decision-making process.
  Factors include long-term monitoring and maintenance, post-closure maintenance, durability, future underground activities, land management issues, long-term contaminant degradation and sustainable use of the soil.

A further criterion should also be included where possible:

- Criterion 5: The decision-making process includes an appropriate level of engagement of all stakeholders.
  This deals with social parameters which have not been considered in this work. However, the physical causes of these impacts have been included as part of the above criteria, and a discussion is included in the appropriate case study. The developed methodology presented here could be expanded to incorporate these areas if required.

The range of technologies that can potentially be used for remediation is necessarily large to address the huge variability in contamination, soil properties (chemical, physical and biological) and groundwater conditions. Therefore any assessment of the impacts of such techniques has to be wide-ranging, and able to consider a wide variety of potential impacts that might arise as a result of their implementation.

A methodology has been developed that incorporates a multi-criteria analysis (MCA) and a detailed impact analysis (DIA), both of which take a life cycle approach (see Harbottle *et al.*, 2005; 2006a; 2006b for more details). This includes impacts not obviously connected with remediation, but occurring as a direct consequence of it (e.g. in raw material excavation), both on immediate and long-term timescales. Applying the assessment methodology to completed remediation projects allows a realistic assessment of technologies and facilitates the identification of particular problems with technologies which can then be used in the design procedure of future remediation projects, as well as identifying areas for improvement.

The MCA takes an overview of a project and allows the inclusion of both quantitative and qualitative effects. It is based on the Environment Agency method for selecting optimal remediation technologies for a particular site (Postle *et al.*, 1999), where a number of categories of information are scored, weighted and summed. The categories and sub-categories that have been included to date in the developed methodology are listed below, although these could be revised and expanded upon as required:

- human health and safety (risks to site users; risks to public)
- local environment (surface water quality and quantity; groundwater quality and quantity; air quality (pollution); quality and structure of soil; habitat and ecology)
- stakeholder concern (acceptability of remediation)
- site use (duration of works; impacts on landscape; future site use; surrounding land use)
- global environment (air quality (greenhouse gases); natural resource use; waste)
- cost (taking into account changes in land values as well as the cost of remediation)

Scores are developed for each sub-category, for both the site itself and any ancillary sites, and for both during and after remediation in both those cases. They can be based on either quantitative or qualitative information. The relative importance of each sub-category and category for a particular site is then incorporated through the use of weights, and these are then combined to give a final overall score.

The DIA compares projects on individual sub-categories to identify where the major impacts are for particular methods. Primarily quantitative data such as emissions, waste generation or material use can then be compared on an individual basis to identify where the major areas of impact arise for particular techniques. In both cases the use of the life cycle approach means that impacts on other sites are also considered, and this means that work on landfills and other ancillary sites is included in the analysis.

The methodology has deliberately not adhered to a rigid format or set of indicators, but is amenable to alteration and to the inclusion of other factors. Therefore, although only mainly technical and environmental impacts have been considered so far, there is considerable scope for expansion, bringing in social and wider economic aspects of a project.

The outcomes of detailed assessment and comparison in two case studies of completed remediation projects are presented here as examples of how the developed methodology can be applied. The first study is the comparison on the same site of the impacts of in situ stabilisation/solidification, which was actually performed, excavation and disposal to landfill and taking no action. The second is the comparison of five different remediation technologies on five different sites based on five completed remediation projects.

9.2.3 Case study 1

The case study presented here compared the advantages and disadvantages of either using in situ stabilisation/solidification (S/S), excavation and disposal to landfill or taking no action on the same contaminated site. The site in question was remediated using S/S, although excavation and disposal was considered and hence there are data available for that scenario also. The 'no action' option assumed that the site conditions remained the same as prior to remediation. Excavation and disposal to landfill was included in this study as it is still a commonly used remediation technique and is usually assumed to be inherently unsustainable; hence it is used as a baseline for comparison. Brief details of the options being considered are given in Table 9.1, and a flow diagram of both remediation processes showing major remediation stages and inputs is shown in Figure 9.3. A common unit of measure was employed; each quantitative measure or score was normalised with respect to the tonnage of soil remediated on the site. The methodology also allows the inclusion of effects on sites other than that being remediated (e.g. landfills or borrow pits). Full details of this work are presented elsewhere (Harbottle *et al.*, 2005; 2006a; 2006b).

The site itself was previously used for industrial purposes, which had contaminated the coarse-grained soil layers present to a depth of ~4 metres with organic contaminants, particularly BTEX (benzene, toluene, ethylbenzene

Table 9.1   Brief details of the options being compared.

Project	Details
In situ stabilisation/ solidification	A cement-based binder material was used to treat contaminated areas. Auger rigs with hollow flight augers were used to deliver and mix the binder with the soil in situ. The in situ nature of the project meant that no waste material was produced and the majority of the work was performed on the site itself.
Excavation and disposal to landfill	All contaminated material (the same volume as was treated with S/S) was assumed to have been removed through excavation followed by disposal at a suitable landfill. The excavated material was replaced with virgin fill.
No action	No contaminant removal or containment was attempted prior to redeveloping the site.

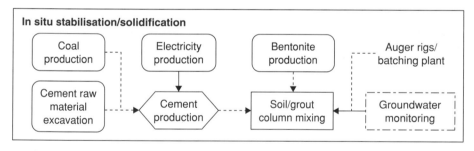

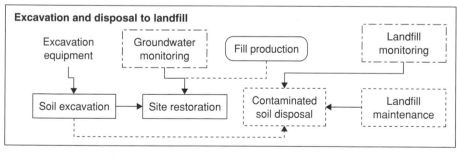

Figure 9.3   Flow diagram of in situ S/S and excavation and disposal to landfill in case study 1 showing major remediation stages and input.

and xylenes). The site was to be redeveloped for residential use, and potential receptors included future site residents and a nearby river.

Tables 9.2, 9.3 and 9.4 present the scores developed for each of the three remediation options considered – no action, S/S and landfilling. Each table gives four individual scores for each of the sub-categories considered, for on site and off site, both during and after remediation. In addition, a brief justification of these scores is given. Table 9.5 presents a summary of the weights (i.e. the relative importance on the site) of each sub-category and category, again with brief justifications. These tables help to identify where the major impacts were considered to occur, as well as illustrating the process through which overall scores were developed.

The outcome of the MCA is presented in Figure 9.4. Scores give indications of the *change* in impact due to remediation (i.e. if there is no change

Table 9.2   Scores and their justifications for MCA of no action (taken from Harbottle *et al.*, 2006a).

Category	Criterion	Scores During On site	Scores During Off site	Scores After On site	Scores After Off site	Justification
Human health and safety	Risks to site users	0	0	5	0	Possible gradual improvement in time due to natural attenuation.
	Risks to public	0	0	0	0	No effect.
Local environment	Surface water quality	0	0	−100	0	Contamination expected to eventually reach river.
	Surface water quantity	0	0	0	0	No effect.
	Groundwater quality	0	0	5	0	Possible gradual improvement in time due to natural attenuation.
	Groundwater quantity	0	0	0	0	No effect.
	Air quality (pollution)	0	0	0	0	No emissions.
	Quality/structure of soil	0	0	5	0	Possible gradual improvement in time due to natural attenuation.
	Habitat/ecology	0	0	0	0	No effect.
Third-party/ stakeholder concern	Third-party/ stakeholder acceptability	0	0	−100	0	Unacceptable to leave contamination in place (both to nearby residents, future site users and developers).
Site use	Duration of remediation	0	0	0	0	N/A
	Impact on landscape	0	0	0	0	No effect.
	Site use	0	0	0	0	No effect.
	Surrounding land use	0	0	−100	0	Continuation of blight effects.
Global environment	Air quality (greenhouse gas)	0	0	−1	0	0.36 kg/t due to natural CO_2 emissions.
	Use of natural resources	0	0	0	0	None
	Non-recyclable waste	0	0	0	0	None

Table 9.3 Scores and their justifications for MCA of stabilisation/solidification (taken from Harbottle *et al.*, 2006a).

Category	Criterion	Scores During On site	Scores During Off site	Scores After On site	Scores After Off site	Justification
Human health and safety	Risks to site users	−30	−5	90	0	Minimal contaminant emission and site work, little offsite work transportation. Long-term reduction in risk due to contamination, no effects off site.
	Risks to public	−2	−6	10	0	Small amount of dust/odours, small number of HGV movements, transportation and dust relatively low off site. Small improvement after due to contamination being stabilised (no offsite effects).
Local environment	Surface water quality	0	0	40	0	Improvement after through prevention of contamination reaching river on site.
	Surface water quantity	0	0	−100	0	Reduction in permeability of site due to solidified material.
	Groundwater quality	0	0	95	0	Contamination reduced by factor of 50.
	Groundwater quantity	0	0	−20	0	Reduction in site permeability.
	Air quality (pollution)	−73	0	0	0	4.11 kg/t non-greenhouse emissions (all assumed on site).
	Quality/structure of soil	−40	0	−40	0	Increase in pH, strength and reduction in permeability on site. Stabilised mass remains for foreseeable future.
	Habitat/ecology	−100	0	−40	0	Effective loss of soil and surface habitats during remediation. Continued loss of soil habitat although risk due to contamination is reduced.
Third-party/ stakeholder concern	Third-party/stakeholder acceptability	−1	−1	90	0	Relatively low noise, dust, odours, transportation on and off site. Majority of risk from contamination removed
Site use	Duration of remediation	−100	0	0	0	Two months.
	Impact on landscape	0	0	0	−5	Possible small impact from extraction of raw materials off site.
	Site use	0	0	83	−8	Potential future onsite use could be in any of five categories (residential, commercial, industrial, non-green and green public open space). Off site, loss of 0.5 due to raw materials extraction.
	Surrounding land use	−1	−1	100	0	Small impact due to congestion on and off site. After: removal of blight, reuse of land.
Global environment	Air quality (greenhouse gas)	−100	0	10	0	42.78 kg/t (all assumed on site) After: −4.26 kg/t (absorption)
	Use of natural resources	−9	0	0	0	89.5 kg/t (during)
	Non-recyclable waste	0	0	0	0	None

Table 9.4 Scores and their justifications for MCA of landfilling (taken from Harbottle et al., 2006a).

Category	Criterion	Scores During On site	Scores During Off site	Scores After On site	Scores After Off site	Justification
Human health and safety	Risks to site users	-100	-100	100	-5	Highest risk due to contaminant emission and site work (on and off site). Reduction in contamination risk (small risk of escape of contaminants off site).
	Risks to public	-100	-100	20	-5	Large effects of dust and transport on and off site. After, improvement due to contamination removal on site, small contaminant escape risk from landfill.
Local environment	Surface water quality	0	0	50	-5	No effect during remediation. Onsite improvement due to source removal, small risk of escape from landfill.
	Surface water quantity	0	0	0	0	No effect.
	Groundwater quality	-5	0	100	-5	Small risk of contaminant escape during excavation. After, removal of all contamination, although small risk of escape from landfill.
	Groundwater quantity	-100	0	0	-5	Disturbance/dewatering during groundworks. Possible small effect due to landfill cap/liner.
	Air quality (pollution)	-100	0	0	0	3.88 kg/t non-greenhouse emissions
	Quality/structure of soil	-100	-40	10	-20	Loss of structure on site due to excavation, contamination all transferred off site.
	Habitat/ecology	-100	-80	50	-20	Loss of surface/soil habitats on site plus surface on landfill site. After, habitat restored on site, although contamination remains in landfill.
Third-party/ stake-holder concern	Third-party/stakeholder acceptability	-40	-20	100	-50	Largest impact from noise, dust and transportation on and off site. After, no risks remain on site but landfill remains unpopular.
Site use	Duration of remediation	-50	-50	0	0	One month (includes work on landfill, borrow pit).
	Impact on landscape	0	0	0	-100	Long-term negative impact of landfill.
	Site use	0	0	100	-67	Potential future onsite use could be in all six categories (residential, industrial, commercial, agricultural, non-green and green public open space). Off site, loss of four due to raw material extraction and landfill.
	Surrounding land use	-20	-20	100	-80	Largest impact due to congestion on and off site. After, removal of blight associated with landfill.
Global environment	Air quality (greenhouse gas)	-31	0	-1	0	12.85 kg/t during and 0.35 kg/t after
	Use of natural resources	-100	0	0	0	1005.6 kg/t
	Non-recyclable waste	-100	0	0	0	1000 kg/t

Table 9.5 Weights and justifications for MCA (taken from Harbottle *et al.*, 2006a).

Category	Criterion	Weights			Justification
		On	Off	Cat	
Human health and safety	Risks to site users	1	0.5	1	On site: future residential use. Off site: few site users, for working hours only.
	Risks to public	1	0.5		On site: high population in local area. Off site: low population near landfill.
Local environment	Surface water quality	1	0.7	0.9	On site: protection of the river is an objective of the project although groundwater flow and expected probability of significant impact are low. Off site: nearby lakes for recreational use.
	Surface water quantity	0.2	0.2		On site: water not used for abstraction, small site. Off site: little effect on nearby lakes.
	Groundwater quality	0.8	0.7		On site: shallow aquifer, not used for drinking but important for river quality. Off site: some importance due to nearby lakes.
	Groundwater quantity	0.2	0.2		On site: not used for abstraction, slight importance for river. Off site: slight importance (nearby lakes).
	Air quality (pollution)	0.7	0.5		On site: high local population. Off site: low local population.
	Quality/structure of soil	0.2	0.1		On site: slight importance due to construction. Off site: little importance as contained in landfill.
	Habitat/ecology	0.1	0.3		On site: low importance. Off site: landfill near to nature reserve.
Third-party/ stakeholder concern	Third-party/stakeholder confidence	0.8	0.6	0.8	On site: local/site population is high and so remediation affects more people. Off site: lower population.
	Third party/stakeholder acceptability	1	0.7		
Site use	Duration of remediation	0.6	0.4	0.9	On site: important due to disturbance and need for development. Off site: less important.
	Impact on landscape	0.5	0.7		On site: some importance – high local population but is in urban area and to be redeveloped. Off site: rural area so some importance.
	Site use	1	0.6		More important in urban area due to pressure for land.
	Surrounding land use	0.8	0.6		
Global environment	Air quality (greenhouse gas)	1	1	0.7	Global importance – equal weights.
	Use of natural resources	1	1		
	Non-recyclable waste	1	1		

before and afterwards then there is a zero score). The scores from six main categories (human health and safety, local environment, stakeholder concern, site use, global environment and cost) are presented. Each is made up of weighted scores from a number of sub-categories. Positive scores can be interpreted as 'good' and negative as 'bad'. It can clearly be seen that excavation and disposal to landfill produced some highly negative scores compared to in situ S/S and no action, with 'cost', 'human health and safety' and 'global environment' providing particular cause for concern. This is perhaps not surprising; the technique has long been considered to be particularly unsustainable. In comparison, S/S performed well. Overall, the 'cost' category scored well, as the benefits of redevelopment outweighed the cost of performing the remediation. The major negative impact of this technique was that of greenhouse gas emissions (as indicated by the performance of the 'global environment' category in Figure 9.4), due to carbon dioxide release in cement production. Otherwise, the in situ nature of the remediation helped to minimise a number of potentially onerous impacts, such as those arising from transportation and waste. The main impacts with the 'no action' option were in the 'stakeholder concern', 'site use' and 'local environment' categories. As there would be little change in both the effects on human health and costs/land values, these have a low score and are not visible on the figure.

The results of the MCA allow the comparison of broad categories of data. They allow both quantitative and qualitative data to be included in a single analysis, although the latter includes the risk of a certain degree of subjectivity. An assessment of individual sub-categories has also been performed to further investigate these broad scores. This covers a number of areas such as emissions, waste production, contamination risks and long-term effects. Examples of the data are given in Figure 9.5 for use of raw and recycled

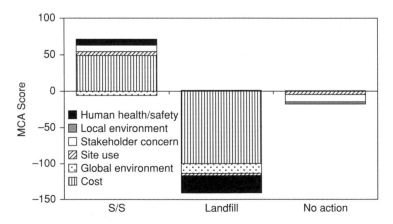

Figure 9.4   Outcome of the MCA for case study 1 (taken from Harbottle *et al.*, 2006a).

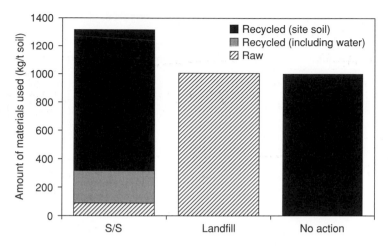

Figure 9.5 Total recycled and raw materials in the DIA for case study 1 (taken from Harbottle *et al.*, 2006a).

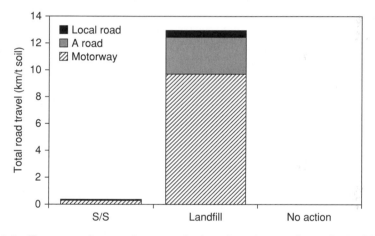

Figure 9.6 Transportation requirements, broken down into road type, in the DIA of case study 1 (taken from Harbottle *et al.*, 2006a).

materials impacts and in Figure 9.6 for transportation impacts. Figure 9.5 shows the relatively small amount of raw material usage with S/S when compared to the fill required in the excavation and disposal to landfill. The larger total material usage with S/S (including reused site soil) indicates the increase in density of the treated soil. Figure 9.6 shows how an in situ remediation technique significantly reduces road usage in comparison to disposal of the material off site. In both examples presented here, taking no action has no impact.

A summary of the outcome from the detailed impact analysis is presented in Table 9.6, which lists the main impacts of each of the three scenarios assessed. Excavation and disposal to landfill had a number of

Table 9.6   Summary of impacts, both positive and negative, of the scenarios used in case study 1.

In situ stabilisation/ solidification	Excavation and disposal to landfill	No action
• Low intensity operations	• High transportation	• Stakeholder concern
• Low waste	• High waste production	• Continued risks to site users
• Low transportation	• High material use	• Impact on river quality
• Low noise	• Impacts on landfill site	
• High CO_2 emissions	• High energy use	
• High energy use	• Long term	
• Contaminants remain	• High disturbance	
• Changes to soil properties		

negative impacts that arose through its intensive onsite work and extensive use of offsite transportation and disposal. These include emissions from transport and site work, use of materials, considerable waste production and potential long-term effects of contamination in the landfill. Despite the ease of use of this remediation technique and the thoroughness with which the remediated site is cleaned, it would still be expected to be of considerable concern to stakeholders, particularly those who live near the landfill site. S/S has a mixture of low and high impacts; emissions are high owing to cement production but this is offset by the reduction in onsite and offsite work required. S/S also has long-term impacts, although this time the contaminants remain on the site itself and as such may pose a future risk to site users. Taking no action has serious impacts with respect to the effects of the contamination on receptors, but of course it involves no work in implementing it and so has no impact on many of the sub-categories considered in this analysis.

Although criterion 5 was not formally assessed in the case studies presented here, a number of qualitative points may be made on the impact of the various site treatment options on stakeholders. In situ S/S, for example, is rapid, relatively cheap and does not lead to significant disturbance either on site or off site, and is therefore likely to be relatively popular with developers and possibly residents near to the site, although the latter group might be concerned by contaminants remaining on site. Excavation and disposal to landfill is also rapid and, importantly, is a tried and tested technique. Developers may be satisfied with this, but there are a number of potential stakeholders who might be adversely affected by various aspects of it, particularly those affected by the landfill itself or transportation of materials to it. 'No action' does not have any problems in terms of implementation, but it would be unacceptable to regulators and probably nearby residents and other local stakeholders. None of these scenarios actually separates contaminant from the soil, and so the actual problem of contamination is not removed, but merely contained. Therefore, each option has unsustainable social aspects associated with it.

9.2.4 Case study 2

This case study considered five separate remediation projects, each undertaken using one main technology (although in most a degree of excavation and disposal to landfill was employed). The five technologies were in situ stabilisation/solidification (S/S), soil washing, ex situ bioremediation, cover system, and excavation and disposal to landfill. Further information is presented by Harbottle *et al.*, 2007a; 2007b. Again, excavation and disposal to landfill was included as a baseline with which to compare the other technologies. Details of the five projects are given briefly in Table 9.7. Each project took place on a different site around the UK, and so comparison between projects was considerably more difficult than in case study 1, where

Table 9.7   Brief details of the five projects in case study 2.

Project	Details
In situ stabilisation/ solidification	Past industrial use contaminated the site with organic pollutants. The site was redeveloped for residential use, using a cement-based binder material to treat contaminated areas. The in situ nature of the project meant that no waste material was produced and the majority of the work was on the site itself. Potential receptors included a nearby river and future residents.
Soil-washing	This contaminated former gasworks site was treated using a soil-washing procedure for removing fine-grained soil constituents and the organic contamination contained therein. Remediation allowed redevelopment for commercial use, as well as protecting a river and groundwater supplies. Cleaned, coarse-grained soil was reused on site. This significantly reduced the amount of material disposed of in landfill, in turn reducing local disturbance through transportation.
Ex situ bioremediation	A range of former industries had left a number of organic contaminated areas on this site and so a combination of disposal to landfill (for very heavily contaminated soil) and bioremediation in windrows on the site itself was utilised. Bioremediated soil was reused on site. Potential receptors included a nearby river, groundwater supplies and future site users on this mixed-use redevelopment.
Cover system	The site had previously been a gasworks, resulting in a range of contaminants. This heavily contaminated site was remediated through hotspot excavation and offsite disposal followed by application of a cover, using recycled material where possible. The site was developed for commercial reuse. Further impact on groundwater was to be avoided although specific treatment was not performed due to the generally degraded nature of the local area.
Excavation and disposal to landfill	Contaminated material containing both organic and inorganic pollutants was excavated and disposed of in landfill. Recycled material was used for backfill. The site was then reused for light industrial purposes.

techniques were compared on the same site. To alleviate this, a common unit of measurement has been employed; each quantitative measure or score has been normalised with respect to the amount of soil remediated on the respective sites. This ensures that the comparison is not unduly affected by project size. As in case study 1, impacts occurring away from the remediated site itself, for example on landfills or borrow pits, have been included, and both MCA and DIA were carried out.

The analysis in this assessment was performed in a similar way to that in case study 1, and hence the tables illustrating how the various scores and weights in the MCA were developed and combined are not presented in this case. The scores from the same five main categories mentioned in case study 1 (human health and safety, local environment, stakeholder concern, site use and global environment) are presented. Each is made up of weighted scores from a number of sub-categories (data not presented). The weightings vary depending on the specific site considered, hence the weighting for the in situ S/S and landfilling in case studies 1 and 2 are not necessarily the same.

The outcome of the MCA is presented in Figure 9.7. It can again be seen that excavation and disposal to landfill produced some highly negative scores in comparison with the other technologies, with 'human health and safety', 'stakeholder concern' and 'site use' providing particular cause for concern. In comparison, the other technologies performed well. This is particularly the case for the cover system, where a large amount of soil was considered to have been remediated and therefore impacts were correspondingly reduced. S/S also had relatively low impacts owing to its

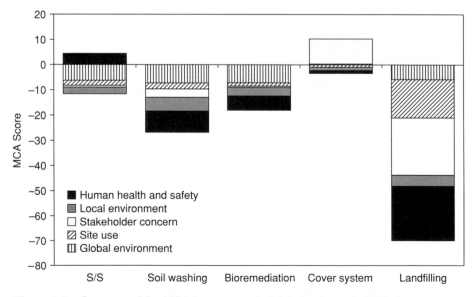

Figure 9.7 Outcome of the MCA for case study 2 (Harbottle *et al.*, 2007a).

in situ nature; a reduction in transportation and offsite work considerably reduced potential impacts. As all projects other than S/S involved a certain degree of excavation and disposal to landfill, it is clear that the impacts of the 'pure' technologies (for soil-washing, bioremediation and cover system) are less onerous than is presented here. It should be noted that these scores cannot be directly compared with the scores from the MCA performed as part of case study 1, and vice versa, as they were developed through comparison within these five projects only. However, despite different sites being used, some correlation may be achieved by taking the excavation and disposal to landfill projects as benchmarks.

Results from the DIA are given in Figure 9.8 for use of raw and recycled materials and in Figure 9.9 for transportation. Figure 9.8 clearly shows the reuse of site soil in all techniques other than excavation and disposal to landfill, although in this case study the use of recycled fill in the project was a positive reuse of material. Transportation per tonne of remediated soil was, as expected, greatest for this project also, while the effect of reducing offsite soil disposal is evident in the remaining four projects. The cover system project imported a large amount of material, accounting for its relatively high value.

A list of the main impacts of each of the remediation projects assessed based on the DIA is presented in Table 9.8. Both the cover system and excavation and disposal to landfill projects have impacts that back up the outcome of the MCA. The former had few highly negative impacts, owing

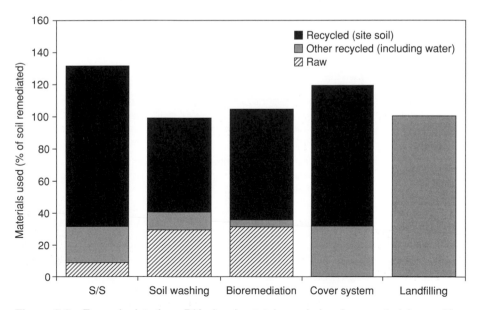

Figure 9.8   Example data from DIA showing total recycled and raw materials used for case study 2 (Harbottle *et al.*, 2007b).

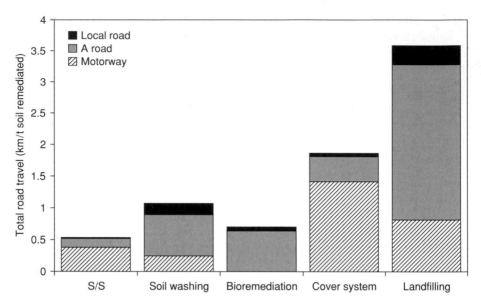

Figure 9.9   Example data from DIA showing transportation requirements, broken down into road type for case study 2 (Harbottle *et al.*, 2007b).

Table 9.8   Summary of impacts from the DIA for the projects used in case study 2.

	Impacts	
In situ stabilisation/ solidification	• Low intensity operations • Low waste production • Low disturbance • Low transportation • Low noise	• High CO_2 emissions and energy use due to cement production • Contaminants remain • Changes to soil properties
Soil-washing	• Reduced transportation • Reduced waste • Short duration • High energy use	• Impacts on landfill site • Fill use • High intensity operation • High noise
Ex situ bioremediation	• Destruction of contaminants • Low transportation • Low disturbance • Fill use	• Impacts on landfill site • High intensity operation • High noise
Cover system	• Low waste • Low emissions • Rapidity • Reduced transportation	• Low energy use • Some contaminants remain • Impacts on landfill site • High noise
Excavation and disposal to landfill	• High transportation • High waste production • High material use • Impacts on landfill site	• High energy use • Long duration • High disturbance

to the relatively low effect of remediation with respect to the amount of soil remediated. Conversely, excavation and disposal had a number of negative impacts and little positive to recommend it, because of intensive onsite work and offsite transportation and dumping. The remaining three projects, in situ S/S, soil-washing and ex situ bioremediation, all have a mixture of areas with low and high impacts. Of particular comment for in situ S/S are again the gaseous emissions, primarily due to the production of Portland cement, and the persistence of contaminants on the site with attendant long-term impacts. Soil-washing reduced the effects of offsite disposal but required intensive onsite operations to bring this about, while bioremediation was the one technique that enabled contaminant destruction but required time and space and also considerable work on the site. A different way in which the DIA data was used was to rank the five remediation projects in order of their performance in each of the areas considered in the DIA. This is useful when attempting to compare specific sustainability issues between remediation projects, and was the approach was adopted in Harbottle *et al.*, 2007a and 2007b.

9.3 Sustainability improvements to remediation techniques

9.3.1 Improvements based on sustainability assessment

As well as being used for the comparative assessment of different remediation techniques and projects, the above sustainability assessment is also useful in quantifying potential ways in which technical and environmental sustainability aspects of remediation techniques can be improved. A brief list of impacts, benefits and potential improvements to certain techniques has been compiled in Table 9.9, giving an indication of problems and possibilities.

9.3.2 Improvements based on experimental investigations

The technique considered experimentally in detail here is stabilisation/solidification (S/S). S/S is a process-based containment method for contaminated soils, whereby binders are mixed with the contaminated soil in order to produce a solidified mass (LaGrega *et al.*, 2001; Evans *et al.*, 2001). Stabilisation is a chemical process to minimise the hazardous nature of contaminants by converting them into less soluble, less mobile or less toxic forms. Solidification is a physical process to convert the contaminant substrate into a durable and dense monolith, reducing the potential for the leaching of the contaminants from it. It is most suitable for the immobilisation of heavy metals and to a lesser extent for organic contaminants.

Table 9.9 Summary of impacts from and potential improvements to certain remediation technologies.

Remediation technique		Description	Environmental impacts		
			Resources used	Energy used	Emissions
Landfill/dig and dump		Excavation/disposal of contaminated soil	Use of land for landfill, fill, cap and liner	Fuel (transport and site work)	Volatile contaminants, emissions from transport, pretreatment, site work, landfill (long term)
Capping/cover system		Cap over contaminated area to prevent access	Cap materials/ geosynthetics	Fuel (transport and site work)	Transport and site work
Stabilisation/ solidification	In situ soil mixing	Use of binders etc. to fix contaminants	Binder material, soil resource lost	Fuel (site work), power for cement production	Contaminant release (long term), cement production
Barriers	Containment	Barrier walls surround site	Barrier material	Fuel (transport and site work)	Escape in long term
	Permeable reactive barrier	Barriers funnel contaminated groundwater to reactive zone	Barrier material	Fuel (transport and site work)	Reduction in barrier efficiency over time
Incineration		Combustion/destruction of soil organic material	Fill	Energy for heating, fuel (transport, site work)	From heating, transport, site work
Thermal desorption		Soil excavated and heated, contaminants volatilised	As above	As above	As above
Pump and treat/flushing		Groundwater extraction, contaminant removal, reintroduction	Surfactants, sorbents, flocculants, etc.	Energy for pumping	Chemical/contaminant loss to groundwater
Soil-washing		Separation of contaminated (fine) and clean (coarse) soil	As for landfill, plus wash additives	Fuel (transport and site work), electricity (washing plant)	Volatile contaminants, from transport
Soil vapour extraction		Vacuum induces air flow in soil, volatilises contaminants	Sorbent etc.	Energy for creating vacuum	Volatilised contaminants, emissions from pumping
Onsite landfilling		Encapsulation of contaminated soils on site	Cap/liner, fill	Fuel (transport and site work)	Volatile contaminants, transport, site work
Phyto- remediation	Accumulation/ stabilisation	Contaminant accumulation in plants, stabilisation in soil	Nutrients/ composts etc.		Possible combustion of harvest to recover metals
	Degradation	Contaminant degradation by plants/associated microbes	Nutrients	Irrigation	
Bioremediation	Natural attenuation	Monitored natural contaminant degradation			Risk of contaminant transport off site
	Biosparging/ bioslurping	In situ soil aeration (pumping/injection)		Energy for pumping	Volatile contaminants
	Enhanced in situ bioremediation	Nutrient introduction via groundwater/injection	Nutrients, bacteria	Pumping	Effect of nutrients (e.g. nitrates) on flora and fauna
	Bioreactors	Soil slurried, treated in reactor, controlled conditions	Nutrients etc.	Transport, equipment fuel	
	Windrows	Soil in rows, additives used e.g. compost, regularly tilled	Nutrients/ composts etc.	Fuel (site work)	Leachate
	Treatment beds	Soil spread on suitable surface, tilled, moisture adjusted, nutrients added	Nutrients/ composts etc.	Fuel (site work)	Leachate

Other	Social impacts	Economic impacts	Limitations	Other impacts/benefits	Possible improvements
Transport, contaminants remain, change to soil conditions/groundwater flow, site disturbance	Transport, stigma of landfill, health effects	Landfill tax, requirement to pretreat, blight near landfill/borrow pit	Durability, lack of landfill space	No residual contamination, rapid, simple, cheap, landfill gas	Reactive cap/liner barriers, construction on landfill, stimulate biodegradation
Transport, contaminants remain	Transport, remaining blight?	Reduced site value	Durability, not for risks to groundwater etc.	Rapid, simple, cheap	Reactive cap material, enhance biodegradation
Contaminants remain	Remaining blight?	Reduced site value	Problems with organics, durability	Rapid, cheap	Degradation (chemical, bio), sustainable binder
Contaminants remain	Remaining blight	Reduced site value	Durability	May hinder development	Reactive barrier, better barrier material
Production methods of barrier material	Long duration	Reduced site value during implementation	Durability	Small footprint, hinders development	Improved/more sustainable material
Excavation, loss of soil structure/organic matter	Incinerator location, transport		Organic pollutants only	Possible reuse of incinerated material	
As above	Transport, location of plant		Volatile organic pollutants only		
Possible effect on water supplies	Concern about chemicals in soil		Potentially long timescale, limited containment	Small footprint	Use gravity to reduce pumping
Loss of soil structure/organic matter, site disturbance	Transport, blight at landfill	Blight at landfill	As for landfill		
			Volatile organics only, duration	Can enhance biodegradation	
Excavation, contaminants remain		Reduction in site value	Limited space, construction problems		As for landfill
Contaminants may remain, use of non-native plants		Recovery of bio-energy and metals	Few capable plants, near surface only		
Introduction of non-native plant species			Organics near surface only		
	Long duration	Continued cost of monitoring	Long duration, prevents development, only hydrocarbons	May not need demolition	
			Hydrocarbons only	May not need demolition	Increase air temperature
			Hydrocarbons only	May not need demolition	
Excavation/disturbance of soil	Possible transport		Hydrocarbons only, limited throughput		
Excavation/disturbance of soil			Hydrocarbons only, large area required		Greenhouse cover, use of earthworms etc.
Excavation/disturbance of soil			Hydrocarbons only, large area required		As above

The technique has been used on a number of projects in the UK in the last decade, owing to its numerous advantages including speed of implementation, use of well-established techniques and materials, elimination of offsite disposal if performed in situ, low risk and relatively low costs. However, a major concern of the technique is that the contaminants are neither removed nor destroyed but remain within the bound material, and may become a future hazard if the containment system breaks down or the environmental conditions change. In addition, the high pH, ~12–13, of the binder systems usually employed, which are often Portland-cement-based, is not a suitable environment for microorganisms. There is also a considerably reduced pore space, as well as reduced pore connectivity, meaning that transport of essential nutrients can be severely limited. In addition, it is likely that the bioavailability of contaminants to any potential degrading organisms will be hindered, owing to their immobilisation or sequestration. Hence these are not appropriate conditions to facilitate the biodegradation of any organic contaminants present. This could prevent or significantly reduce any natural attenuation with the stabilised/solidified soil matrix that might otherwise occur naturally within the untreated soil. Hence, the inclusion of a biodegradation mechanism in the system would eliminate this negative aspect of this technique and would make it a more sustainable remediation option for the treatment of contaminated land. In addition, the development and application of a suitable low-pH cement could also lead to appropriate conditions for combined S/S and biodegradation. The following two sections provide an overview of the work carried out in those two novel areas of research.

9.3.3　*A novel low-pH phosphate cement binder system*

The high alkalinity of Portland-cement-based S/S binders is ideal for the immobilisation of heavy metals which precipitate as insoluble hydroxides, with minimum solubility within a pH range of 8 to 11. Nevertheless, such systems have found limited suitability in fixing organic contaminants, since some organics have been reported to interfere with the hydration processes of Portland cement (PC), resulting in reduced final strength and impaired stabilisation (LaGrega *et al.*, 2001). One way of facilitating the biodegradation of organic contaminants within a cement-based system is to investigate suitably low-pH cement systems. A pH range of 6 to 8 is favourable for the growth and survival of microorganisms. However, at this near neutral pH there is a potential for the leaching of cement-stabilised metals, owing to increased metal solubility. Magnesium phosphate cements (MPCs) formed at room temperature by acid-base reaction, between magnesium oxide (MgO) and an acid phosphate source in solution, have been reported to develop much lower initial pH than that of PC. In addition, phosphates of metals have a very low solubility. Hence MPCs are a promising candidate for the investigation in hand.

Based on the phosphate source used, different variants of MPCs can be formed. One common phosphate source is potassium dihydrogen orthophosphate (KH_2PO_4) which forms magnesium potassium phosphate cements. The main reaction product of MgO with KH_2PO_4 is a hard, dense ceramic of magnesium potassium phosphate hexahydrate ($MgKPO_4.6H_2O$) (Wagh, 2004). For small volume works, in order to reduce the reaction rate, the magnesia used is generally dead burned magnesia, which is a low reactivity magnesia form produced by calcining of magnesite ($MgCO_3$) at around 1300°C. However, the resulting reduced solubility of the calcined magnesia would still be too high for mixing larger volumes (Wagh, 2004).

Therefore pretreatment of MgO by retarder additions becomes necessary in order to decelerate the acid-base reaction and to lower the consequent heat generated. While a reorganisation of the MgO surface has been reported to occur as a result of calcination lowering its reactivity (Soudée and Péra, 2002), the addition of a retarder in the form of boric acid has been reported to form a polymeric coating of low-solubility magnesium-boron-phosphate compound, called 'Lünebergite' ($Mg_3B_2(PO_4)_2.(OH)_6.6H_2O$) on the MgO particles, prolonging the reaction of MgO with the acid phosphates in solution (Wagh, 2004).

MPCs have been reported to possess improved traits compared to PC, such as higher early-age and long-term strengths (Abdelrazig *et al.*, 1989), higher freeze–thaw and wet–dry durability (Li *et al.*, 2004), lower permeability (El-Jazairi, 1987) and higher resistance to sulphate attack (Li *et al.*, 2004), which have made MPCs an attractive option for special applications. All these advantages are beneficial in the development of a more sustainable containment system and in particular in response to aggressive environmental conditions such as those likely to be imposed by future climate change scenarios.

MPCs are used for a wide range of applications including refractory materials, dental cements, structural repair works and waste management. In waste management, the phosphate reactions have been reported to convert most of the hazardous contaminants into non-leachable phosphate reaction products, and the phosphate cement itself encapsulates these insoluble reaction products into a dense and durable matrix (Singh *et al.*, 1998). Phosphates of these contaminants have a much lower solubility than their corresponding oxides or other salts (Rao *et al.*, 2000). Thus MPCs provide a very effective chemical immobilisation and physical encapsulation of contaminants. Furthermore, unlike PC, MPCs have also been reported to set even in the presence of hydrocarbons (Wagh, 2004).

In the work reported here, potassium dihydrogen orthophosphate, denoted as P, was used as the phosphate source. A much more cost-effective source of phosphate was also investigated for comparison. This is triple super-phosphate, denoted as TSP, which is predominantly calcium dihydrogen phosphate $(Ca(H_2PO_4)_2.H_2O)$; it also contains some gypsum $(CaSO_4.2H_2O)$, and is a common fertiliser. One type of dead burned magnesia, DB10 (Richard Baker Harrison, UK) with a citric acid reactivity of 1170 mins, denoted as M, was used throughout and boric acid (H_3BO_3), denoted as B, was used as the retarder. Following some trials, ratios of P:M or TSP:M ranging from 8:1 and 1:5 were tested with B contents ranging from 0 to 1.5%. A range of cement paste mixes at their standard consistence, where appropriate (water content 11–23%), were tested. Where it was not possible to test the standard consistence a water content of 50% was used, which was the value used when applying those cement mixes as grouts to the soil. The mixes ranged from dense solid monoliths at the high M content end to highly fractured structures at the high P content end.

The pH of the cement pastes for some of the mixes at 1 and 28 days is shown in Figure 9.10. The figure shows clearly the lower pH produced by the MPC mixes compared to PC mixes. The mixes differed in the way their pH changed over time, many increasing but some also decreasing at 28 days compared to 1 day. The figure also shows the effect of the two different P sources used and the effect of the different P:M ratios used.

A number of the cement paste mixes were then mixed with a sand con-taminated with zinc chloride and lead nitrate each in concentration of

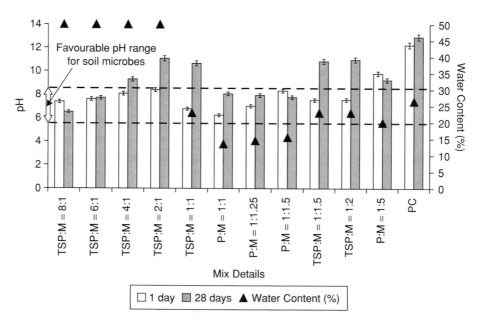

Figure 9.10 The 1-day and 28-day pH of a range of the MPC paste mixes (all with B = 1.5%), as well as the PC paste mix.

3000 mg/kg soil with a total water content of 10%. The cements were prepared as a grout with a water:cement ratio of 0.5:1 and mixed with the soil in a ratio of 3:7. The S/S-treated contaminated sands were tested in batch leaching at 28 days (BSI, 2002). The leached concentrations of Zn and Pb for all the MPC mixes were below their detection limit of 0.013 mg/L and 0.1 mg/L respectively, while PC produced concentrations of 0.087 mg/L and 1.97 mg/L respectively. The effect of the MPC mixes on reducing the leachability of those heavy metals despite the lower pH is clear. This work is still ongoing and is now investigating the effectiveness of these MPCs in facilitating biodegradation within their matrix and their durability in the longer term under aggressive climate conditions.

9.3.4 Combined S/S and biodegradation for Portland cement binder systems

Very few relevant studies on the modification of S/S systems to include degradation mechanisms have been reported. An abiotic method was proposed for chlorinated organic contaminants, whereby added iron created a reducing environment leading to contaminant dechlorination (Hwang and Batchelor, 2000; 2001; 2002). In addition, the biological clean-up of contaminated concrete surfaces has been investigated (Beklemishev and Kozliak, 2003), although this was in relation to the removal of contaminants applied to an existing surface rather than stabilised within it. Substantial reduction in organic contamination was achieved when surfaces covered with microbial biomass were applied to contaminated concrete surfaces. It was suggested that the diffusion of contaminants from the concrete pore structure was important in determining the degradation rate.

The work presented here focused on investigating ways to amend the properties of PC-based stabilised/solidified soil to encourage microbial survival, and hence biodegradation of contaminants. These included studies on additives to the systems, such as green waste compost, an oxygen-releasing compound (sodium oleate), nutrient agar (plate count agar), a water-retaining agent (polyacrylamide) and a nitrogen source (ammonium sulphate). An example of the effects of the different additives (in isolation) and in different addition rates on the dehydrogenase activity after 14 days in uncontaminated clayey silt model soil stabilised with PC grout is given in Figure 9.11. The effect of the cement on significantly reducing the enzyme activity in the soil is clear when compared to the soil alone. Compost was found to have a particularly strong effect on the enzyme activity, which increased with compost content, returning the enzyme activities to their level in the soil alone. A similar increasing trend, although at much lower enzyme activity, with the ammonium sulphate (AS) and the plate count agar (PCA) was observed. The enzyme activity also increased at higher concentrations

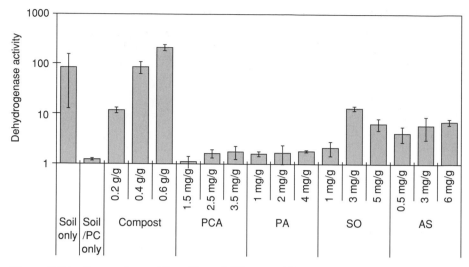

Figure 9.11   Comparison of the effect of different additives on enzyme (dehydrogenase) activity within clayey silt/PC system microcosms after 14 days' incubation at 37°C. PCA – plate count agar; PA – polyacrylamide (water retaining agent); SO – sodium oleate (air entraining agent); AS – ammonium sulphate (nitrogen source).

of the sodium oleate (SO). These results and others led to the use of compost only as an additive in later experiments.

Compost addition was expected to provide suitable refuges for microorganisms within the soil/cement matrix, as a method of protection from the harsh environment. This was achieved by amending the soil with green waste compost prior to mixing with the cement. Compost has the additional advantages of containing a large number of microorganisms as well as a supply of nutrients, and aids in water retention. The impact of compost addition at different rates to a silty sand model soil, spiked with 2-chlorobenzoic acid (2CBA) as the organic contaminant and stabilised/solidified with PC, is presented as an illustration. 2CBA is a non-volatile organic compound and is commonly referred to in the literature as a degradation product of polychlorinated biphenyls (Niedan and Schöler, 1997). It was chosen for this work because of its relative ease of degradation in the soils considered, and its relatively low toxicity, minimising the impact on microorganisms that perhaps are already stressed owing to the harsh environment within soil/cement systems. All experiments were performed at 37°C and approximately 95% relative humidity. This example is taken from Harbottle and Al-Tabbaa (2007).

Figure 9.12 illustrates, on a logarithmic scale, how the amendment of soil/PC systems with compost can affect microbial activity, in terms of dehydrogenase enzyme activity. It can be seen that the inclusion of 0.2 g/g

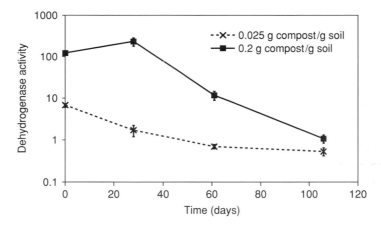

Figure 9.12   Comparison of the effect of two different rates of green waste compost addition on enzyme (dehydrogenase) activity within the silty sand/PC system microcosms.

compost considerably improved the levels of activity in the soil/PC systems. However, the activity decreased with time in such a way that at 106 days the level is similar in both systems. The fate of the contaminant within the soil/PC microcosms is shown in Figure 9.13, in terms of percentage contaminant recovery, with gradual contaminant loss up to 61 days, followed by considerable decreases in contaminant recovery between 61 and 106 days, particularly with 0.2 g/g added compost. The pH in the soil/PC microcosms for the same time duration (Figure 9.14) shows the initial high-pH environment provided by the PC, which decreased over time

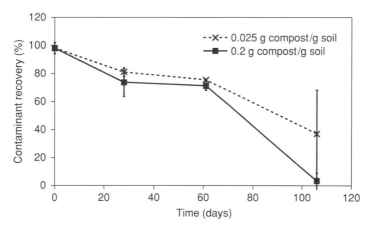

Figure 9.13   Comparison of the effect of two different rates of green waste compost addition on 2CBA recovery within silty sand/PC cement system microcosms.

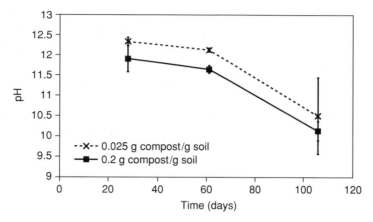

Figure 9.14 Comparison of the pH of silty sand/PC system microcosms with two different rates of green waste compost addition.

in the warm, humid environment in which they were stored. The 2CBA concentration in stabilised/solidified soil has, therefore, been shown to decrease over time, corresponding to a decrease in pH of the system. The major decreases do not correspond to the highest levels of enzyme activity, which may imply that the latter are due to a stress response upon addition of the high-pH grout. The subsequent lack of activity over the period in which the contaminant disappeared may be due to low survival rates of microorganisms in the systems tested.

9.4 Conclusions

The objective of the first part of the work presented here was to identify, compare and address the wider impacts of a range of remediation technologies in use in the UK. An assessment methodology was developed based on multi-criteria analysis and detailed impact assessment, both of which incorporated life cycle approaches. This was used to compare three options on one particular site and in a separate study compare five technologies on five different sites. The use of data from past projects has allowed the inclusion of true information from both implementation and post-remediation stages that is not normally available or is predicted. These studies highlighted the major impacts from each, with in both cases the excavation and disposal to landfill project performing poorly. Both excavation and offsite disposal had significant separate impacts when all parts of remediation, on and off site and during and after remediation, were taken into account. In contrast, many of the other technologies included in this study had low scores relative to this when the impacts of excavation and disposal in each project were neglected. This highlights the particular benefits of methods that minimise any offsite impacts and those that require relatively little intervention on the site itself, particularly if remediation can be performed in situ. The two best-performing techniques in case study 2, S/S and cover system, both used rapid, onsite methods that minimised both excavation and offsite disposal.

In case study 1, the inclusion of the 'no action' option, whereby no remediation would be performed prior to redevelopment, performed rather better than excavation and disposal to landfill. This illustrates the possibility that certain remediation projects might actually be doing overall harm environmentally, socially or economically compared to the initial state of affairs, and highlights the importance of performing a thorough assessment of the true impacts.

The current methodology only addresses the physical impacts that lead to social and/or economic impacts, but the flexibility of the developed methodology would allow the effects of these to be incorporated relatively easily.

Full assessment and consideration of sustainability or of wider impacts of remediation is not currently performed in practice, although the results from the survey show that awareness of sustainability issues exists and that certain aspects are being implemented. It is hoped that the analysis presented here will assist in informing the selection of remediation technologies through knowledge of their true impacts in tandem with their efficacy and cost. It also would allow the design of particular projects to be tailored specifically to minimise impacts for the area in which they take place.

The objective of the second part of the work was to identify possibilities for improvement of currently used remediation technologies. Improvements based on the sustainability assessment developed were tabulated. Improvements based on experimental investigations concentrated on the remediation technique of stabilisation/solidification. Although stabilisation/solidification is a fledgling promising solution to contaminated land remediation in the UK, it suffers from the fact that while it effectively deals with most inorganics, it hinders the degradation of organic contamination. Hence two specific areas were addressed. The first was the investigation of the development of a low-pH phosphate-based cement binder system which would be capable of facilitating biodegradation of organics as well as the stabilisation/solidification of inorganics. The second was the investigation of a range of additives that would facilitate biodegradation of organic compounds within a standard Portland cement binder system. This work presents an example of how significant advances can be made in modifying existing remediation methods to address specific aspects of their negative sustainability impacts.

Acknowledgements

Data for the two of the case studies were kindly provided by May Gurney Ltd and Delta Simons Ltd. The authors would like to thank those who responded to the questionnaire and to the collaborators on the questionnaire, Dr G. Sellers and Dr R. van Herwijnen, also of SUBR:IM, and also to Srinath Iyengar for his work on the low-pH phosphate cements.

References

Abdelrazig, B.E.I., Sharp, J.H. and El-Jazairi, B. (1989) The microstructure and mechanical properties of mortars made from magnesia-phosphate cement. *Cement and Concrete Research*, **19**, 247–58.

Andersson, J. (2003) *Environmental impacts of contaminated site remediation: a comparison of two life cycle assessments*. MSc thesis, Environmental Science Programme, Linköpings Universitet, Sweden (ISRN LIU-ITUF/MV-D—03/02—SE).

Bardos, R.P., Nathanail, C.P. and Weenk, A. (2000a) *Assessing the Wider Environmental Value of Contaminated Land Remediation: a review*. R&D Technical Report P238. Environment Agency (England and Wales), Bristol.

Bardos, R.P., Nathanail, C.P. and Weenk, A. (2000b) *Added Environmental Value: A Tool to Help Understand the Effects of Remediation of Land Contamination within the Context of Sustainable Development*. Workshop Report R&D Project Record P5/023/01. Environment Agency (England and Wales), Bristol.

Bardos, R.P., Mariotti, C., Marot, F. and Sullivan, T. (2001) Framework for decision support used in contaminated land management in Europe and North America. *Land Contamination and Reclamation*, **9**, 149–63.

Bardos, R.P., Lewis, A., Nortcliff, S., Matiotti, C., Marot, F. and Sullivan, T. (2002a) *Review of Decision Support Tools for Contaminated Land Management, and their Use in Europe*. CLARINET Report. Umweltbundesamt GmbH, Germany.

Bardos, R.P., Nathanail, J. and Pope, B. (2002b) General principles for remedial approach selection. *Land Contamination and Reclamation*, **10**, 137–60.

Bayer, P. and Finkel, M. (2006) Life cycle assessment of active and passive groundwater remediation technologies. *Journal of Contaminant Hydrology*, **83**, 171–99.

Beklemishev, M.K. and Kozliak, E.I. (2003) Bioremediation of concrete contaminated with n-hexadecane and naphthalene. *Acta Biotechnologica*, **23**, 197–210.

Blanc, A., Métivier-Pignon, H., Gourdon, R. and Rousseaux, P. (2004) Life cycle assessment as a tool for controlling the development of technical activities: application to the remediation of a site contaminated by sulfur. *Advances in Environmental Research*, **8**, 613–27.

BSI (2002) *Characterisation of Waste Leaching: Compliance Test for Leaching of Granular Waste Materials and Sludges*. BS EN 12457: Parts 1 to 4. British Standards Institution, London.

Diamond, M.L., Page, C.A., Campbell, M., McKenna, S. and Lall, R. (1999) Life cycle framework for assessment of site remediation options: method and generic survey. *Environmental Toxicology and Chemistry*, **18**, 788–800.

El-Jazairi, B. (1987) The properties of hardened MPC mortar and concrete relevant to the requirements of rapid repair of concrete pavements. *Concrete*, **21**, 25–31.

Evans, D., Jefferis, S.A., Thomas, A.O. and Cui, S. (2001) *Remedial Process for Contaminated Land: Principles and Practices.* CIRIA Report C549, Construction Industry Research and Information Association, London.

Harbottle, M.J. and Al-Tabbaa, A. (2007) Degradation of 2-chlorobenzoic acid in stabilised/ solidified soil systems. *Submitted to the Journal of Environmental Management.*

Harbottle, M.J., Al-Tabbaa, A. and Evans, C.W. (2005) The technical sustainability of in-situ stabilisation/solidification. In: *Proceedings of the International Conference on Stabilisation/Solidification Treatment and Remediation* (eds A. Al-Tabbaa and J.A. Stegemann), Cambridge, 13–14 April, pp. 159–170. Balkema, London.

Harbottle, M.J., Al-Tabbaa, A. and Evans, C.W. (2006a) Assessing the true technical/ environmental impacts of contaminated land remediation – a case study of containment, disposal and no action. *Land Contamination and Reclamation,* **14**, 85–99.

Harbottle, M.J., Al-Tabbaa, A. and Evans, C.W. (2006b) A comparison of the technical sustainability of in situ stabilisation/solidification with disposal to landfill. *Journal of Hazardous Materials,* **141**, 430–40.

Harbottle, M.J., Al-Tabbaa, A. and Evans, C.W. (2007a) A comparison of the technical/ environmental sustainability of five remediation projects. Part I: Multi-criteria analysis. *Submitted to the ICE Journal of Geotechnical Engineering.*

Harbottle, M.J., Al-Tabbaa, A. and Evans, C.W. (2007b) A comparison of the technical/ environmental sustainability of five remediation projects. Part II: Details impact assessment. *Submitted to the ICE Journal of Geotechnical Engineering.*

Hwang, I. and Batchelor, B. (2000) Reductive dechlorination of tetrachloroethylene by Fe(II) in cement slurries. *Environmental Science and Technology,* **34**, 5017–22.

Hwang, I. and Batchelor, B. (2001) Reductive dechlorination of tetrachloroethylene in soils by Fe(II)-based degradative solidification/stabilisation. *Environmental Science and Technology,* **35**, 3792–7.

Hwang, I. and Batchelor, B. (2002) Reductive dechlorination of chlorinated methanes in cement slurries containing Fe(II). *Chemosphere,* **48**, 1019–27.

LaGrega, M.D., Buckingham, P.L. and Evans, J.C. (2001) *Hazardous Waste Management.* McGraw-Hill Publishers, New York.

Li, Z., Zhu, D. and Zhang, Y. (2004) Development of sustainable cementitious materials. *Proceedings of the International Workshop on Sustainable Development and Concrete Technology,* Beijing, China, pp. 55–76.

Niedan, V. and Schöler, H.F. (1997) Natural formation of chlorobenzoic acids (CBA) and distinction between PCB-degraded CBA. *Chemosphere,* **35**, 1233–41.

Page, C.A., Diamond, M.L., Campbell, M. and McKenna, S. (1999) Life-cycle framework for assessment of site remediation options: case study. *Environmental Toxicology and Chemistry,* **18**, 801–10.

Postle, M., Fenn, T., Grosso, A. and Steeds, J. (1999) *Cost-Benefit Analysis for Remediation of Land Contamination.* R&D Technical Report P316, Scottish Environmental Protection Agency/Environment Agency, Bristol.

Rao, A.J., Pagilla, K.R. and Wagh, A.S. (2000) Stabilization and solidification of metal-laden wastes by compaction and magnesium phosphate-based binder. *Journal of Air & Waste Management Association,* **50**, 1623–31.

Singh, D., Wagh, S., Tlustochowicz, M. and Jeong, S.Y. (1998) Phosphate ceramic process for macroencapsulation and stabilization of low-level debris wastes. *Waste Management,* **18**, 135–43.

Soudée, E. and Péra, J. (2002) Influence of magnesia surface on the setting time of magnesia-phosphate cement. *Cement and Concrete Research* **32**, 153–7.

Suèr, P., Nilsson-Påledal, S. and Norrman, J. (2004) LCA for site remediation: a literature review. *Soil and Sediment Contamination*, **13**, 415–25.

Vegter, J., Lowe, J. and Kasamas, H. (eds) (2002) *Sustainable Management of Contaminated Land: An Overview*. CLARINET Report. Umweltbundesamt GmbH, Germany.

Volkwein, S., Hurtig, H.-W. and Klöpffer, W. (1999) Life cycle assessment of contaminated sites remediation. *International Journal of Life Cycle Assessment*, **4**, 263–74.

Wagh, A.S. (2004) *Chemically Bonded Phosphate Ceramics*. Elsevier Ltd, UK.

10

'The Creature Lurks Within?' Restoring Acid Tar Lagoons

Simon Talbot, Nigel Lawson and Colin Smith

10.1 Introduction

Although not the most common form of land contamination, acid tar lagoons (ATLs) pose considerable hazards to human health, controlled waters and wider ecosystems. Their complex and highly toxic properties present considerable challenges to conventional remediation techniques which seek to break pollutant linkages; for example, capping systems have often failed owing to the mobility of the acid tars, excavation is increasingly viewed as environmentally unsustainable and total encapsulation and destructive techniques are financially prohibitive. In short, there is a paucity of robust and durable long-term remediation options for reclaiming these sites. It is because of this complexity that ATLs are classified as 'Special Sites' under the Environmental Protection Act 1990, which sets out the legal framework for regulating contaminated sites (see Chapter 3).

While extant literature on ATLs has focused on the technical issues (for example, Nancarrow *et al.*, 2001; Reynolds, 2002), little attention has been afforded to the multifarious human health, socio-economic and political issues they raise for decision makers and affected communities. This chapter examines ATLs from these two broad perspectives (technical and social scientific) and highlights the way that social and political acceptability is crucial in determining the appropriate remediation technology adopted for cleaning up particular sites. We investigate these propositions through a case study of an ATL in north-west England. This site contains several of the potentially most problematic threats arising from ATLs: potential development and property price blight, severe contamination, access restrictions, multiple nearby risk receptors, and so on.

This chapter also argues that the current state of knowledge on how to remediate ATLs in the UK is extremely limited. For this reason, the chapter also presents lesson-drawing research undertaken by the authors in mainland Europe where ATLs have been successfully remediated following key stakeholder engagement and risk communication. We assess the prospects for incorporating key lessons into UK practice.

The chapter is organised into three parts. The first part presents a technical overview of acid tars and acid tar lagoons, their properties and characteristics. In order to further understanding of the 'real world' dynamics of ATLs, the second section presents a case study of the various technical and social issues affecting a particular site in north-west England. The final part looks beyond the UK for potential solutions to remediating ATLs. To do this, it focuses on an exercise in 'lesson-drawing' from Germany, where considerable experience and expertise in remediating ATLs exists. Overall, we are concerned in this chapter with developing a better characterisation of the various potential remedial strategies that could be adopted on ATLs and understanding the way in which socio-political issues affect the choice of action implemented.

10.2 Acid tar lagoons: a technical introduction

Acid tar is a waste residue of petrochemical processes, which are now mostly abandoned. Its production can be traced from the end of the nineteenth century (Milne *et al.*, 1986). There are three main processes that produce acid tars: benzole refining, white oil production and oil re-refining (Nancarrow *et al.*, 2001). Each involves the use of concentrated sulphuric acid as a washing liquid to purify an organic material, which results in a residual tar containing a high proportion of sulphuric acid compared with other acid tars from coal carbonisation processes.

Benzole refining is a set of processes which extract purified fractions of benzene, toluene and xylene from crude benzole, a by-product of coal carbonisation. Washing with concentrated sulphuric acid removes two major impurities: sulphur-containing compounds and unsaturated hydrocarbons (Claxton, 1961). After the washing is complete, the purified benzole is decanted for further treatment and the bottom acid tar is run off from the base of the washer. Oil re-refining is a process that regenerates spent lubricants. The oil is fed into a contact tank where it is mixed with concentrated sulphuric acid and fuller's earth to remove non-hydrocarbon material, unsaturated hydrocarbons and sulphur-containing compounds (Milne *et al.*, 1986). White oils are highly purified compounds used for medicinal, cosmetic and specialised lubrication purposes. Again, sulphuric acid is used for removal of unsaturated and sulphur-containing compounds.

It is hard to be precise about the scale of the problem that ATLs pose as historical information about benzole and oil-refining is extremely limited. However, there were approximately 140 benzole-refining plants in the United Kingdom with an estimated total production of 2.5 million tonnes during the period 1930–1980. White oil production generated an estimated 2 million tonnes of acid tar across 10 to 12 sites in the UK. There were about 30 to 40 small plants typically producing 500 tonnes of acid tar per year over an average 40-year operation (Nancarrow *et al.*, 2001). Historically, the methods used for acid tar disposal were mostly by landfill into existing holes or lined lagoons, usually near the former chemical plants. The acid tar sometimes underwent a limited pre-treatment, and was often co-disposed with other materials such as drums of various chemicals, sugar waste, sand, ash, clinker, vegetation, PCBs and so on. Typical lagoon depths reported in the literature vary from 4–10 metres for open lagoons, and volumes vary from ~3 000–60 000 cubic metres (Banks *et al.*, 1998; Nichol, 2000; Chambers, 2001).

Before the 1970s, the disposal of acid tar was usually not an environmentally friendly process. In the benzole-refining industry, the acid tar produced was often pre-treated by being diluted with creosote oil and then steamed to recover any entrained benzole and some of the sulphuric acid (Claxton, 1961). Some efforts were made to neutralise the acid by mixing the acid tar with lime and other alkaline materials. However, because of poor mixing with the thick acid tar, this usually proved to be inefficient.

10.2.1 Characteristics of acid tar

10.2.1.1 Composition

The chemical composition of acid tar from each individual process differs significantly owing to the different starting material and final product. Therefore the chemical characteristics of acid tar cannot be closely specified. However, the basic characteristics of acid tar remain similar across the three major processes. Acid tar is a complex mixture of hydrocarbon, sulphuric acid, water and a various range of co-disposed materials. The chemicals inside acid tar can be grouped into the following categories: aliphatic hydrocarbons, aromatic hydrocarbons, phenols, metals, organic acids, sulphonated hydrocarbons, and gases such as hydrogen sulphide, sulphur dioxide and methane (Nancarrow *et al.*, 2001). Disturbed acid tars may thus give rise to significant odour problems. Some acid tars were pre-treated to reduce their acid content before disposal. Nichol (2000) reports a typical composite breakdown of acid tar as 44% sulphuric acid, 42% oil residues, 8% sulphated oil residues and 6% water, though sulphuric acid content can vary significantly in different tars.

Table 10.1   Physical properties of acid tars.

Origin	Viscosity	pH	Colour	Odour
Benzole refining	Thin to fairly viscous	pH 2 or lower	Generally black	Strongly aromatic
White oil production	Very viscous	The most acidic tars, pH below 1	Generally black	Oily
Oil re-refining	Variable	pH 2 or lower	Browner than others	Oily

Source: Nancarrow *et al.* (2001)

Physically, acid tars consist of three phases: a free acid tar phase, free oil and a sulphate-rich acidic aqueous phase. The latter will not dissolve in the acid tar or oil and may exist as an emulsion similar to those found in coal tars (Payne and Charles, 1987) or as pockets of clear liquid (Nichol, 2000). Table 10.1 shows some of physical properties of acid tar from different processes.

The viscosity of acid tar is highly sensitive to temperature. At high temperatures the acid tar will become fairly mobile and tends to be fluid, but at lower temperatures the acid tar solidifies to a variable degree, depending on its composition. The variation in viscosity is apparent within the normal range of ambient temperature, that is 0 to 25°C. At exposed surfaces oxidation will occur, resulting in an anticipated increase of viscosity within the lagoon surface layers, in some cases leading to surface cracking. The density of the acid tar is reported at between 1 020 and 1 430 kg/m^3 (Hao and Smith, 2005) and 1 200 and 1 400 kg/m^3 (Nichol, 2000), which are both higher than typical coal tars at 1 060 kg/m^3 (Oudijk and Coler, 1995); it may therefore be regarded as a 'dense non-aqueous phase liquid' (DNAPL). It is assumed that the higher acid tar density is due in part to the high sulphuric acid content (density 1 960 kg/m^3).

The typical major contaminants found within acid tar are polycyclic aromatic hydrocarbons (PAHs), petroleum hydrocarbons, phenols, benzene, toluene, xylene (BTEX), acid, heavy metals and sulphate, all of which can cause environmental problems if they migrate into the surrounding soil and groundwater environment. Air pollution may be caused by both volatile organic compounds (VOCs), sulphur dioxide, and dusts from the weathering of exposed acid tar surfaces (Nancarrow *et al.*, 2001).

10.2.1.2   Physical and chemical mobility of acid tar

In any consideration of the environmental impact of an acid tar lagoon, it is necessary to examine both the physical and chemical stability of the acid tar, which are interdependent and must be controlled. Physical barriers to contaminant migration may be mechanically disrupted and/or chemically attacked. Figure 10.1 depicts a conceptual model of a range of processes and

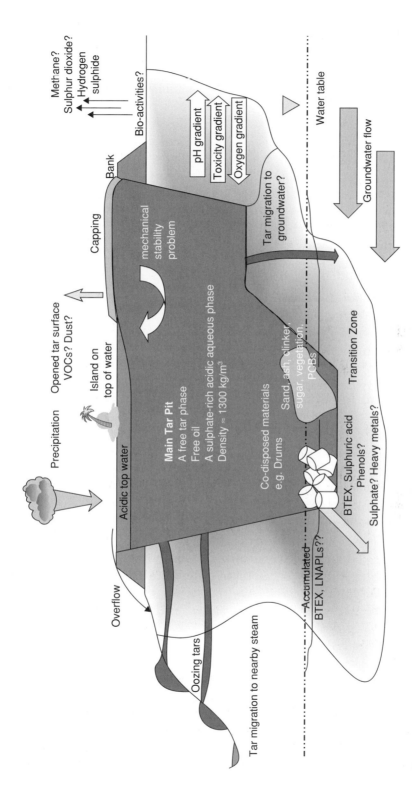

Figure 10.1 Conceptual model of generic acid tar lagoon processes (Talbot *et al.*, 2004).

potential migration pathways that have been observed or are inferred for a general acid tar lagoon.

Nichol (2000) reports results of probings of an acid tar lagoon that indicate stratification of the acid tar into semi-solid layers separated by acid tar bands of softer consistency. Reynolds (2002) reports seismic survey data indicating softer surface acid tars underlain by more viscous acid tar. It is not clear whether this is simply due to differing batch properties as acid tar was placed in the lagoon, or due to long-term separation or weathering processes, or due to a combination of all three processes.

A challenging characteristic of acid tars is that in their semi-fluid state they can be driven, by the pressure head of the overlying acid tar and any capping material, through any cracks and other pathways in the soil, for example along tree roots. Flowing acid tar may also dissolve components of the mineral matrix, helping to widen fissures, increasing mass permeability and easing flow. In the upper soil zone above the water table the tar is likely to weather to a more mechanically stable form, particularly when it reaches the surface. However, while the outer layers may weather, this protects the inner tar which can continue to flow. On surface slopes this can lead to slow surface flows of tar down gradients (of the order of m/year, depending on ambient temperature), with the weathered material 'riding' on the underlying fresher unweathered material. Leaching tests carried out by the authors indicate that where the tar is in contact with groundwater it tends give rise to sustained long-term leaching of sulphates, acid and organics, while physically the tar breaks down into fine particulates. In highly permeable soil these particulates may migrate for some distance. It is not known as yet whether there is continued washing away of tar as particulates or whether the process ultimately self-seals. However, it should be emphasised that these results were drawn from one acid tar type and should not necessarily be generalised.

Acid tar is generally lighter than soil, and therefore typically heavier soil-based capping materials will tend to sink into it. This may result in gross instability of a cap, or alternatively pressurisation of the underlying tar which can lead to localised tar 'eruptions' with tar emerging at point locations and subsequently spreading out over the surface.

10.2.1.3 Environmental impact
While it is possible to infer and describe likely processes occurring in and around acid tar lagoons, there exists little scientific data to quantify many aspects of these processes or to establish their significance. Because of the inherent variability of acid tars, it may also not be possible to generalise site and acid-tar-specific results in all cases. The available literature currently indicates that mechanical stability is a significant issue, particularly with respect to capping layers but also with the ability of the acid tar to

flow through fissures in the ground and emerge some distance away. Nichol (2000) reports virtually no noteworthy transfer of contaminants from acid tar to any contacting water in laboratory leaching tests. Available field evidence in the literature also indicates no significant offsite migration of contaminants in ground or surface waters (Banks *et al.*, 1998; Nichol, 2000). This may be because of a combination of low leaching and natural attenuation; however, it may also only be specific to these particular lagoons and should not at present be generalised. Banks *et al.* (1998) indicate that contamination from acid tar pits does have the effect of lowering the local groundwater pH and increasing concentration of chloride, sulphate and some metals (detailed organic analyses were not reported).

Reynolds (2002) highlights the fire risk at acid tar lagoons. In 1980, the Llwyneinion lagoon (near Rhos, Wrexham, North Wales) had a layer of volatile hydrocarbon floating on the water overlying the surface acid tar. The volatile hydrocarbon ignited and burnt off, in the process evaporating the acid water and burning some of the acid tar beneath. The resulting smoke plume necessitated the evacuation of a nearby town. However, it is not clear whether the volatile hydrocarbon arose from the particular type of acid tar at that site itself or arose from co-disposed chemicals. The fire risk may be reduced by maintaining a water cover at all times. In addition, maintaining a water cover reduces the quantity of VOCs and sulphur dioxide escaping to the atmosphere.

Table 10.2 lists an extensive but not necessarily exhaustive set of potential linkages and environmental impacts of acid tar lagoons. Any restoration will seek to manage all of these linkages and reduce them to acceptable levels.

10.3 Regulating risk on an acid tar lagoon

Our concern in this section is briefly to outline the regulatory framework which guides the remediation of contaminated sites, specifically Special Sites, and then to examine the implementation of the regulatory regime on an ATL in north-west England. We explore the social, political and technical factors involved in attempting to determine risk and select remedial options. The research upon which this case study is based was derived from semi-structured interviews with 20 key 'actors' associated with this site which were conducted by the authors from May to August 2005. These actors included key stakeholders involved in deliberating on the future of the site – past and current local authority environmental health officers, the Health Protection Agency, the site owners – and others excluded from internal decision-making (local councillors, neighbourhood representatives and so on). In addition to this, over a three-year period, the authors convened, participated in and observed 'internal' stakeholder meetings. Notes of each

Table 10.2 Acid tar lagoons: potential linkages.

Source	Pathway	Receptor
BTEX, PAHs, VOCs	Volatilisation in upper soil surface	Site users
BTEX, PAHs, heavy metals, phenols and sulphate	Inhalation of dust and vapours	Site users, residents of adjacent properties
BTEX, PAHs, heavy metals, phenols and sulphate	Ingestion of surface soils	Site users, residents of adjacent properties
BTEX, PAHs, heavy metals, phenols and sulphate	Groundwater discharge	Nearby water course
BTEX, PAHs, heavy metals, phenols and sulphate	Surface run off	Nearby water course
BTEX, PAHs, heavy metals, phenols and sulphate	Migration along engineered structures (outfall)	Nearby water course
BTEX, PAHs, heavy metals, phenols and sulphate	Leaching/migration of contaminants through the soil and unsaturated zone	Major/minor aquifer
Methane, sulphur dioxide and hydrogen sulphide	Migration through permeable soil	Site users, residents of adjacent properties
BTEX, PAHs, heavy metals, phenols and sulphate. Acid (burns)	Bulk migration of acid tar. Direct contact	Site users, residents of adjacent properties
Toxic combustion products	Smoke from combustion of acid tar in lagoon	Site users, residents of adjacent properties

Source: Talbot *et al.* (2004)

meeting were recorded by the authors and were drawn upon in this analysis. Other documentation came from internal local authority files from previous meetings related to the site. Owing to the sensitivity of the site under investigation, confidentiality has been maintained throughout.

10.3.1 Regulatory policies

Part IIA of the Environmental Protection Act 1990 and its related statutory guidance requires local authorities to identify contaminated land in their areas and ensure that it is managed in an appropriate manner. Part IIA specifies that local authorities are also required to create and maintain registers that contain information on remediation notices, which the public have a legal right to access (see Chapter 3).

In order for local authorities to make a determination that a particular site meets the statutory definition of contaminated land, it must be demonstrated that a pollutant linkage exists: that is, a source of contamination must be present along with a pathway and a receptor. Part IIA creates a sub-set of contaminated land called 'Special Sites'. Acid tar lagoons are one of a number of industrial activities that fall within this definition (Regulation 2(1)(b) Contaminated Land (England) Regulations 2000). All Special Sites, once identified and designated by the local authority, are regulated by the Environment Agency (EA). It is the EA's role to carry out inspections on Special Sites and act as the enforcing authority to ensure that suitable remediation is undertaken to break the pollutant linkages. Local authorities remain the enforcing authority for all other statutorily contaminated land (Catney *et al.*, 2006). In the UK the principal tool for determining significant human health pollutant linkages is the Contaminated Land Exposure Assessment (CLEA) model, published jointly by the Department for Environment, Food and Rural Affairs (Defra) and the EA.

10.3.2 Local implementation

One of the study sites investigated by SUBR:IM researchers was a capped ATL site located in north-west England. The site investigated is located within a former industrial area adjacent to housing and a brook. Ownership is currently divided between the local authority and a private company. Historical records and anecdotal evidence suggest that the site was previously a sandpit which was bunded and infilled with waste acid tars and other wastes such as foundry sand between the mid 1930s and the early 1960s, forming an acid tar lagoon. In 1970 part of the site was purchased, levelled and capped by the local authority so that it could be put back into beneficial use as 'public open space'. However, the weight of the cap displaced the acid tar laterally and by the mid 1990s acid tars had broken through to the surface, into the private-sector-owned parts of the site. Intrusive site investigations were carried out in 1994 and 1997 by the local authority and a detailed quantitative risk assessment (DQRA) of the site undertaken in 1998 'produced' Site-Specific Target Levels (SSTL) for a number of PAHs above what was considered to be an acceptable risk for users.

By 2001 acid tar had started to become visible at the surface within the area of public open space. As a stopgap, an additional temporary soil cap was placed over the affected part of the site. It was only at this stage that the local authority decided to inform home owners adjacent to the site by letter that this work was being undertaken, although they stated that the work was only of a temporary nature. The regulatory process was further delayed when it became clear that the Part IIA legislation and guidelines would require another DQRA to be undertaken on the site because the introduction of more Soil Guideline Values (SGVs) by the EA meant that the previous assessment was not compatible with the regulatory guidelines post 2000. Hence, in April 2003 the local authority reassessed the levels of contamination in the surface soils. This assessment revealed high levels of PAH contamination indicating, on the balance of probabilities, a risk to human health across most of the site, with the notable exception of the area of public open space covered by the temporary cap.

In early 2004, the Environment Agency commissioned a Human Health Risk Assessment covering the part of the site open to public access. The conclusion was that 'even by adopting less conservative modelling assumptions, there remain one or more pollutant linkages that would still appear to present a significant risk of significant harm to members of the public using the site', with the recommendation that 'exposure of the local population to such contaminants from the site should be reduced to as low a level as reasonably practicable'. The local authority is required by the regulatory guidelines to act upon these conclusions and recommendations and designate the part of the site open to public access as statutory contaminated land under Part IIA. However, there have been a number of delays in the regulatory process on this site which has meant that, three years on from the findings of the EA risk assessment and more than ten years after acid tars were first discovered at the surface by members of the public, the site still remains undesignated.

The first problem was of an external nature. During the late 1990s and early 2000s, central government was formalising the policy regime for contaminated land. This created considerable uncertainty for the local authority as to whether it would be responsible for the regulation of the site, what standard of clean-up would be required and what precise steps needed to be taken to ensure that the regulatory process fell into line with the emergent guidelines. The second problem was internal. There was considerable turnover in the staff of the local authority's environmental health department; this was compounded by the lack of specialist contaminated land officers. (It was common practice until relatively recent times for contaminated land issues to be dealt with within local authorities by generalist environmental health officers with a minimum of contaminated land training.) These factors created considerable discontinuities in the regulatory process and delayed the local authority's response to an identified hazard as new staff needed considerable time to be trained and familiarised with the complex nature of the problems on the site. (With the advent of the 2000 regulatory guidelines there has been a progressive increase in the resources allocated within environmental health departments to contaminated land issues, with an increasing number of specialist contaminated land officers now being employed.)

With respect to the state of the soil, the question confronting local contaminated land officers was the extent of the health risk to the general public if they used the site. While it was known that the site was contaminated by acid tar near the surface, there was less certainty over the precise exposure levels of the constituent key contaminants hazardous to human health if inhaled, touched or ingested. Although there were no reports of health problems that could be linked directly to people living near or using the site, there is a history of public health issues arising from other acid tar sites. Site owners are compelled to act if there is a possibility that their land could pose a danger to human health as they have a statutory duty of care towards the public using their land under the Occupiers' Liability Acts of 1957 and 1984.

However, some local authority officers felt that the case study site should not have been as high a priority for remedial action as it was:

[T]here was no resident pressure, but it is our longest running case and we do have the risk assessment that suggests that there is a source-pathway-receptor, but it's very obscure. We have to take some action at some point . . . We also have had two relatively high-profile sites that we have worked on that have been quite pressing. (Interview with local authority contaminated land officer, May 2005)

Despite the lack of action, deliberation on remediating the site retained its importance because the 'internal' stakeholders in the consultation process lobbied to keep it so. For example, the local office of the Health Protection Agency (HPA) for the area covering the site expressed concerns with heavy metals from the site leaching into groundwater and a local surface water course, and hence posing a long-term public health risk. In addition, the HPA expressed concern over PAHs contaminating the site, but it did argue that more evidence was needed of exposure to PAHs from other sources outside the site. This raised the problem of the need to measure for site-specific exposure to both PAHs and heavy metals.

Preliminary results of the EA's risk assessment on the site suggested that the greatest risk to the public was from potentially contaminated soils in the gardens of houses built adjacent to the site before a bund was installed. The local authority, in contrast, felt that the risk from this pathway was low. A number of local authority contaminated land officers interviewed also questioned the quality of the EA's risk assessment undertaken on the site. At one level, they questioned the basic assumptions of and parameters set by the EA's consultants. For example, it was argued that they failed to include the private-sector-owned parts of the site in the analysis. While not open to the public, this area contains acid tar at the surface which has weathered down to a dust that is easily picked up and blown by wind into the public open space and adjacent housing. At a more general level, it was argued that the CLEA threshold guidelines for Benzo(a)pyrene were unrealistically low. Concerns over the CLEA threshold for this contaminant led the local contaminated land officer, who believed that this particular contaminant's threshold was going to be raised in the near future, to consider postponing designation (or designating and then 'de-designating' the site) in anticipation of a change. In recent years an ongoing debate in the UK has focused on the levels at which SGV thresholds have been set. The development industry has argued that the current thresholds are set too low and that local authority contaminated land officers have treated SGVs as standards, rather than as guidance. This, it is claimed, has led to the over-remediation of sites. These problems with SGVs led the government in 2005 to place a moratorium on developing new SGVs while a task force deliberated their future. One reason given by the local authority for delays in the process of designating the site and carrying out remediation was the frequent changes to the government's regulatory policies over the past decade.

From a technological perspective, the local contaminated land officers needed to find a method of remediation that would not increase the risk of public exposure. One 'remedial' proposal considered by the local authority was to exclude the public from the contaminated area by erecting fencing and installing warning signs to notify the public of the general risk the site posed. The local authority argued that this action could be deemed acceptable under the Part IIA guidance as it would break the pathway of people coming into physical contact with the dangerous parts of the site. However, the proposal was dependent on a limited reading of the potential pathways of the site's contamination and the risk it posed to the wider environment, such as the pollution to local groundwater and the adjoining river.[1] A 3-D model of the site is presented in Figure 10.2, and illustrates a range of environmental interactions of concern.

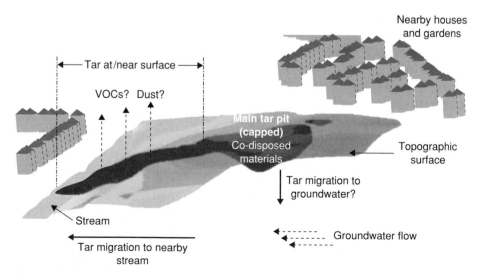

Figure 10.2   Study site acid tar body.

A further hurdle to the remediation of the site was the shared ownership. While the local authority owned one half of the site, the other half was owned by a private company. This left questions of liability. Under Part IIA there is a need to identify the appropriate person who is responsible for remediating the site, for example the original polluter(s) or current owners. However, in this case the original polluter who operated the tip before it was bought by the local authority could not be found as the company had long since gone out of business. The private company that now owned half the site stated (interview, 2005) that it did not welcome designation of the whole site as it did not wish to be seen to own identifiable contamin-ated land. It did state that it did not object to the local authority's half being designated but that it would oppose its half being included. With limited resources local authorities are tempted to follow the line of least resistance. In this case this, the authority was minded to designate only the half of the site it owned. However, it could be argued that the only permanent and cost-effective remedial solution would involve treating the whole site.

10.3.3 Stakeholder considerations

The location, the dual ownership, the regulatory scenario and the technical challenges demonstrate that remediating this site in a sustainable manner while also addressing the potential risk to human health and respecting the concerns of the affected stakeholder groups is extremely complex.

Discovering a potential risk to human health due to contaminated land, and the subsequent remediation process, can impact on a large and varied group of interested parties ('stakeholders'). In general, stakeholders in the deliberative process fell into two distinct categories:

- *internal stakeholders* – actors 'in the know' and able directly to influence the process
- *external stakeholders* – invariably not fully informed, and with less direct influence on the process, yet subjected to both its short- and long-term direct effects

While the risk to both groups of stakeholders will vary, the concerns of all stakeholders must be considered. Under Part IIA of the Environmental Protection Act 1990 the public have a legal right to all information on re-mediation notices. Furthermore, the UN Economic Commission for Europe (UNECE) Convention on Access to Information, Public Participation in Decision-Making and Access to Justice in Environmental Matters (The Århus Convention) substantially increases the rights of the public to receive all environmental information held by public authorities. The convention

Table 10.3   Internal-external interests and concerns.

Stakeholders – internal	Primary interest and concerns
Site owner/s	Financial; cost/time benefit.
Local Authority	Legal obligations, including environmental protection; income beneficial use of land; local amenities.
Environment Agency	Protection of groundwater and surface water quality
National legislators/policy makers	Use of brownfield land to meet development targets
Health Protection Agency/Primary Care Trust	Minimising risk to general public
Health and safety officers	Protection of remediation contractors
Remediation consultants	Clients' interests
Remediation contractors	Narrow contractual obligations
Stakeholders – external	
Local politicians	Local electorate
Media	Story, sensationalising issues
Local community	Property values; safety; local amenity value of site
Wider community	Area amenity value

also includes the rights of citizens and environmental organisations to participate in environmental decision-making from an early stage (Europa Environment, 2006). Table 10.3 describes the specific interests of various stakeholders affected by contaminated land.

Different stakeholders on this site held differing perceptions of risk, and correspondingly varied in their responses to the risk assessment on the site. One-on-one interviews of key stakeholders were undertaken and reveal clearly their differing interests and concerns:

- The local authority, which is also the owner of the public open space, considered the risk assessment as lacking in clarity because it failed to qualify remarks such as 'significant severity', 'significant harm', 'significant possibility'. More importantly, the authority felt compelled to react to the assessment, despite its apparent flaws.
- A former pollution control officer for the local authority with first-hand knowledge of both the site and the risk assessment ventured: 'I would take my kids to play (on this particular site).'
- The Environment Agency advocated a precautionary approach to the site: 'Our preferred option [in response to the risk assessment] would be urgent action such as fencing to secure the site followed by remediation.

Blight is inevitable, but with designation it would look as if something is being done' (interview, Environment Agency Officer, May 2005).

- A representative from the private company that owned the fenced-off part of the site expressed little concern about potential liabilities or political fall-out because the risk assessment did not cover their side of the site.
- The local politician (the Lead Member for the Environment) saw a need to take action based on the risk assessment for the site 'because it is in the public domain'. However, financial resources are of prime consideration: 'contaminated land has a low priority in budgetary terms'.
- Two local ward councillors considered it a desirable place to live, and that residents' only concern appeared to be the possibility of the site being developed with a consequent loss of their recreational area. Local residents do not consider themselves to be at risk despite an acute awareness of the history of the site.
- The neighbourhood manager for the local authority, responsible for coordination between various actors and agencies operating in the area and liaison with the public 'to keep a finger on the pulse of issues in the area', was unaware of any particular concerns about the site.

The 'internal' and 'external' distinctions between stakeholders reveals the exclusionary way – for local residents at least – in which consultation was undertaken on this site. It is common to argue that best practice in communicating risks is to be open and transparent with the general public at an early stage (see Chapter 3). An open process of communication can help to generate trust between decision makers and affected communities. For example, on the case study site the public were informed at a relatively early stage by letter that the site posed a potential risk to human health but were given no further information, nor were they consulted on the future of the site. There was concern within the local authority that the local public would become unnecessarily alarmed if informed that their local environment posed a risk that had been tolerated for many years. Furthermore, local authority officers expressed concern over the reaction to the possibly 'sensational' manner in which the local media might approach the issue.

One conclusion to be drawn from a consideration of the interactions of the stakeholders at the case study site is that despite a quantified risk being identified, inaction has occurred because none of the stakeholders' interests were aligned. Furthermore, despite a carefully constructed regulatory system being in place to protect human health, the 'expert' internal stakeholders were unable to agree whether the quantitative risk assessment overestimated or underestimated the human health risks. However, it would be unwise to draw any firmer conclusions at this time as the local authority is required to manage potentially contaminated land in a prioritised manner and information on other sites was not made available to the SUBR:IM researchers. It is therefore quite possible that the local authority's resources have been directed to more pressing sites.

10.4 Lesson-drawing from Germany: an appraisal of the state of the art in remediation

10.4.1 Introduction

As noted above, the UK past experience of ATL-capping has generally been one of failure owing to the mobility of the acid tars. Long-term containment, although ostensibly technically straightforward, has been found to be very difficult in practice. A second remediation methodology that has been considered has been stabilisation. Acid tar lagoons have been remediated through excavation, neutralisation using lime and subsequent landfilling. However, this is unlikely to be acceptable under current environmental regulations in the UK. In Europe and the USA the dispersing by chemical reaction (DCR) method (Grajczak and McManus, 1995) has been successfully used; this process involves mixing the acid tar with calcium oxide. However, there are health and safety concerns relating to the exothermic nature of the reactions and the consequential release of VOCs, SO_2 and other fumes, making in situ remediation extremely challenging and effectively requiring the DCR only to be carried out in closed plants (Boelsing, 1988). More recently ex situ stabilisation technologies have been investigated in the UK (Slater, 2003); successful trials have been carried out in Belgium (Pensaert, 2005) and they have been successfully used on specific small-scale lagoons in Germany (Gruss, 2005).

A third approach to remediating ATLs that was considered to be very attractive is their excavation and incineration in cement kilns. However, this raises further issues relating to emissions control. Cement manufacturers are increasingly wary of the impact that the raised sulphate content of the acid tar has on their product. Whether this is actually a technical issue is not clear; however, customer perceptions are a strong factor.

For some of the larger sites in the UK (for example, Llwyneinion, Hoole Bank and Cinderhills), significant investigation and assessment work has been carried out but no clearly viable remediation option acceptable to the various stakeholders has emerged. In part this is attributable to no clearly robust and demonstrated technologies being available at the time. In addition, any remediation technique that involves excavation has to control VOCs and sulphur dioxide (SO_2) emissions from the exposed surface of the acid tar. A successful remediation solution must include identifying both environmentally acceptable methodologies for excavating acid tars and demonstratably robust technologies for treating them. In this case it is useful to examine experience in Germany (Gruss, 2005), where several lagoons have recently been successfully remediated under modern strict environmental regulations, and remediation is ongoing with several other lagoons. This section draws upon international lessons to examine technical applications and regulatory/social issues in the German context with an eye to their application in the UK.

First, however, a short note on the process of 'lesson-drawing' may be useful. Lesson-drawing is an approach that seeks to explain the processes by which 'knowledge about institutions, policies or delivery systems at one sector or level of governance is used in the development of institutions, policies or delivery systems at another sector or level of governance' (Evans, 2006, p. 480). Rose (1993, p. 27) defined a lesson as 'a detailed cause-and-effect description of a set of actions that government can consider in the light of experience elsewhere, including a prospective evaluation of whether what is done elsewhere could someday become effective here'. The lessons drawn may or may not lead to the transfer of policies or practices, depending on whether the lessons are deemed positive or negative (Bache and Taylor, 2003, p. 280).

One of the main benefits of lesson-drawing is that the costs of developing innovative programmes are incurred by the first nation. Other countries can learn lessons from the experience of this nation without having to go through a process of trial and error, although there are often costs involved with the process of adaptation (Rose, 2001, p. 4). An analysis of policy transfer is a useful approach when examining how lessons can be drawn from different countries and applied in others; it can also help us understand why these transfers are often not successful.

10.4.2 Background to German experience

Acid tar production was ongoing in the former East Germany until relatively recently (1991), prior to reunification, resulting in a legacy of acid tar disposal to lagoons. This was carried out under fairly unrestrictive environmental regulations. Following unification, Germany pressed ahead with a policy framework effectively providing up to 100% funding to companies/enterprises for remediation designed to clean up contaminated land sites. However, the enterprises usually had to carry an appropriate portion of the remediation costs themselves. This policy was designed to remove obstacles to inward investment to eastern Germany.

The former nationalised Motimol DDR waste oil re-refining plant at Chemnitz was one of two such sites in the former East Germany. It was denationalised in 1991 and purchased by the Baufeld company. Baufeld took on ownership and responsibility for five acid tar lagoons at two

Figure 10.3   Two of the Chemnitz lagoons prior to remediation. (Courtesy of Baufeld-Umwelt-Engineering GmbH, Chemnitz, Germany)

locations with a total volume of 120 000 m³ arising from refining operations before 1991. Figure 10.3 depicts two of the lagoons prior to remediation (Gruss, 2005).

Since 1997, Baufeld has trialled a range of remediation approaches to acid tar lagoons and has developed considerable experience. Currently three lagoons, with a total quantity of nearly 95 000 tons of acid tar residues, have been successfully remediated. Remediation has occurred using a variety of methods primarily based around conversion of the acid tars to a 'substitute fuel' for energy recovery (energetic utilisation).

The Chemnitz lagoons presented many of the remediation challenges common to UK lagoons, namely proximity to housing, variable consistency, variable size and contaminated surface water ponds. Prior to remediation a number of possible methods were developed and evaluated. These included a variety of stabilisation techniques, liquefaction and solidification of acid tars for utilisation in the production of sulphur chemicals or for incineration in cement kilns and power plants. Excavation and conversion to a solid substitute fuel for use in a modern coal-fired power station was selected as the best environmental option. By using excavation techniques that minimised atmospheric emissions and efficient onsite processing, this approach not only permanently removed a contaminating waste but converted it to a beneficial use. Specifically, the acid tar's high calorific content was used for power generation which replaced the burning of an equivalent calorific content of primary fossil fuels. The consequential reduction in the consumption of primary fossil fuels is a significant positive factor in a 'life cycle analysis' of this remediation process (Smith *et al.*, in review).

However, alternative methods were retained as options for specific circumstances. These included using treated acid tars as 'daily cover' on landfill sites. In the following sections the regulatory, social science, financial and technical issues associated with the Chemnitz experience are discussed in greater detail.

10.4.3 German regulatory, social, and financial issues

While the financial provisions available for remediation differ between the western and eastern parts of Germany, the regulatory framework is the same. Germany's political structure is a federal state. Under the Federal Soil Protection Act 1998 and the subsequent Soil Protection and Contaminated Sites Ordinance 1999, sites with historic contamination (Altlasten) that pose a hazard to the environment or to human health are required to be remediated to a condition that is suitable for the current or future use of the land. Clean-up standards are determined by trigger and action threshold values of contamination. Although the German legislation is operated to strict standards, it is a flexible and pragmatic system with different threshold values applied to different land uses. This is similar to the UK's 'suitable for use' approach. In practice, the federal law sets the overall legislative framework and the trigger and action thresholds while the federal states (Länder) are responsible for administering the system and compiling inventories of contaminated sites. Under this legislative framework licences are issued for all of the different elements of a remediation scheme.

The German legislative framework is similar to that of England and Wales as embodied within the Environmental Protection Act 1990, the Environment Act 1995, the Pollution Prevention and Control Regulations 2000 and the adopted Landfill Directive 2005.

While the German authorities have discretion over which party is responsible for remedial actions, it is generally the site owners who must remediate their sites, with the actual remediation controlled through a system of licences for the constituent activities. The regulatory and licensing functions are carried out at the three administrative levels: the Länder, county and district levels. However, the principal environmental regulatory expertise is maintained at the county level.

For each remediation project a detailed 'remediation plan' is required to be submitted to the local district authority. Remediation plans include all the desk study and intrusive investigation information and details of the remediation actions that will be carried out. The final remediation plan is lodged with the local district authority for public inspection and, if required, a public inquiry is held.

The two acid tar remediation projects currently being carried out by Baufeld are regulated by the Free State of Saxony and the county authority of Chemnitz, while the original 'remediation plans' were submitted to the Stollberg and Chemnitz district councils.

Because of the close proximity of residential properties to the acid tar lagoons, the works were strictly monitored. This involved continuous monitoring of emissions from the exposed acid tars and the air quality at the site boundary adjacent to the residential properties. As the principal constituent of the emissions was sulphur dioxide, this was taken as the indicator for all of the airborne contaminants. In addition, continuous recordings of temperatures, wind speeds and wind direction were taken via semi-permanent weather stations immediately adjacent to the sites. The specialist environmental monitoring work and supervision were carried out by independent consultants.

Owing to the particularly aggressive nature of the surface emissions, in addition to the automated air-sampling equipment, the site boundary was randomly patrolled by specialist consultants employed to detect adverse odours. The results of this monitoring were reported directly to the regulators. It was also explained to the residents via information meetings and public presentations that there would be periods of smell and noise from the site, which were unavoidable consequences of the permanent removal of the acid tars.

The first acid tar lagoon to be treated was located 50 metres away from residential properties. Through detailed discussions with the regulators and the local community, a fully enclosed tent with negative air pressure and active air scrubbing was operated for the excavation and removal of the acid tar. This material was subsequently treated off site.

The system of regulatory control that was set up for the Chemnitz acid tar remediation projects was open and transparent and designed to provide both protection and reassurance for the local communities immediately affected by the site works. Weekly project management meetings were attended by the Baufeld project management team, the environmental monitoring consultants and specialists from the regulator's contaminated land unit.

In respect of the remediation costs, the regulating authorities took into account the cost-effectiveness of the substitute fuel approach. This was found to be an approach with similar costs but lower risks and more security than other tested techniques. This financial advantage was particularly significant owing to the proximity of a modern brown-coal-fired power station at Swartz Pumpe. Owing to the high sulphur content of brown coal, this power plant has very efficient flue gas scrubbers. However, the financial advantage of this approach is likely to be further enhanced in the future by the increase in landfill gate prices following the implementation of the Landfill Directive and a substantial increase in 'waste' volume associated with alternative solidification processes (Smith *et al.*, in review).

10.4.4 Technical issues

The history, layout and size of the Chemnitz acid tar lagoons have been described by Gruss (2005). The main techniques trialled at Chemnitz may be summarised as follows:

- excavation of the acid tar with associated emission mitigation procedures
- onsite or offsite processing of the acid tar into a substitute fuel suitable for burning in a brown coal power station (majority of acid tar)
- onsite and offsite processing of the acid tar to form a mechanically and chemically stable material for subsequent reburial in a containment cell (minority of acid tar)
- onsite and offsite treatment of the acid-tar-contaminated soils around the perimeter of the lagoon for subsequent disposal to landfill
- pumping out of the oil–water–sulphuric acid emulsion (OWSE) for offsite treatment in a refinery
- liquefaction and removal by pumping of acid tar for offsite use as a liquid substitute fuel

The technique or techniques chosen must be appropriate to the acid tar lagoon type, location and proximity of available treatment works/power stations. In the geographical and regulatory environment of the Chemnitz lagoons, the most economical and environmentally preferable solution was to utilise the energy inherent in the acid tar through conversion to a viable substitute fuel for use in power stations. This solution also provides a complete removal of the contaminants (Smith *et al.*, in review).

10.4.5 Excavation with emission control

Emissions from acid tars may be generated at relatively low volumes from the surface of an undisturbed lagoon, particularly where a layer of overlying water acts as an absorbent barrier. However, disturbance of the acid tars and the exposure of fresh surfaces to the atmosphere can lead to unacceptably high gas emissions. The key pollutant is sulphur dioxide, but associated emissions include volatile hydrocarbons, resulting in both hazardous air pollution and strong, unpleasant smells.

At Chemnitz, excavation of the acid tar was carried out by long arm excavator followed by immediate transfer to a wheel loader with a sealed shovel. The loader was able to weigh the acid tar, which allowed close control of the subsequent treatment process. The acid tar was then transferred to a pneumatically sealed processing unit (or, where logistics required offsite treatment, the tar was placed in lockable sealed transfer containers).

The cabins of all operating plant (excavators, front-end wheeled loaders) were supplied with clean air from activated-carbon filtration units.

In this way emissions from acid tar being moved were minimised. However, it was more challenging to control emissions from the disturbed lagoon itself. As mentioned earlier, for one small lagoon it was necessary to enclose the lagoon in a tent as it lay in close proximity to housing. This solution is clearly only practicable for the smaller lagoons; for the larger lagoons, emissions to the air were unavoidable. Unfavourable atmospheric conditions could lead to unacceptable concentrations at nearby housing and strict emission monitoring was necessary to permit excavation work to continue and/or to trigger the use of high-volume air blowers to dilute emissions (Smith *et al.*, in review).

Recent work by Pensaert (2005) has also reported success in reducing emissions from a lagoon surface by covering the tar surfaces with a thin layer of lime slurry.

10.4.6 Reprocessing as fuel

The onsite processing unit at Chemnitz was constructed of durable low-maintenance materials to keep interruptions to the remediation processes to a minimum. After delivery of the pre-weighed acid tars into the processing unit, the first stage of the treatment process was the addition of wood chips to absorb liquids and to minimise the 'sticky' character for better handling and conveying to the mixing unit. The mixture was then neutralised through the addition of a tightly specified type and mass of brown coal fly ash which contained free lime and calcium oxide. Following a defined time interval for reaction and consolidation processes to complete, the resulting substitute fuel was acceptable to the strict quality standards of the brown coal power station. The plant was capable of processing up to 3000 t/month.

The entire process was carried out in a closed system. Emissions released in the reaction processes (dust, volatiles and pollutants) were passed through an air-cleaning scrubber. This essentially consisted of a neutralisation washing process to remove the sulphur dioxide, and an oil-fired after-burner unit for the hydrocarbon vapours. Continuous emission-monitoring guaranteed compliance with legally permitted levels.

The substitute fuel was temporarily stored in an enclosed storage bay before being transported to the power station. In order to feed the substitute fuel into the fuel stream of the power station (up to 5% replacement of brown coal), a substitute fuel acceptance station was constructed on the power station site.

The acid-tar-contaminated soils from the dams and bases of the lagoons, which had negligible caloric value, were also neutralised and treated for subsequent disposal to landfill. The uncontaminated soils were stored for refilling and profile after excavation of the contaminated materials.

10.4.7 Summary of German experience

The Chemnitz project was planned and implemented in a transparent manner that involved both internal and external stakeholders. All parties recognised that there would be some short-term 'loss of amenity' associated with odours, noise and vehicle movements while the site work was being carried out. However, in the long term, through the conversion of the acid tars to a substitute fuel for use in a modern power station, the environmental and human health risks associated with the ATLs were permanently removed.

10.4.8 Application in the UK

The Chemnitz experience emphasises three key issues that should be considered in any remediation scheme involving acid tars:

(1) The treatment process must be sufficiently robust to cope with variations in composition of the acid tars.
(2) Any process that disturbs the acid tar has a strong likelihood of giving rise to significant gas and odour problems.
(3) Strict control and monitoring measures must be put in place. These will require significant liaison with local residents and regulatory authorities.

From a technical point of view the Chemnitz approach is perfectly applicable in the UK. Additional challenges, however, will be present in terms of regulatory and financial/logistical issues. Of the large UK lagoons in the public domain (e.g. Hoole Bank, Llwyneinion and Cinderhills), all are of a magnitude that would make excavation and processing a viable approach. Broadly they share the same issues of proximity to housing encountered at Chemnitz. The three key challenges are therefore:

- mobilisation cost and licensing of an onsite processing plant (adequate space and access to erect a processing plant adjacent to the lagoon will also be required)
- identification of a source of suitable fly ash or equivalent additives (e.g. lime products)

- location of a suitable power-generating plant willing to accept the processed fuel

Finding a suitable power station in the UK with adequate flue gas scrubbing capabilities, will have proved very difficult up to now. However, with the advent of the next generation of refuse-derived fuel power stations this may no longer pose a problem. It is likely over the next few years that a network of such installations will be constructed sustainably across the UK to manage waste and divert it away from landfill. Such plants will have efficient and effective emission control and monitoring systems.

Smaller lagoons in the UK, such as at the case study site, are unlikely to justify the expense of an onsite processing plant and will therefore require offsite processing coordinated with the remediation of other larger lagoons.

10.5 Conclusions

This chapter has sought to highlight the complex physical, chemical and toxic properties of acid tar lagoons. Conventional capping techniques have proven inadequate at breaking pollutant linkages. Unfortunately there currently exists limited experience of successfully treating ATLs in the UK. For these reasons, this chapter has sought to do two things from a technological perspective. The first is to characterise acid tars, to develop a greater understanding of their properties and then to explore the characteristics and environmental impacts of acid tar lagoons. Developing durable and sustainable technological solutions requires such research. However, this chapter has also sought to draw upon the experience of remediation engineers in Germany who have had considerable experience of dealing with highly toxic and large-scale ATLs.

In line with the philosophy of the SUBR:IM consortium, this chapter has also sought to develop an understanding of the broader human health, socio-economic and political issues that ATL sites can raise for decision makers and affected communities. The UK case study site utilised in our research demonstrated some of the threats arising from ATLs: potential development and property price blight, political controversy, loss of local amenity and so on. It showed how the implementation of regulatory policies can be mediated by the interests and concerns of local stakeholders and the context within which such processes take place (see also Chapter 3).

The ATL remediation at Chemnitz has demonstrated a viable approach to acid tar remediation involving complete removal of the contamination and beneficial utilisation of the acid tar as fuel. The project has clearly demonstrated that solutions to key technical problems are available, namely robust processes able to handle the inherent variability typical of lagoon-deposited

acid tars and effective control of gas emissions generated by lagoon disturbance. However, just as importantly, the Chemnitz experience also clearly
demonstrates that through an open and transparent regulatory system all
of the other human health, socio-economic and political issues can be
resolved to the satisfaction of both the internal and external stakeholders.

The 'lesson-drawing' from Germany could be distilled into the following
observations. Sustainable ATL remediation projects can be only be considered when the interests of all the 'stakeholders' can be demonstrated
broadly to coincide. Furthermore, such projects can only be successfully
completed when all the necessary conditions are present. These include
positive owner involvement, robust technologies, regulatory consent, clear
policy alignment, local community agreement, political approval and, importantly, appropriate finance. Up to now, it would appear that in the UK
not all of these conditions have been available at the same time.

Note

1. Early evidence suggested that an aqueous contamination plume was spreading down-
    gradient from the original tar lagoon, although later evidence suggested that pollution
    to the brook might have come from further upstream of the site.

References

Bache, I. and Taylor, A.J. (2003) The politics of policy resistance: reconstructing higher
    education in Kosovo. *Journal of Public Policy*, **23** (3), 279–00.
Banks, D., Nesbit, N.L., Firth, T. and Power, S. (1998) Contaminant migration from
    disposal of acid tar wastes in fractured coal measures strata, South Derbyshire.
    In: *Groundwater Contaminants and their Migration* (eds J.D. Mather, D. Banks,
    S. Dumpleton and M. Fermor), pp. 283–311. Geological Society Special Publication.
Boelsing, F. (1988) *Remediation of Toxic Waste Sites: DCR Technology in the Field
    of Immobilisation and Fixation of Hazardous Compounds.* Ministry of Economics,
    Technology and Traffic, Hannover, Germany.
Catney, P., Henneberry, J., Meadowcroft, J. and Eiser, J.R. (2006) Regulating contamin-
    ated land through 'development managerialism'. *Journal of Environmental Policy and
    Planning*, **8** (4), 331–56.
Chambers, J.E. (2001) *The Application of 3D Elecrical Tomography to the Investigation
    of Brownfield Sites*, Vol. 1. Department of Civil and Structural Engineering, University
    of Sheffield, Sheffield.
Claxton, G. (1961) *Benzoles: Production and uses.* National Benzole and Allied Products
    Association, London.
Evans, M. (2006) At the interface between theory and practice: policy transfer and lesson
    drawing. *Public Administration*, **84** (2), 479–89.
Grajczak, P. and McManus, R.W. (1995) Remediation of acid tar sludge at a superfund
    site. *Superfund Proceedings*, **1**, 243–4.

Gruss, D. (2005) Säureharzaltlasten, Innovative Technologien zur Sanierung und energetischen Verwertung. *Terratech*, **3–4**, 15–18.

Hao, X. and Smith, C.C. (2005) Physical and chemical properties of acid tars. *9th International FZK/TNO Conference on Soil-Water Systems*, Bordeaux Convention Centre, Bordeaux, France, 3–7 October 2005.

Milne, D.D., Clark, A.I. and Perry, R. (1986) Acid tars: their production, treatment and disposal in the UK. *Waste Management & Research*, **4**, 407–18.

Nancarrow, D.J., Slade, N.J. and Steeds, J.E. (2001) *Land Contamination: Technical Guidance on Special Sites: Acid Tar Lagoons.* Environment Agency R&D Technical Report P5-042/TR/04.

Nichol, D. (2000) Geo-engineering problems at Hoole Bank acid tar lagoon, Cheshire, UK. *Land Contamination & Reclamation*, **8** (3), 167–73.

Oudijk, G. and Coler, M. (1995) Beneficial use of the upwelling phenomenon in coal-tar remediation efforts. *International Symposium and Trade Fair on the Clean-up of Manufactured Gas Plants*, 19–21 September, 1995. Prague, Czech Republic.

Payne, J.R. and Charles, R. (1987) Petroleum spills in the marine environment, the chemistry and formation of water-in-oil emulsions and tar balls. *Marine Chemistry*, **20** (3), 297.

Pensaert, S. (2005) The remediation of the acid tar lagoons, Rieme, Belgium. *International Conference on Stabilisation/Solidification Treatment and Remediation, Advances in S/S for Waste and Contaminated Land*, 12–13 April 2005. Cambridge University, Cambridge.

Reynolds, J.M. (2002) The role of environmental geophysics in the investigation of an acid tar lagoon, Llwyneinion, North Wales, UK. *First Break*, **20** (10), 630–36.

Rose, R. (1993) *Lesson Drawing in Public Policy: A Guide to Learning Across Space and Time.* Chatham House, Chatham, NJ.

Rose, R. (2001) *Ten Steps in Learning Lessons from Abroad.* ESRC Future Governance Programme. Available at: http://www.hull.ac.uk/futgov/. Last accessed: 22 April 2006.

Slater, D. (2003) Boys from the black stuff. *SUSTAIN Built Environment Matters*, **4** (6), 39–40.

Talbot, S., Smith, C., Lawson, N., Shaw, S. and Hao, X. (2004) Restoration of acid tar lagoons. *SCI Conference: Contaminated Land: Achievements and Aspirations*, Loughborough, UK.

Part 4
Joined-up solutions

11

Climate Change, Pollutant Linkage and Brownfield Regeneration

Abir Al-Tabbaa, Sinead Smith, Cécile De Munck,
Tim Dixon, Joe Doak, Stephen Garvin and
Mike Raco

11.1 Introduction

Although there is considerable uncertainty in predicting future impacts of climate change, there is global and national evidence to suggest that the UK will be subjected to warmer and wetter winters, hotter and drier summers, rising air temperatures, increased storminess and heavier rainfall. These factors may contribute to an increase in the risk of significant pollutant linkages forming; thus sources of contamination, which currently pose little risk to the environment, are likely to become significant in the future. Climate change may therefore require different adaptation strategies for contaminated brownfield sites.

The potential impact on contaminated land containment systems, such as landfills, barriers, cover and stabilisation/solidification systems, is large. This also applies to any ground contamination, particularly at shallow depths: for example, many of the untreated contaminated brownfield sites, spillage of underground storage tanks, or contamination which has been placed in unengineered excavations (e.g. foot and mouth carcasses). Ground temperatures and evaporation losses from the land surface will increase, causing the soil to crack and resulting in upward capillary suction of water from depth and an increasing risk of exposure of contaminated materials at the ground surface. Higher ground temperatures may also increase the mobility and volatility of certain organic contaminants in the ground. Higher intensity rainfall will challenge soil infiltration capacity and increase the risk of soil erosion and particulate spread of contamination. In addition, there may be a seasonal rise in groundwater level which may bring clean groundwater in contact with ground contaminants. It has also been suggested that the weather will become more cyclic, in particular around freezing point. This could impact on the durability of containment systems at shallow depths. The impact on these pathways will, in part, be affected by the vegetation on the site, itself subject to potential ecological adaptation as a result of climate change. Such changes need to be considered in the technical design of future containment systems or in the future management of contaminated land. They also need to be addressed in any adaptive stakeholder responses. This chapter is an example of an integrated technical and social science approach to the analysis of these issues.

This chapter concentrates on the following:

- experimental quantification of potential impacts of climate change predictions on pollutant linkages
- numerical modelling of potential impacts of climate change predictions on pollutant linkages
- assessment of appropriate technical adaptation strategies to account for climate change
- examination of the adaptive response of key brownfield stakeholders to climate change

This chapter shows that although there is clear evidence of potential significant impacts on contaminated land and containment systems, such evidence is still currently not considered in decision-making in relation to the remediation and redevelopment of contaminated land.

11.2 Evidence of impacts of climate change on contaminated land systems

11.2.1 Potential impacts of climate change: literature review

Most studies on contaminant fate and transport in soil have generally been limited to relationships between contaminant behaviour and intrinsic soil chemical properties (soil organic matter, nutrient, cation exchange capacity, etc.) and biological properties (microbial activity, population, etc.). Some studies focusing on different techniques for soil contamination management have tried to relate contaminant behaviour to variations in different aspects of the climate (temperature, precipitation, humidity, etc.) without drawing any direct linkage between observations and the general notion of 'climate change'. Hence currently these are the main literature studies available for use in developing an understanding of the potential impacts of climate change on contaminated land systems and its implications in terms of pollutant linkages.

Generally, studies on the effect of climate parameters on soil biological properties such as soil respiration, microbial enzyme systems and microbial number, which have implications for contaminant behaviour, show that these effects are dependent on the scale of the changes as well as the original conditions of the soils. However, a plausible consensus points to increases in soil biological properties due to increased temperature, while drought conditions would lead to an inverse effect on these properties. These changes in soil properties as a result of climate change on an average annual scale would, however, be small compared to changes due to seasonal fluctuations (Papatheodorou, 2004; Sowerby *et al.*, 2005). This would have implications for soil contaminants such as hydrocarbons whose degradation and fate are dependent on the rate of soil biological processes.

Studies relating soil chemical properties such as cation exchange capacity (CEC) and pH to climate parameters have compared soils across topographical transects as analogous to different climate conditions (Chadwick *et al.*, 2003). Increased temperature and lower precipitation were found to lead to higher CEC and pH values, as would be accounted for by reduced leaching. However, these effects, especially on CEC, appear to be only significant after a long period of sustained climate conditions coupled with the absence of additional sources of the leached materials. Other chemical properties such as redox potential have been shown to respond to precipitation levels, with increased precipitation leading to soil saturation condition and reduced redox potential while drier conditions would lead to more oxidised conditions (Twining *et al.*, 2004). These conditions would affect heavy metal soil contaminants whose solubility and toxicity are dependent on those soil chemical properties.

It can hence be concluded from studies such as those presented above that climate changes towards warmer conditions would favour biologically driven degradation of compounds amenable to degradation while drier conditions would have the opposite effect. On the other hand, heavy metal soil contaminants, which are more related to issues of leachability, would feature less risk in higher temperatures and a drier climate due to increased soil CEC and pH. However, this may be counteracted by increased redox potentials associated with the drier conditions leading to increased solubility. Reduced forms of heavy metals are reported to be generally less soluble and thus less toxic. Drier climate conditions would present high risks due to increased metal solubility in oxidised conditions. Ageing also generally affects the mobility of heavy metal contaminants. Contaminant mobility is thought to be determined by sorption kinetics, where contaminants become sequestered over time within inaccessible micro-sites in the soil matrix (Alexander, 1995).

Soil organic matter, which depletes over time in bare soils owing to degradation by the soil biota as well as loss resulting from surface runoff, is also likely to respond to different climate conditions. It has been shown that hot regions are not able to retain as high a level of soil organic carbon as cold regions, largely because of high annual decomposition rates as a result of high temperatures (Franzluebbers *et al.*, 2001). However, Bazzaz and Sombroek (1996) suggested that this depletion of soil organic matter might be more than compensated for by the greater organic matter supply from vegetation and crops growing more vigorously because of higher photo-synthesis, greater evapotranspiration and higher water use efficiency under high-CO_2 atmosphere. Changes in soil organic matter also lead to changes in soil water retention parameters at extreme matric suction: for example, wilting point and total porosity (Korodjouma *et al.*, 2006).

In terms of the impact of climatic changes on remediated contaminated land, relevant studies are available on two commonly used containment systems, namely stabilisation/solidification (S/S), with the addition of binders, (see section 11.2.2) and engineered cover systems. Mechanical properties of S/S materials, such as strength and permeability, change with time owing to continued hydration reactions and contaminant interactions which are affected by temperature and moisture content changes (Perera *et al.*, 2005). Severe wet–dry and freeze–thaw cycles have been shown to increase con-taminant leachability from those systems (Boes and Al-Tabbaa, 2001).

Engineered cover systems have been extensively used and studied and a range of field- and laboratory-scale tests have been used to monitor climate-related impacts on cover systems. Cover system failures due to desiccation caused by climate effects, such as wet–dry cycles, have been reported (Wagner and Schnatmeyer, 2002). As a result, the permeability of the cover material can increase, reducing its effectiveness. Freeze–thaw cycles are also reported to cause similar damage when a freezing front advances downwards form-ing ice lenses which form a network of cracks. Figure 11.1 shows a typical cover system damaged by desiccation (Benson, 1999).

It is clear from the above that a wider range of interrelated impacts could be expected depending on the climate change conditions, their magnitude and the nature of the soil system in question.

11.2.2 Potential impacts of climate change: experimental observations

11.2.2.1 Imposed climate change scenarios and design considerations
In order to provide some relevant direct evidence of the impact of climate-related effects on contaminated land systems, an extensive programme of experimental laboratory work was conducted. Highlights from this work and its findings are presented here.

Figure 11.1 Cracks in a compacted clay cover system in southern Wisconsin caused by desiccation (Benson, 1999).

A number of soil systems were tested, some of which are summarised in Table 11.1. Predicted average and extreme seasonal temperatures for 2050 and 2080, from UKCIP (Hulme *et al.*, 2002) under the high emissions scenario for the south-east of England were used to select the temperature conditions to be used in the tests. Extreme summer temperatures are defined as 1 in 10 day extreme values and these were the values used in the tests as detailed in Table 11.2. In the absence of predicted extreme winter temperatures, values of −2°C for 2050 and either 0°C or +2°C for 2080 were used. These are mainly based on the prediction that there will be considerable temperature fluctuation around freezing point, and this zone of temperature is usually the most damaging. In order to compare those climate scenarios with a current baseline scenario, a constant temperature of 20°C (room temperature) was used throughout the year for soil systems 1–5 in Table 11.1. Although this value is not strictly representative, it was used mainly because of the lack of environmental cabinet space for such a large size and number of samples. For soil systems 6–9, for which the samples were much smaller, a winter temperature of 0°C was used, as detailed in Table 11.2. All systems were subjected to ~100% relative humidity (RH) throughout the period of testing. A temperature of 20°C and relative humidity of ~100% were also used for the spring and autumn seasons for all soil systems, during which soil systems were also recharged to their original water content.

Three different summer precipitation scenarios and two different winter precipitation scenarios were imposed as detailed in Table 11.3. Two years

Table 11.1 Details of the different soil systems tested and tests conducted.

Soil system	Soil system symbol	Climate scenarios applied (see Tables 11.2 and 11.3)	Soil system details	Properties tested
Stabilised/solidified contaminated soil	SSCS	B (2050 and 2080)	Silty clayey made ground contaminated with Pb, Cu, Zn, Ni, Cd (up to 3000 mg/kg) and mineral oil (up to 9700 mg/kg). Soil modelled on soil from a chemical works site. Treated with cement-bentonite grout (10:1 ratio) with water:solids ratio of 1.6:1 and applied in grout:soil ratio of 1:3.7. Soils tested at 28 days	Water content, unconfined compressive strength (UCS), permeability, Pb and Cu leachability, leachate pH
Aged stabilised/solidified contaminated soil	ASSCS	D (2080)	Same soil system and grout treatment as above. Accelerated ageing techniques were used to age the soil system by up to 15 years	Water content, UCS, permeability, Pb and Cu leachability, leachate pH
Stabilised/solidified uncontaminated soil	SSUS	A, B, C, D, E, F (2050 and 2080)	Same soil as above but uncontaminated. Same grout used as above. Soil tested at 28 days	Water content, UCS, permeability
Aged stabilised/solidified uncontaminated soil	ASSUS	D (2080)	Same soil as above. Same grout used as above. Accelerated ageing techniques were used to age the soil system by up to 15 years	As above
Compacted clay cover	CCC	A, B (2050 and 2080), C, D, E, F (2050)	Silty clay soil, with 22% water content and compacted to wetter than optimum	Water content, permeability, sorptive capacity
Amended contaminated site soil	ACSS	B, D (2080)	Silty clay (water content 20%) contaminated with heavy metals mainly Cu, Zn and Cd (up to 9580 mg/kg) from a zinc mining site. Soil is amended with compost and zeolite additives in 79:14:7 ratio, the compost was a mixture of green waste compost and sewage sludge compost in 3:1 ratio (see further details in Chapter 8)	Water content, Cd and Cu leachability, leachate pH, soil redox state, CEC
Amended spiked contaminated site soil	ASCSS	B, D (2080)	Same soil above spiked with an organic compound (2-chlorophenol, 260 mg/kg soil)	As above
Amended, spiked and bioaugmented contaminated site soil	ASBCSS	B, D (2080)	Same soil above but bioaugmented with commercially available microbial consortium (42 g/kg soil) commonly used for bioremediation purposes	As above plus extractable hydrocarbon content, soil biological activity, dissolved organic carbon
Bioremediated site soil	BSS	B, D (2080)	Silty sand, site soil formerly contaminated with hydrocarbons (with little heavy metal contamination) and bioremediated	As above

Table 11.2 The actual/predicted and applied temperatures (°C) where the applied temperature has a margin of error of ±2°C.

Year/period	Annual average	Actual/ predicted summer average	Predicted extreme summer	Selected summer average	Actual/ predicted winter average	Selected winter average for soil systems 1–5 in Table 11.1	Selected winter average for soil systems 6–9 in Table 11.1
1961–1990 (baseline)	9.5	15	23	20	4	20	0
2050	12	18	27	27	5.5	−2	–
2080	13.5	19.5	31	31	7	0	2

Table 11.3 The different precipitation scenarios applied.

Scenario symbol	Scenario	Summer conditions	Winter conditions
A/B	No summer rainfall and flooded winter conditions or frequent winter rainfall	Dry with no water recharge	A: Fully immersed in water, referred to as 'flooded' B: Maintained at its natural water content, referred to as 'saturated'
C/D	Intermittent summer rainfall and flooded winter conditions or frequent winter rainfall	Completely dry for one week (with no recharge) and then recharged to its initial water content for another week (under 100% RH conditions)	C: Flooded D: Saturated
E/F	Frequent summer rainfall and flooded winter conditions or frequent winter rainfall	Maintained at its natural water content (under 100% RH conditions), referred to as saturated	E: Flooded F: Saturated

of climate change conditions as predicted for 2050 and 2080 were modelled. Summer and winter were modelled for a three-month period each while spring and autumn were each modelled for a one-month period only.

11.2.2.2 Evidence of impacts on soil containment systems

The results presented here are those of the first four S/S soil containment systems in Table 11.1 (SSCS, ASSCS, SSUS and ASUS), including contaminated and uncontaminated soils, both fresh and aged. The results generally showed that at the end of the first summer, despite various levels of water content changes, there was little physically observed change or damage in the containment systems except for the uncontaminated S/S system (SSUS) in scenario 2080 C/D (intermittent summer rainfall), and the contaminated S/S system (SSCS) in scenario 2080 B (no rainfall), where in both cases some cracking was observed, as can be seen in Figures 11.2 (a) and (b) respectively. The wet–dry intermittent summer rainfall precipitation scenario was found to be the most damaging. Damage in the contaminated S/S system in the no-rainfall scenario is due to the lower degree of cement hydration at 28 days. After the first winter the same uncontaminated S/S soil system in Figure 11.2 (a) suffered severe cracking and deterioration after the flooded scenario (Figure 11.3 (a)) while very little additional damage was observed after the saturated scenario (Figure 11.3 (b)). Figures 11.4 (a) and (b) show the uncontaminated S/S system in the 2050 A/B scenarios which initially suffered no damage in the summer but was then damaged in the more severe 2050 winter conditions, again more so under the flooded than the saturated conditions.

As a consequence of the above physical damage observations, the permeability after the first summer was found to increase in the damaged samples (by up to two orders of magnitude compared to the baseline). After the first winter, the saturated conditions were least damaging (least increase in permeability), and the flooded conditions combined with intermittent summer rainfall were the most damaging, increasing the permeability by up to four orders of magnitude. The second year was found to produce far smaller permeability changes following the changes produced in the first year. Similar trends to those found by physical observations were noted in the permeability of the contaminated samples, with the presence of contaminants causing insignificant changes although the permeabilities were in

(a)   (b)

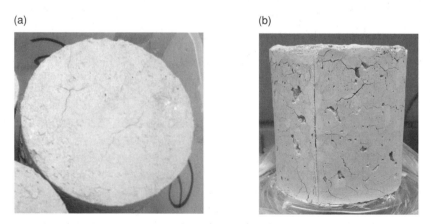

Figure 11.2   Typical S/S soil samples at the end of the first 2080 summer: (a) uncontaminated sample (SSUS) in scenarios C/D; (b) contaminated sample (SSCS) in scenario B.

(a)   (b)

Figure 11.3 Typical S/S uncontaminated soil samples (SSUS) at the end of the first 2080 winter in (a) scenario C and (b) scenario D.

(a)   (b)

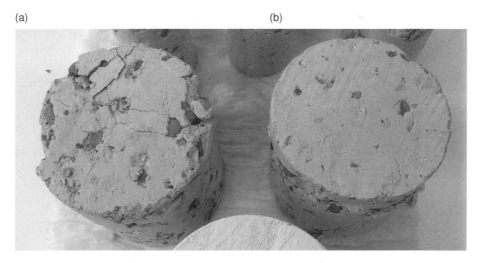

Figure 11.4 Typical S/S uncontaminated soil samples (SSUS) at the end of the first 2050 winter in (a) scenario A and (b) scenario B.

general an order of magnitude higher than those of the uncontaminated soil system samples.

The trends in the unconfined compressive strength (UCS) were comparable with the permeability trends, with the least damaged systems producing the highest UCS (close to those of the baseline response) and the most damaged systems producing the lowest UCS values (up to three times lower than the baseline response). The UCS values of the contaminated samples were generally lower than those of the uncontaminated samples (around half) and similar trends owing to the climate conditions imposed were observed.

The trends in the leachate pH and leachability of Pb and Cu were less obvious. From summer to winter there was little change in leachate pH but from one summer to the next there was a decrease in leachate pH (of around 1 unit) compared to the baseline response, which showed little change. There was less leaching of copper and more leaching of lead after the first winter compared to after the first summer, and this was followed by very little leaching of either after the second winter and less leaching compared to the baseline response.

The aged S/S samples (ASSCS, ASSUS), after the first summer, showed the same level of cracking as observed with the unaged samples (scenario D), while after the first 2080 winter the cracking did not increase. Hence there was far less cracking observed than in the unaged samples, suggesting that age improves the performance of S/S systems against freeze–thaw conditions.

For the compacted clay cover system, system CCC in Table 11.1, considerable variations in water content were observed for scenarios A/B and C/D between summer and winter, with up to 64% water content loss in the dry summers. Despite this, generally very little physical damage or cracking was observed in the samples after the first summer while cracking was observed after the winters owing to the formation of ice lenses. Permeability was found to slightly decrease following the first summer (expected to be due to shrinkage and resulting densification of the samples – an effect which is only observed in small laboratory tests (Albrecht and Benson, 2001)) and increase following each winter (by up to one order of magnitude). It is likely that increases in the permeability in the cracked samples were actually even higher than measured because the cracks close to some degree during laboratory testing under the application of a confining pressure.

The results for the compacted clay cover system were input into the model HELP (Hydrologic Evaluation of Landfill Performance) (Berger, 2000), which calculates the water balance in a cover system, in order to validate the model for the soil cover system studied. The model was then used to extend the experimental work to much longer-term severe climate change conditions by inputting annual long-term changes in soil properties (soil organic content, wilting point, field capacity, total porosity) from the literature together with annual long-term changes in climate conditions (temperature, precipitation, solar radiation, relative humidity). This was used to assess the annual performance of the clay cover system over a 50–80-year period of the same climate change scenarios applied experimentally. The results clearly show the interaction between those two sets of parameters and the impact of long-term exposure to such severe climate conditions.

11.2.2.3 Evidence of impacts on amended or bioremediated contaminated soils

These results relate to contaminated soil systems 6–9 in Table 11.1 (ACSS, ASCSS, ASBCSS and BSS). Systems 6–8 are based on the same site soil, which was subjected to different amendments, and system 9 which was a bioremediated site soil. At the end of the dry summers, the water content of these systems was found to reduce by ~60%. The soil redox potential was found to be higher in the summers than in the winters, and was highest in the driest summer conditions (scenario B). The CEC was found not to change with time, season or climate scenario.

The trend of the leachate pH was such that the values were slightly lower in the summer than in the winter (by ~0.5 units) with the bioaugmented site soil (ASBCSS) having a lower pH throughout (by ~0.5 units) caused by the bioaugmentation process itself. The bioremediated site soil (BSS) showed the same trend but with higher pH values throughout. Insignificant changes due to the two different summer precipitation scenarios were observed. The leachability of Cu and Cd showed slight fluctuation between seasons which had an inverse relation to the leachate pH. The leachability of the metals in the presence of the organic contaminant produced no differences (ASCSS compared to ACSS). The bioaugmentation process (ASBCSS) caused increases in the leachability of the metals. With respect to differences in summer precipitation scenarios these changes were only evident in the bioaugmented soil (ASBCSS), with slightly higher leachability in the wetter scenario (D) than in the drier scenario (B), and highest in the baseline scenario. Changes in the copper leachability were also observed between the summers and winters in the same bioaugmented soil system (ASBCSS). Increases in leachability in the summers were observed, followed by slowing down and some reduction in the winters. The BSS soil system did not show such a trend because the bioremediation treatment was performed some time ago. The Cu leachability results for the three scenarios tested on the four soil systems are shown in Figure 11.5.

The effect of the climate scenarios employed on soil hydrocarbon contaminants was observed by measuring changes in the amount of solvent-extractable hydrocarbon from the soil. This gives a measure of hydrocarbon loss due to both biological and physical processes as well as other contaminant–soil interactions such as adsorption. There was less extractable hydrocarbon content in the bioaugmented soil (ASBCSS), hence potentially more degradation had taken place compared to the other two

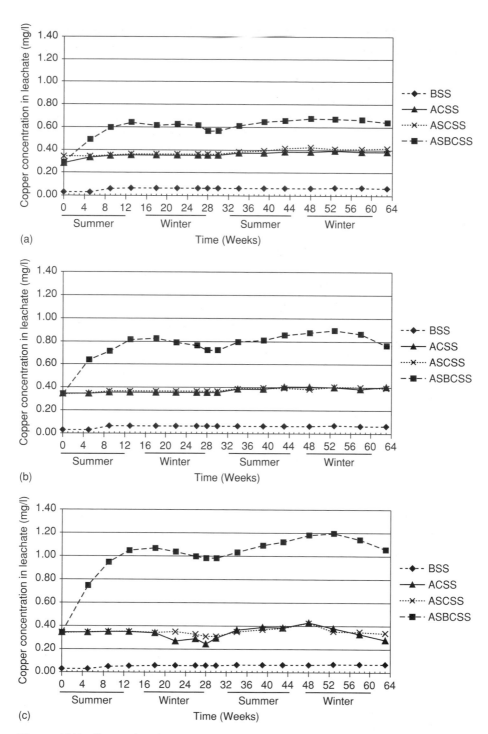

Figure 11.5   Copper leachate concentrations in the four soil systems (ACSS, ASCSS, ASBCSS and BSS) over the two-year period tested for (a) scenario 2080 B, (b) scenario 2080 D and (c) the baseline scenario.

systems (ACSS and ASCSS). Within each season the extractable quantity reduced with time, the rate of reduction being much less in winter, with the second winter showing little change in the quantity of extractable hydrocarbon. This suggests that prolonged conditions resulted in the hydrocarbon being less available to processes such as biodegradation and evaporation. The lowest extractable hydrocarbon content was observed in the baseline scenario and the highest in the worst climate scenario imposed, suggesting that the climate scenarios imposed reduced the hydrocarbon loss. The profile of the extractable hydrocarbon content is shown in Figure 11.6.

In terms of soil biological activity, this was much higher in the first summer than in the first winter and reduced in the second year. The baseline scenario had the highest activity and the B scenario the lowest. The higher biological activities in the summer correlate well with the higher rates of loss of extractable hydrocarbon in the summers and the trend of lower biological activity in the second year also correlates well with the trend of lower rates of change in extractable hydrocarbon in the second year. The trends in the soil biological activities observed are presented in Figure 11.7.

The experimental observations reported here show that certain climate change scenarios or combinations of scenarios do give rise to potentially significant impacts on the four different soil systems investigated, namely S/S contaminated and uncontaminated soils, compacted clay cover systems, amended soils with compost and mineral additives and bioremediated soils. For the S/S soil systems, wet–dry conditions were found the most damaging, the damage was most severe in the first season of severe climate conditions, and ageing of the system was found to be an advantage. In the compacted clay cover system, more damage was observed after the winters than the summers with an increase in permeability of one order of magnitude. Combining the results with model predictions enabled the assessment of the longer-term impact of exposure to severe climate conditions. For the amended or bioremediated contaminated site soils, the results show that the changes were more severe between seasons and between different soil systems compared to between climate change scenarios, and that the overall changes are a combination of both sets of changes together with long-term natural changes in soil conditions.

Such observations will have an impact on the management of contaminated land and existing containment systems as well as the management and design of future systems.

For example, a permeability value of 10^{-9} m/s is usually required for both engineered cover and S/S systems. Hence the design of future systems might require much lower initial permeability to be achieved to allow for orders of magnitude of potential increases over time due to climate change

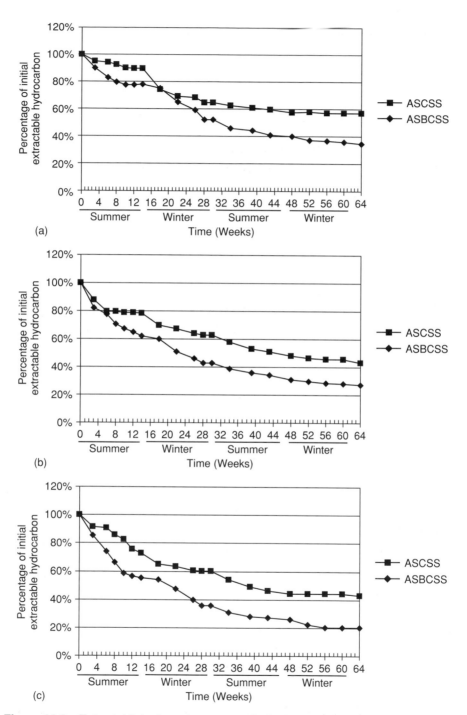

Figure 11.6   Extractable hydrocarbon content in the two relevant soil systems (ASCSS and ASBCSS) over the two-year period tested for (a) scenario 2080 B, (b) scenario 2080 D and (c) the baseline scenario.

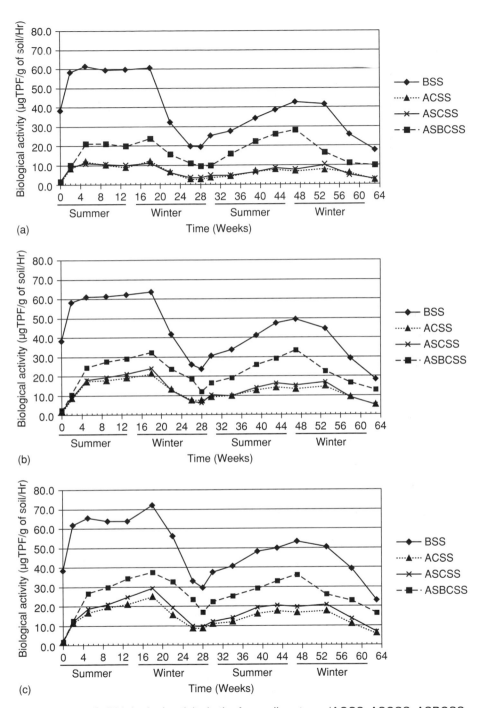

Figure 11.7 Soil biological activity in the four soil systems (ACSS, ASCSS, ASBCSS and BSS) over the two-year period tested for (a) scenario 2080 B, (b) scenario 2080 D and (c) the baseline scenario.

conditions. Containment systems with improved technical performance, and which are more durable and sustainable and hence likely to offer an improved resistance to climate change conditions, are being investigated and developed as detailed in section 11.2.2.

11.3 Modelling potential impacts of climate change and the creation of greenspace on contaminated land

11.3.1 Introduction

Modelling studies can examine the effects of predicted climate and atmospheric carbon dioxide changes on contaminant availability and transport without the timescale associated with laboratory simulations, especially if these are not practically achievable. Whenever possible, results of predictions in a current climate should be compared with onsite or laboratory measurements to ensure their validity. If the predictions are sufficiently accurate, then such validation can provide confidence for the use of modelling methods for simulations in other geographical areas under ranges of predicted future environmental conditions.

Modelling tools can be used in two ways: first, for studying how climate change may affect pollutant linkage on contaminated sites; and second, for simulating remediation options prior to conducting the remediation at the field scale. In this way modelling can be used as a risk-forecasting tool in retrospective or future-planning processes.

When choosing or defining a model it is important that decisions are made on a suitable level of process complexity or simplicity (i.e. which processes are taken into account) that will be relevant to the end user's objectives. A good balance must be found between simple predictive tools that aid planners in decision-making and rather complex models (and associated data complexity required for operation and validation purposes) used by researchers.

Understanding the processes and environmental parameters that control pollutant linkage is a first step in modelling them and a prerequisite for defining appropriate remedial responses. Examples of such links are shown schematically in Figure 11.8. Moreover, seasonal influences should be considered because climate-induced changes of environmental parameters controlling pollutant linkages may increase or decrease over the course of the year (as shown in section 11.2). Many processes are also spatially and temporally sensitive. Consideration of such influences on processes will need to be accounted for when modelling potential impacts of climate change.

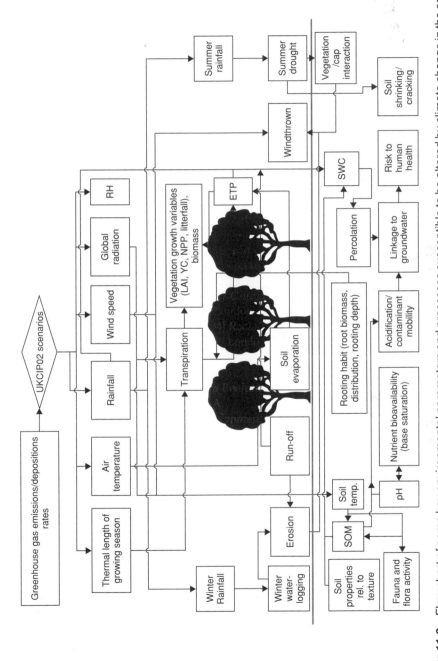

Figure 11.8 Flow chart of some key environmental parameters and processes susceptible to be altered by climate change in the context of pollutant linkage. RH (Relative Humidity), LAI (Leaf Area Index), YC (Yield Class), NPP (Net Primary Productivity), ETP (EvapoTranspiration Potential), SWC (Soil Water Content), SOM (Soil Organic Matter). For example, vegetation characteristics are key parameters with regard to soil erosion as vegetation can protect the soil from raindrops' impact (litterfall, vegetation/crown characteristics and growth) but can also contribute to soil dryness by consuming water and nutrients from the soil.

11.3.2 UKCIP02 climate change scenarios: technical implications with regard to modelling environmental processes

The Hadley Centre Global Circulation Models predictions for the 2020s, 2050s and 2080s offer various spatial (50 km or 5 km) and temporal (annual, seasonal, daily or monthly) resolutions for predicted changes. In order to scope out the effects of potential changes it is recommended that a minimum of two contrasting scenarios should be examined (Hulme *et al.*, 2002).

The main challenge when modelling environmental processes in a climate change context is to match temporal and spatial data resolution between the processes being studied under the scenarios being investigated. A good example of this is when modelling the effects of climate change on soil erosion processes using the water erosion model RUSLE2 (Foster, 2004) and UKCIP02 scenario data. This example illustrates an inconsistency between the required input parameters of the water erosion model and the climate change data available from UKCIP.

The average annual rainfall erosivity factor (MJ.mm/ha.h.yr) as defined within RUSLE2 is

$$R = \frac{1}{n} \sum_{j=1}^{n} \left[\sum_{k=1}^{m} (E)(I_{30})_k \right] \tag{11.1}$$

where E represent the total storm kinetic energy of a given storm (MJ/ha), I_{30} is the maximum 30-minute rainfall intensity during the storm (mm/h), j is the index for the number of years used to produce the average, k the index for the number of storms in each year, n the number of years used to obtain average R and m the number of storms in each year.

Although the model requires an *annual* average of storm erosivity, closer examination of Equation 11.1 shows that *sub-hourly* rain gauge records would be needed across several years to compute this value. In contrast, the UKCIP02 scenario format best suited to input the model is at a *monthly* resolution.

In order to address inadequacies between the temporal resolution of the climate change scenario data and the data requirements of the chosen model, modellers often have to either find published techniques or set up new methodologies that will enable them to use available climate change scenario data.

11.3.3 Models available for modelling risk to surface water by water erosion

Several parametric models have been developed to predict soil erosion at drainage basin, hill slope and field scales. With few exceptions, these models are based on soil type, land use, landform, and climatic and topographic information. Erosion models are often classified in two categories, physically (or process) based models and empirical models. More and more models, however, include both empirical and physically based equations, depending on the data and scientific knowledge available. Tables 11.4 and 11.5 present an appraisal of the most commonly used process and empirical models for predicting changes in water erosion due to climate change using data available.

The Revised Universal Soil Loss Equation, RUSLE2, (Renard *et al.*, 1997; Toy and Foster, 1998; Foster *et al.*, 2003a,b; Foster, 2004; and Figure 11.9) was chosen as the most appropriate model for studying climate change impacts on soil erosion processes. Although developed and validated for the USA, RUSLE2 can be applied to other countries where the information is available for local evaluation of the erosivity factor (Equation 11.1). The model evaluates soil loss from a hillslope profile caused by raindrop impact and overland flow (collectively referred to as interrill erosion) plus rill erosion, but does not account for gully erosion. Using topography and derived

Table 11.4 Characteristics for process models.

Name of model	Name stands for	Source of information	Spatial scale	Soil loss computed	Climate data source
PESERA	Pan European Soil Erosion Risk Assessment	(Van Rompaey *et al.*, 2003)	Continental (European) – no hillslope profile[1]	annual	User
WEPP	Water Erosion Prediction Project	(Morgan, 2005)	Hillslope profile or watershed or grid	continuous simulation – can also be operated for a single event	USA climate generator 'CLINGEN'[2]
GUESS	Griffith University Erosion Sedimentation System	(Morgan, 2005)	Hillslope	single event[3]	User (with 3 parameters needing calibrating)
EUROSEM	EUROpean Soil Erosion Model	(Morgan, 2005)	Individual fields or small catchments	single event[3]	User

General note: Greyed cells highlight the reason why a model may not be appropriate for modelling water erosion processes on contaminated land.
Note 1: PESERA deals with continental erosion and is therefore not adapted to modelling erosion at the scale of a contaminated site.
Note 2: WEPP works with a built-in climate generator for the USA, making it cost-effective to adapt to a UK climate.
Note 3: In the context of climate change, annual soil loss modelling is required.

Table 11.5   Characteristics for empirical models.

Name of model	Name stands for	Source of information	Spatial scale	Soil loss computed	Climate data source
USLE[1] (Obsolete)	Universal Soil Loss Equation	(Morgan, 2005) (Foster *et al.*, 2003a)	Field on hillslope	Mean annual	User
RUSLE[1] (Obsolete)	Revised Universal Soil Loss Equation 1	(Toy and Foster, 1998) (Renard *et al.*, 1997) (Morgan, 2005) USDA-ARS	Field on hillslope	Mean annual	User
RUSLE2	Revised Universal Soil Loss Equation 2	USDA-ARS USDA-NRCS (Foster *et al.*, 2003b) (Foster, 2004)	Field on hillslope	Mean annual (and mean daily values[2])	User
SLEMSA[3]	Soil Loss EstiMator for Southern Africa	(Morgan, 2005)	Field on hillslope	Mean annual	User (mean annual rainfall)
Morgan, Morgan and Finney method	–	(Morgan, 2005) (Shrestha *et al.*, 2004)	Field on hillslope	Mean annual	User (mean annual rainfall, number of wet days)

General note: Greyed cells highlight why a model may not be appropriate for modelling water erosion processes on contaminated land.

Note 1: Foster *et al.* (2003a) describe the reasons why users should not be using USLE and RUSLE1 as the RUSLE2 was designed to more accurately predict erosion.

Note 2: RUSLE2 computes average daily erosion, which represents the average erosion that would be observed if erosion was measured on that day for a sufficiently long period.

Note 3: SLEMSA was developed and validated for Southern Africa only (specific rainfall regime, very different erosion processes than the UK) and its approach is too simplistic compared to RUSLE (using classes instead of specific values for soil, vegetation, and using mean annual rainfall distribution).

soil erosion rate estimates, the model evaluates sediment delivery both on the slope and to watercourses.

11.3.4 *Example of how modelling can predict climate change effects on pollutant linkages induced by water erosion*

11.3.4.1 Background

UKCIP02 scenarios of monthly precipitation and air temperature generated from The Hadley Centre Global Circulation Models were used for two regions of the UK (one in the South West (Cornwall) and one in the South East (East Sussex)) that have contrasting climates to model the impact of differences in rainfall regimes on soil erosion rates and subsequent movement of contaminants. Changes in rainfall erosivity were estimated based

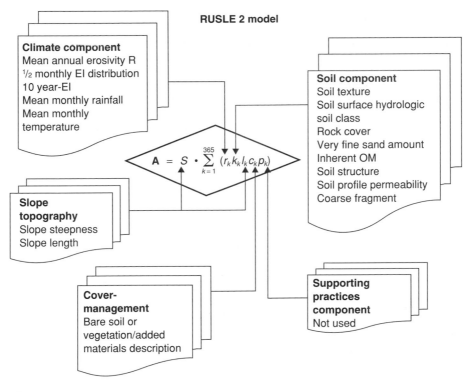

where:

A = average annual soil loss (t/ha/yr)

k = index for day of the year

S = slope steepness factor (the same for every day) (fraction, unit less)

r_k = rainfall and run-off erosivity factor (MJ.mm/ha.hr)

k_k = soil erodibility factor (t.hr/MJ.mm)

l_k = soil length factor (fraction, unit less)

c_k = cover-management factor (added soil, vegetation and vegetation residue) (fraction, unit less)

p_k = supporting practices factor (practices aiming at reducing erosion) (fraction, unit less)

Figure 11.9   Components of the RUSLE2 model (Foster, 2004).

on its relationship with the Modified Fournier Index (Renard and Freimund, 1994). Rainfall erosivities were then calibrated for the UK regions using two published techniques (Renard and Freimund 1994; Renard *et al.*, 1997) with sub-hourly rain gauge records from two weather stations located within the regions studied. In order to scope out the potential impacts of climate change on erosion rates for the two regions, two contrasting emissions scenarios (low and high emissions) were used to run RUSLE2 using topography and soil characteristics for a single case study site. This was a spoil heap on a disused tin-mining site located in the southern part of the Tamar Valley in south-west England (Figure 11.10). This represents a highly erodible coarse (1–2mm) sandy material on a steep-sided slope (54%) (Table 11.6).

Figure 11.10   Case study site main spoil.

Table 11.6   Values for the soil description used in the RUSLE 2 model to simulate the case study site spoil.

Variable	Value
Slope length	46 m
Slope steepness	54%
Soil texture	Sand (4% clay, 6% silt, 90% sand)
Very fine sand amount	16%
Inherent organic matter	2.5%
Soil structure	Coarse (1–2 mm)
Permeability	Moderate
Hydrologic class	Lowest run-off
Rock cover	0%

The site is completely devoid of vegetation and has significantly elevated levels of arsenic, lead and cadmium in the spoil material. In order to simulate regional differences in climate-induced changes to erosivity, climatic data for the South East region were applied to the same case study site.

11.3.4.2 UKCIP02 precipitation predictions

The UKCIP02 scenarios (Hulme *et al.*, 2002) predict that a large proportion of the UK will experience a small reduction in annual average precipitation of between 0 and 15% by the 2080s, depending on scenario. There are likely to be larger seasonal and regional differences compared to current weather. The seasonal distribution of precipitation is predicted to change, with winters becoming up to 30% wetter and summers up to 50% drier for some regions

Table 11.7 Annual and seasonal precipitation totals for the SW and the SE.

		DJF total[1] (mm)	MAM total[2] (mm)	JJA total[3] (mm)	SON total[4] (mm)	Annual total (mm)
Baseline	SW	412.2	298.8	220.3	324.7	1255.9
	SE	241.1	190.0	151.0	204.4	786.4
l 2020	SW	430.7	294.5	197.7	317.8	1240.7
	SE	253.4	189.6	135.8	198.2	777.1
l 2050	SW	445.2	291.2	179.9	312.5	1228.8
	SE	263.1	189.3	124.0	193.4	769.8
l 2080	SW	459.0	288.0	163.1	307.4	1217.5
	SE	286.3	188.9	101.4	183.8	760.5
h 2020	SW	434.2	293.7	193.4	316.5	1237.9
	SE	255.8	189.5	133.0	197.1	775.3
h 2050	SW	464.6	286.7	156.2	305.3	1212.9
	SE	276.1	188.8	108.0	187.0	760.0
h 2080	SW	503.0	277.8	109.3	291.2	1181.4
	SE	301.8	188.0	76.6	174.2	740.6

Note 1: Precipitation total for December, January and February.
Note 2: Precipitation total for March, April and May.
Note 3: Precipitation total for June, July and August.
Note 4: Precipitation total for September, October and November.

by the 2080s. Wetter winters are the result of an increase in both the frequency and the intensity of precipitation events while drier summers are partly the result of less frequent intense precipitation events. Contrasts between winter and summer in both the precipitation seasonal totals and the frequency of intense precipitation events are consistent with the observed trends in the last few decades (Hulme *et al.*, 2002). Storm erosivities are expected to increase in winter as a result of an increase in winter intense precipitation events. It is predicted that eastern and southern parts of the UK will experience the largest percentage changes in precipitation in both winter and summer.

Table 11.7 presents seasonal precipitation totals for the baseline and the low and high emission scenarios for the two UK regions considered. Figure 11.11 illustrates the impact of these precipitation changes on mean annual erosivity for the two regions.

11.3.4.3 Erosion rates results
The results for the two emission scenarios (Figure 11.12) showed a significant but gradual increase in erosion rates with time as a result of climate change predictions. Erosion rates were strongly correlated to mean annual erosivity values. Erosion rates for the South West region were between one and a half times and double those for the South East region, demonstrating how contrasting regional variation in predicted changes in climate could significantly influence soil erosion processes and highlighting the need for

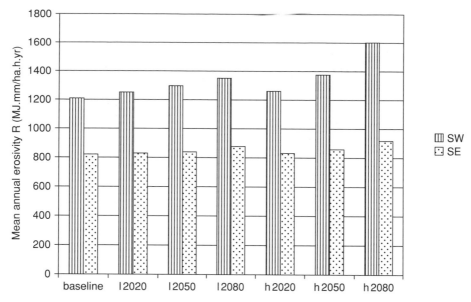

Figure 11.11   Mean annual erosivities for the SW and the SE.

modelling at a localised scale. Taking the average across the low and high emission scenarios, results for the case study site suggested that soil erosion rates could increase by up to 21.9% for the South West region and 4.2% for the South East region by the 2080s, showing a significant regional variation.

11.3.4.4 Role of greenspace for mitigation against pollutant linkage through soil erosion

Modelling the impact of vegetation establishment for a specific site requires that the vegetation type is consistent with the location's climate, irrigation, soil type and fertility, pest control and other management conditions. Key parameters when modelling soil erosion include plant height, rooting habit, canopy cover, live surface cover (the portion of the soil surface covered by live plant parts that touch the soil surface) and long-term biomass yield. Simulated vegetation establishment on the case study site using RUSLE2 (Figures 11.9 and 11.10; Table 11.8) showed that for a well-established vegetation cover there would be a highly significant decrease in erosion rates, equivalent to two orders of magnitude compared to those for the bare spoil (Figure 11.13). Using actual contaminant concentrations for the case study site spoil, estimates of the mass of contaminants which would create a pollutant linkage via sediment production can be calculated using a simple mass balance:

$$M = C \times A \times 10^{-3}$$

$$(11.2)$$

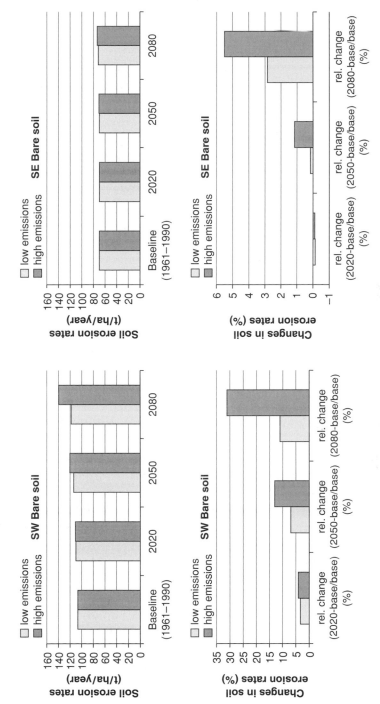

Figure 11.12 Bare soil absolute and relative erosion rates for the SW and the SE. Soil erosion rates refer to parameter A in Figure 11.9.

Table 11.8   Values for the vegetation description used in the RUSLE 2 model to simulate the greening characteristics.

Variable	Value
Residue type	Permanent vegetation/fescue, from RUSLE2 core database
Biomass to yield ratio	1
Above ground biomass at max. canopy cover (kg/ha): • for standard yield: • for ½ standard yield:	 3900 (Ditsch and Collin, 2000) 1950
Root mass in top 100 mm of spoil (kg/ha)	Computed by RUSLE2 growth chart, adjusted to plant characteristics (for yield, fall height, senescence and retardance level)
Canopy cover (%)	Computed by RUSLE2 growth chart
Fall height (m)	Computed by RUSLE2 growth chart, adjusted to plant characteristics
Live surface cover (%)	Computed by RUSLE2 growth chart
Senescence	RUSLE2 core database values, adjusted for canopy cover
Retardance level	RUSLE2 core database values, adjusted for yield
Long-term natural roughness (mm)	6 (Foster, 2004)

where M represents the mass of contaminants released via sediment production (kg/ha/yr), C the contaminant concentration in spoil material (mg/kg), A the average annual soil loss estimated by the RUSLE2 model (t/ha/yr) (see equation in Fig. 11.9) and 10^{-3} a conversion factor.

Owing to the homogeneous nature of the spoil material (mine wastes), contaminants were considered uniformly distributed throughout it. Accordingly, contaminant concentrations were assumed to be homogeneous within the spoil material. This assumption was useful as RUSLE2 did not estimate the depth of soil erosion. Concentrations used for estimating mass of contaminants released via sediment production were onsite concentrations measured in the surface material (0 to 10 cm) of the spoil. The worst case climate scenario (high emissions) for the 2080s showed for the SW a 31% increase in arsenic mobilisation from 3.6 to 4.8 t.ha^{-1}.yr^{-1} for a bare spoil heap. For the same scenario, revegetation of the heap showed a dramatic reduction (by two orders of magnitude) in the amount of metals mobilised in the sediments, reducing the mobilisation well below existing levels (Figure 11.14).

As a remedial option for sites without organic-matter-rich soil materials, species should be chosen which are most suited to low spoil-fertility and

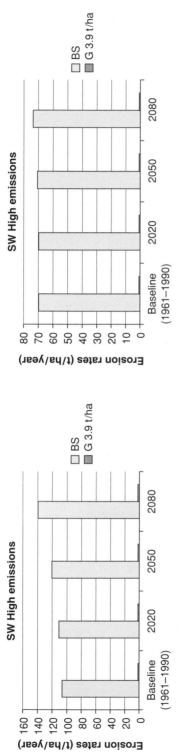

Figure 11.13 Impact of soil cover management options for high emissions scenarios.

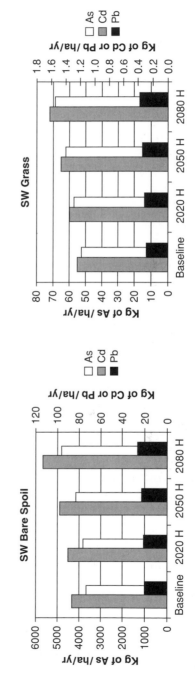

Figure 11.14 Metals mobilisation in the sediments before and after vegetation establishment.

water-holding capacity (Moffat and McNeill, 1994). Modelling demonstrated that erosion rates took several years to stabilise during the establishment phase of revegetation until the maximum grass cover was stable at 83% ground cover. Establishment of perennial grasses for which winter die-back is minimal, as well as adoption of best practice for timing of site operations (tillage, addition of soil-forming materials, sowing, installation of irrigation systems, etc.) will maximise the rate of vegetation coverage over the bare soil/spoil. Some options, such as choosing optimal plant varieties, may be inexpensive while others, such as introducing irrigation, involve major investment. However, it is clear that choice of vegetation must, itself, consider likely effects of climate change on species suitability. In the South East especially, UKCIP02 predictions identify a severe reduction in summer rainfall, so perennial species should be chosen that are comparatively drought-tolerant. Deep-rooting species are better adapted to summer drought, and this habit will also aid in soil erosion protection. In summary, vegetation establishment may represent a cost-effective remedial action to lessen or overcome the adverse effects of climate change on pollutant linkage through soil erosion.

11.4 Climate change mitigation and adaptation

In the previous part of this chapter it was seen that certain climate change scenarios are expected to have significant impacts on current and future contaminated land and containment systems. The next section will initially further explore the differences between mitigation and adaptation, but focuses first on how an adaptive strategy may be developed for remediation in the context of climate change, and second on how key stakeholders (developers and planners) in the UK have reacted to climate change in the development process, which includes brownfield regeneration.

11.4.1 What is mitigation and adaptation?

The IPCC (IPCC, 2001, Annex B) defines **mitigation** as 'an anthropogenic intervention to reduce the *sources* or enhance the *sinks* of *greenhouse gases*'. This emphasises the essentially reactive role of human society to the problems of climate change and an acknowledgement that human actions are a major cause of the current pace and direction of climate change. The same report has a more extensive definition of **adaptation**, starting with the statement that adaptation involves 'adjustment in natural or human systems in response to actual or expected climatic *stimuli* or their effects, which moderates harm or exploits beneficial opportunities', and going on to identify various types of adaptation (i.e. anticipatory and reactive adaptation, private and public adaptation, and autonomous and planned adaptation).

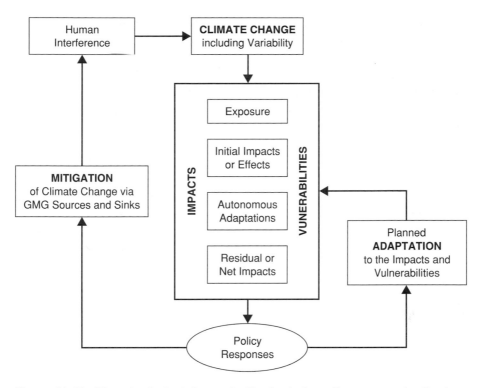

Figure 11.15   The role of adaptation and mitigation in the policy response to climate change (Smit *et al.*, 1999).

In the governmental and academic literature, it is the concept, policy and practice of adaptation that has increasingly taken centre field. While mitigation is important, it is adaptation that is being addressed here. As Smit *et al.* (1999, p. 224) suggest, 'adaptation to climate change and its impacts is receiving increasing attention as an alternative or complementary response strategy to reducing net emissions of greenhouse gases (termed "mitigation" in the climate change community)'. Adaptation focuses on planned, long-term and preparatory changes to socio-economic systems in order to respond to the impacts and implications of climate change. However, both the IPCC and its academic adherents are quick to point out that both concepts are part and parcel of an 'integrated' approach to climate change, as illustrated in Figure 11.15.

11.4.2 Who needs to mitigate and adapt?

Given the clear messages that came out of the Rio Earth Summit in 1992 (UNCED, 1992) and subsequent governmental and non-governmental literature on the topic, it is not surprising that the call to action on climate change mitigation and adaptation is broad in its range of targets. The recent

Stern report (Stern, 2006) sent out a wake-up call to government organisations (at national and international levels) but many 'implementation plans' at the regional or local level specify particular roles and responsibilities that go well beyond the government sector. The implementation plan for south-east England (SEERA 2006), for instance, includes mitigation and adaptation measures to be carried out by

- national policy makers/central government (e.g. Defra, ODPM, DTI, HM Treasury, DfES)
- national regulators/advisors (e.g. Environment Agency, Natural England, English Heritage, HSE, Ofwat, Met. Office)
- regional planners and policy makers (e.g. the Regional Assembly/GOSE/ SEEDA)
- local authorities (specifying a range of service areas)
- health service providers (e.g. Primary Care Trusts)
- business and commerce (including representative organisations, e.g. CBI, chambers of commerce)
- insurance industry
- construction, development and design industry (e.g. architects, landscape architects, engineers, home builders, trade and professional bodies)
- agricultural and forestry industry and landowners (e.g. NFU, farmers, Forestry Commission)
- land and heritage managers
- utilities – water industry, energy providers, telecommunications industry

In the remaining sections of this chapter the current and future adaptation strategies of different stakeholder groups is given some attention, both in terms of specific technical measures that can be developed and in broader strategic terms. In order to address the research findings from the scientific work reported so far, certain organisational and policy issues need to be tackled. Section 11.5 focuses on the appropriate technical and organisational responses relevant to the range of stakeholders involved in remediating contaminated land while section 11.6 concentrates on the particular role of the development industry and local authorities.

11.5   Technical adaptation and risk management strategies

11.5.1   Adaptation strategy

The current approach to risk assessment and management for brownfield sites is based upon the *source–pathway–receptor* model of pollutant linkages (Defra and EA, 2002). By definition there is no risk if there is no pathway for contamination to reach a receptor, even if the contamination exists on the

Table 11.9   Conceptual model of pollutant linkages.

Sources	Pathways	Receptors
Polyaromatic hydrocarbons Located predominantly in the central area at former storage tanks	Direct contact Groundwater Perched water	Humans Canal water
Phenols and organics containing nitrogen, oxygen and sulphur Mainly near storage tanks, but also dispersed	Direct contact Groundwater Perched water	Humans Canal water Building materials
Arsenic and zinc Dispersed in varying concentrations	Direct contact Groundwater Perched water	Humans Canal water Building materials
Sulphate and sulphide Dispersed over the site, variation in levels	Direct contact Groundwater Perched water	Humans Canal water Building materials
Cyanide Mainly near to storage tanks, but dispersed across the site	Direct contact Groundwater Perched water	Humans Canal water

site. Adaptation strategies need to modify and develop the site-specific approach in order to address the climate change issues discussed earlier. The key items to include are as follows:

- The impact of climate change should be based upon site-specific situations as opposed to defining the same response for all brownfield sites. This approach is in keeping with the UK's approach to assessing and managing the risks on any brownfield site.
- The adaptation strategy for each site will take as its starting point the current situation based upon good practice for risk assessment and management as outlined in CLR 11 (Defra and EA, 2002).
- The impact of climate change should be addressed through the use of a conceptual model of pollutant linkages (see Table 11.9) at each stage.
- The impact of climate change should then use the UKCIP scenarios and focus on the actual location of the site and its climate factors.

The adaptation strategy requires a detailed adaptation methodology, as set out in the following section.

11.5.2 Adaptation methodology

The methodology for adaptation to climate change for brownfield sites is set out in four stages. Stage 1 represents the current situation with regard to risk assessment and stage 3 represents the current approach to risk management including remediation of the site. It is expected that experienced consultants will be involved in the risk assessment and management process, and they will be familiar with and able to apply best practice. Stages 2 and 4 represent the additional issues that need to be considered to manage climate change impacts. It is important that the methodology works sequentially through from stage 1 to stage 4, but, as experienced practitioners will be aware, there is often a need for iteration between the various components.

11.5.2.1 Stage 1: risk assessment based on current situation – key requirements
The following steps need to be completed at this stage:

- Carry out a risk assessment, use CLR 11, using specific models or guidance to determine risks to humans (e.g. CLEA models), plants, water and property (e.g. BRE Special Digest 1 (BRE and The Concrete Centre, 2005)).
- Develop a conceptual model of pollutant linkages as part of risk assessment, including determining all sources, pathways and receptors.
- Determine and report the risks of historic contamination on the site, including the conceptual model of pollutant linkages.

This stage is based upon the current approach to risk assessment; experienced and fully qualified practitioners will need to be involved at this stage.

11.5.2.2 Stage 2: risk assessment based on climate change
The following steps need to be completed at this stage:

- Determine UKCIP climate change data, using all four scenarios for a particular location and considering the periods for 2020, 2050 and 2080 using the scenario data.
- Make a qualitative assessment of the impact on sources, pathways and receptors, and determine potential impacts on the pollutant linkages in the conceptual model.
- Re-address the quantitative risk assessment through, for example, changing input parameters to human health risk assessment, or address guidance on the impact of contamination on building foundation materials.
- Redefine the conceptual model of pollutant linkages based upon the periods 2020, 2050 and 2080. Redefine soil levels on the site based on the climate change risk factors from UKCIP and adjust site trigger levels as appropriate. Compare revised values with soil guideline values for human health, water, buildings and other receptors.

- Report the results and the revised conceptual model of pollutant linkages.

11.5.2.3 Stage 3: risk management current position

The current risk management needs to be based on a thorough risk assessment (stage 1). The following steps need to be completed at this stage:

- Use a technology-based approach, either excavation and removal, containment or treatment for the source and pathways. For property, address the materials used to ensure that receptors are resistant to the contamination.
- Non-technical measures for managing sites may be used, or ongoing monitoring or maintenance may be required.
- Follow best practice in risk management and determine verification requirements for the remedial work.

11.5.2.4 Stage 4: risk management based on climate change

The starting point is risk management using the current approach presented in stage 3 and all other information gathered during stages 1 and 2. The following steps will then need to be considered at this stage:

- If stage 2 has demonstrated a potential increase in the risk then reassess the risk management options in stage 3. If no greater risk is perceived, no action is required.
- If the risk is determined as being greater, undertake further assessment of the remediation options. Use modelling, if possible, of the additional risk from climate change over 2020, 2050 and 2080. Alternatively use qualitative judgement to address the requirements of remediation options.
- Any remediation technology that removes, destroys or permanently changes the contamination, without a time-dependent factor, will remove the climate change risk. The use of such technology will not therefore be subject to the impact of climate change.
- Containment and some treatment technologies will need to be addressed as to the impact of climate change. Over time, changes to the materials used in cover systems (as shown in section 11.2), slurry walls and geo-membranes from temperature and moisture changes are likely to be significant. Design changes to the technologies or a change of technology may be required; for example, excavation and removal may be required in place of the use of containment in some situations.

11.5.3 Case study: Greater Manchester site

This case study illustrates the methodology for the technical response to climate change for brownfield land. This site, which the owners requested should remain anonymous, was a former fuel distribution depot, and previous to that a tar works and brickworks in the Greater Manchester area, but had been derelict for a number of years. It is approximately 2 hectares and is generally flat but slopes slightly down to a canal, approximately 150 metres away. The site is located within an area of mixed industrial and commercial land use, and is surrounded by roads on three sides and a railway line on the fourth side, with the canal beyond.

The site was considered for remediation and redevelopment. The proposed remediation method involved placing a cover system over the contaminated soil in situ on the site. The proposed redevelopment of the site included construction of a small industrial building, a car park and some landscaping.

11.5.3.1 Stage 1: current risk assessment: case study

11.5.3.1.1 Hazard identification
The hazard identification stage demonstrated that the following contamination was present:

- hydrocarbons, including polycyclic aromatic hydrocarbons (PAHs)
- phenols
- nitrogen-, oxygen- and sulphur-containing organic compounds
- heavy metal ions (arsenic and zinc)
- sulphate and sulphide
- cyanide

11.5.3.1.2 Hazard assessment
Site investigations were undertaken to determine the extent and nature of contamination present in the groundwater and soils on the site.

11.5.3.1.3 Risk estimation
The conceptual model for the site shows the pollutant linkages as set out in Table 11.9.

11.5.3.1.4 Risk evaluation
Table 11.10 shows the summarised test results from the site and the assessment criteria for the three identified receptors.

The risk evaluation criteria demonstrate that there is a risk to human health and to the canal water course. Both of these risks are based on sufficient exposure of the receptor to the source of contamination.

For building materials the main contaminant of concern was sulphate. However, the levels in the groundwater were found to be below 1200 mg/l and therefore not a risk to concrete.

Table 11.10   Test results and site assessment criteria.

Contaminant	Assessment criteria (mg/kg) Human health		Assessment criteria (mg/l) Water	Assessment criteria (mg/kg) Building materials	Recorded groundwater concentrations across sampled areas (mg/kg)		
	CLEA SGVs[1]	RBCA[2]		SD1[3] BR255[4] EA[5]	Highest value	Mean	True mean US95
Inorganic							
Arsenic	20	Nc	0.01	Nr	30	20	27
Cadmium	1–8	Nc	0.005	Nr	40	30	35
Chromium	130	Nc	0.1	Nr	200	150	185
Lead	450	Nc	0.015	1000	500	450	490
Mercury	8	Nc	0.002	Nr	15	10	13
Selenium	35	Nc	0.05	Nr	50	40	47
Nickel	50	Nc		Nr	60	55	58
Asbestos	Nc	Nc	7×10^6 fibres/l	Nr			
Cyanide	Nc	6.5×10^{-5}	0.2	Nr			
Sulphur	Nc	Nc		Nr			
Sulphate	Nc	Nc	250	1200	936	815	902
pH	Nc	Nc	6.5–8.5	<7	4.5	4.7	4.6
Organic							
PAH Total	Nc	Nc		Nr			
Individual PAHs							
Naphthalene	Nc	48	0.002		60	50	57
Acenaphthylene	Nc	26	0.002		30	20	27
Acenaphthene	Nc	280	0.002		300	200	270
Fuorene	Nc	>120	0.002		150	123	142
Phenanthrene	Nc	>230	0.002		250	224	242
Anthracene	Nc	>6.4	0.002		10	5	8
Fluorathene	Nc	>78	0.002		90	70	83
Pyrene	Nc	>61	0.002		70	50	63
Benzoanthracene	Nc	4.3	0.002		10	7	9
Chrysene	Nc	2.7	0.002		5	3	4
Benzo(b)fluorathene	Nc	4.3	0.002		10	7	9
Benzo(f)fluorathene	Nc	43	0.002		50	35	43
Indeno1,2,3-cdpyrene	Nc	4.3	0.002		10	7	9
Dibenz(a,h)anthracene	Nc	>11	0.002		17	10	15
Benzo(g,h,i)perylene	Nc	>11	0.002		20	17	19

Nc – not considered; nr – not relevant

[1] Defra and EA, [2] EA, [3] BRE and The Concrete Centre, [4] BRE, [5] BRE
Nc: not considered, Nr: not relevant. BR255: Performance of building materials in contaminated land, CLEA: Contaminated Land Exposure Assessment, EA: Environment Agency, RBCA: Risk Based Corrective Action, SGV: Soil Guideline Values

11.5.3.2 Stage 2: climate change risk assessment – case study

The key impacts from climate change on the risk assessment are likely to be experienced by greater loading of groundwater by increased rainfall. In addition, an increased risk of flooding is likely to surcharge the groundwater on sites at a more frequent interval.

Table 11.11 Climate change conceptual model of pollutant linkages.

Sources	Pathways	Receptors
Polyaromatic hydrocarbons Located predominantly in the central area at former storage tanks	Direct contact – increased risk due to soil drying and cracking; critical by 2050 Groundwater – increased movement due to rain and flood; critical by 2050 Perched water – increased movement due to rain and flood; critical by 2050	Humans – greater exposure potential Canal water – increased rate of leaching
Phenols and organics containing nitrogen oxygen and sulphur Mainly near storage tanks, but also dispersed	Direct contact – increased risk due to soil drying and cracking; critical by 2050 Groundwater – increased movement due to rain and flood; critical by 2050 Perched water – increased movement due to rain and flood; critical by 2050	Humans Canal water Building materials
Arsenic and zinc Dispersed in varying concentrations	Direct contact – no change Groundwater – increased movement due to rain and flood; critical by 2080 Perched water – increased movement due to rain and flood; critical by 2080	Humans Canal water
Sulphate and sulphide Dispersed over the site, variation in levels	Direct contact – no change Groundwater – increased movement due to rain and flood; critical by 2080 Perched water – increased movement due to rain and flood; critical by 2080	Humans Canal water
Cyanide Mainly near to storage tanks, but dispersed across the site	Direct contact – no change Groundwater – increased movement due to rain and flood; critical by 2080 Perched water – increased movement due to rain and flood; critical by 2080	Humans Canal water

Higher summer temperatures and a greater risk of prolonged drought give rise to potential for release of PAHs from the soil and direct contact with human receptors.

The sources, pathways and receptors of contamination remain the same as at present. However, there is greater risk to the receptors as set out in Table 11.11.

The revised conceptual model demonstrates that there are increased risks. The risks are indicated as critical by either 2050 or 2080; however, some impact can also be expected by 2020 in all identified instances.

11.5.3.3 Stage 3: current risk management – case study
It was decided to leave the contamination in situ, placing a cover system over it based on the results of stage 1. This would allow redevelopment of the site to be undertaken above the cover system.

The cover system would break the pathway for direct contact between the contaminated soil and human receptors. It would also act to prevent increasing surcharging of the perched water or groundwater from heavy winter rain by 2050 and increased flood incidents. The design of the cover system conformed to the requirements of BRE Report 465 (BRE *et al.*, 2004).

The design considerations of the cover system had indicated that a cover layer thickness of at least 600 mm should be used across the site, as follows:

- a compacted clay soil layer of thickness 400 mm, covering:
- a capillary break layer or clean uncontaminated broken concrete or masonry to a depth of 200 mm, covering:
- the contaminated soil

The small industrial building would be built directly on the cover system with the direct incorporation of a geomembrane across the concrete floor slab to prevent ingress of soil vapours or gases. The car park and access road areas would be covered with tarmacadam, and drainage used to prevent flooding of the site in heavy rain and to remove the run-off to storm drains and sewers.

The landscaped areas of the site would have topsoil applied to green the area, which should help to minimise erosion to the cover system (section 11.3). The landscaped areas would also serve to remove water in times of high rainfall, but this would be retained within the upper layer of the cover system rather than the contaminated layer.

11.5.3.4 Stage 4: risk management based on climate change – case study

Table 11.12 shows low and high emissions climate change scenarios for 2080 and how they may impact on the use of the cover system. Each of the factors in Table 11.12 is likely to become more significant at 2050 and 2080, although some impact will be experienced as early as 2020. The impact of increased rainfall on run-off and erosion and the drying of the cover system in the drought conditions could be modelled using models of erosion (see section 11.3) and soil-drying (see section 11.2). The qualitative assessment of climate change risk (Table 11.12) shows that there is additional risk by the erosion of the cover system and potential opening of the pathway for contamination to cause harm to receptors.

For this site the increased risk is particularly associated, however, with the landscaped areas as opposed to the built-up areas or car-parking areas.

Potential responses to the impact of climate change on the cover system are as follows:

- Excavate and remove contamination instead of using a cover system. However, as there is an increased flood risk the use of the 600 mm or

Table 11.12 Climate change risk management for the case study.

	2080 low emissions scenario	2080 high emissions scenario	Impact on the cover system
Mean annual temperature	+2.5°C	+4°C	Increased drying potential and erosion
Winter temperature	+2°C	+3.5°C	Increased drying potential and erosion
Summer temperature	+3°C	+4.5°C	Increased drying potential and erosion
Mean annual precipitation	+3%	+5%	Increased erosion due to runoff
Winter precipitation	+15%	+30%	Increased erosion due to runoff
Summer precipitation	−15%	−45%	Increased drying potential and erosion
Mean winter wind speed	+6% possibly	+6%	Increased drying potential and erosion
Sea level	+18 to +990 mm (Irish Sea)	+18 to +990 mm (Irish Sea)	Increased erosion and mixing potential with contaminated ground
Flood risk	1 in 25 year flood risk	1 in 5 year flood risk	Increased erosion and mixing potential with contaminated ground

greater cover will help to alleviate potential flood impact on the contaminated soil and the new developments on the site.

- At the building and the car-parking areas the 600 mm depth would be sufficient as drying and erosion will not be an issue. For the landscaped areas the potential impact of drying and erosion will be greater but vegetation should provide additional protection to the cover system. Additionally the depth of the cover system could be increased.
- Use a suitable treatment technology (such as bioremediation) to remove the PAHs prior to applying the cover system. This will reduce the concentrations below guideline values for risks to humans and the water course.

11.5.4 Implications for brownfield development

There is uncertainty in both brownfield land assessment and management and also in climate change scenarios. Putting both these aspects together will result in potentially increased uncertainty. In the absence of a technical response strategy and methodology, the response of assessors may either be an overly conservative approach or may ignore the issue on the grounds of uncertainty. A conservative approach may be to use dig and dump rather than containment, which will have implications for the sustainability of the site redevelopment. Alternatively, ignorance of the risks will leave an ongoing liability for the site and its users. The strategy and methodology set out here has the potential to be used to reduce risks and to improve communication between stakeholders.

It should be remembered that the costs of dealing with contamination and other ground issues in the redevelopment of a brownfield site are estimated to be approximately 10% of the total costs. Of course costs are highly site- and project-specific and will depend on a range of factors. Costs will clearly be involved as a consideration; however, a whole life costing approach is likely to be more useful in justifying any additional costs.

From a development perspective, as a rule of thumb the costs of the remediation or the costs of the redevelopment project should not increase by more than 10% and 2% respectively after climate change adaptation is taken into account. If one of these figures is exceeded then the assessor should address again the management of risk and the precise remediation technology being used. There may be situations where it is considered that cost increases are necessary, but they should only exceptionally exceed the development values created.

11.6 Stakeholder adaptation key issues and findings

Two stakeholder groups given prominence in government reports on adaptation to climate change are local authorities and the development industry. These two groups are crucial to the production and management of the built environment and to the take-up and implementation of climate change mitigation and adaptation measures. This is especially true in relation to strategies and measures that might be applied to the remediation of contaminated land, and they were the focus of the SUBR:IM survey work in this area.

The research on stakeholder adaptation policy and practice involved a questionnaire survey of local authorities in England and Wales and national surveys of all relevant sectors of the development industry (residential and commercial developers, lenders, investors and agents). With 103 and 121 responses respectively, the surveys provide a reasonably robust information base from which to explore how climate change mitigation and adaptation is being dealt with by these key stakeholders, especially in relation to the under-researched areas of land contamination and remediation.

11.6.1 Development industry response to climate change

As suggested above, most commentators now acknowledge that climate change will impose new stresses on both natural and socio-economic systems and that these systems will need to adjust through a process of 'adaptation' (Berkhout *et al.*, 2004). It will be important for organisations to achieve changes to routines if they are to survive the impacts of climate change (Nelson and Winter, 1982). Indeed, Berkhout *et al.* (2004) suggest that in a study of nine companies in the UK (including the water sector and the housebuilding industry), business faces key problems in recognising signals relating to climate change impacts and evaluating the costs and benefits of particular adaptation strategies. As the authors point out, there is often a tendency in commercially driven organisations to 'wait and see' and react to events rather than change practices or cultures. This is also true of the development industry in relation to climate change and its effect on remediation, as will be seen.

In terms of the broader picture of the use of remediation techniques by developers, related SUBR:IM survey work suggested that dig and dump, containment and stabilisation/solidification continue to predominate. In the initial survey of developers conducted during 2004, it was found that dig and dump and barrier/containment methods of remediation were still the most common (see Chapter 5). This is illustrated in Figure 11.16. 'Dig and dump' is certainly the most frequently used method of clean-up, with an average score of 3.6. In fact 31% of commercial developers and 67% of housebuilders 'always' or 'often' use this method. This figure is much higher than that for the next most popular method, containment, where the equivalent figures are 15% for commercial developers and 28% for housebuilders. Commercial developers are more likely than housebuilders to have experimented with alternatives to these two processes and generally appear to have a wider awareness of the different technologies available (Figure 11.16). However, even among commercial developers awareness of thermal processes and, perhaps less surprisingly, emerging technologies such as electrokinetic extraction was low. Although the impact of the EU Landfill Directive is now shifting the emphasis towards in situ methods (see Chapter 5), there is still a continued focus on stabilisation/solidification and containment, which has significant implications in terms of climate change, as we have already seen in the technical part of this chapter.

There are some signs of technical innovation, however. In the housebuilding industry, company size clearly makes a difference in the likelihood of having used alternatives to 'dig and dump', although this was not apparent among commercial developers. The volume housebuilders (producing over 2 000 units per year) are much more likely than smaller operators to use alternative solutions and on average claim to use barrier methods, bioremediation and stabilisation/solidification (S/S) slightly more frequently

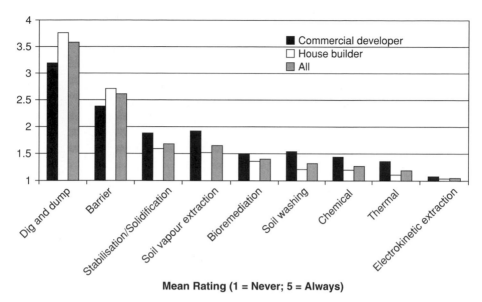

Figure 11.16 Frequency of use of different remediation technologies (by type of company).

than 'dig and dump'. This differential by company size can probably be attributed to resources, access to expertise and the type of development undertaken. A survey of local authorities and consultants on the subject of land remediation has also been undertaken (see Chapter 9). This also revealed that 'dig and dump' was the most frequently used method of clean-up, although barrier/capping came a close second. As found in the survey of developers, these two methods were the most widely used, although local authorities and consultants appear to have greater experience in the use of bio/phyto-remediation than developers.

Given this knowledge of remediation techniques, a further survey of developers was carried out in 2005. This was part of a larger project on flooding (Kenney *et al.*, 2007). A total of 1231 questionnaires was sent out, with 121 responses being received back, which represents a response rate of 9.8%. The relative perceptions of the respondents towards environmental risk suggest that contamination is more important than flood risk, subsidence or storm damage.

When the responses of the 'developers' group (some 43 responses) towards climate change are examined, it suggests that the potential impact of climate change on site remediation is not considered to be as substantive an issue as in the masterplan/site layout, building design, construction and choice of materials phases of the building life cycle (Figure 11.17). However, there were some group differences, with residential developers placing more importance on remediation impacts than choice of materials or the construction process.

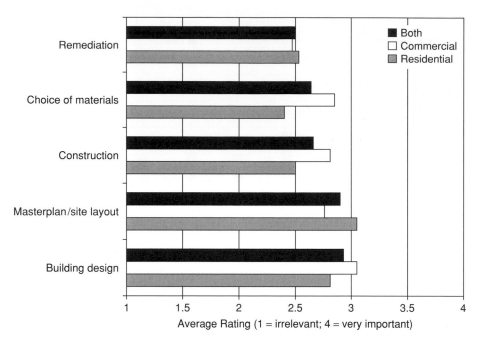

Figure 11.17   Climate change and impact on stages of the development process.

Similarly, most developers believed that subsidence, flooding and storms were more important than either higher temperatures or the increased risk of remediation schemes failing (Figure 11.18).

Finally, in relation to remediation options, developers suggested that there was still some concern over the issue of future climate change. They would therefore be more likely to either reject a particular option and use an alternative, or switch, if there were no additional costs. This suggests that developers are currently cost-driven in this respect.

To follow up the survey work, six interviews were also conducted with three practitioners and three developers. Generally, and unsurprisingly, the level of knowledge regarding the impact of climate change on remediation was greater among the first group than among the latter. As one consultant put it:

My own personal view is that there are still a lot of question marks about the use of cover systems and the retention of contaminated materials on site and I suspect that in a number of cases not adequate consideration is given to the real cost of ensuring long-term durability of those systems.

Several developers were aware of potential future problems, but tended to treat the issue as connected to wider concerns over flooding. As one developer suggested:

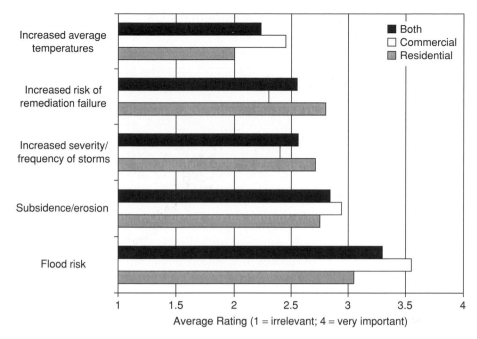

Figure 11.18   Type of climate change impact.

No, we would certainly be concerned . . . [about] the impact . . . climate change would have, if it's going to have an impact on the methodology for carrying out remediation, it is probably going to have a more direct impact on the development itself anyway. So you know so we would be planning for whatever the impact of it was: it has a secondary impact on the methodology for remediating the site – well therefore you know the methodology's got to be changed as well.

The same developer continued:

for instance, you know most of the issues to do with climate change are obviously to do with flood risk and flood risk assessments, so therefore . . . you'd be looking at all of the problems you have and then it would be an holistic approach to the design solution.

11.6.2 *The response of local authorities to climate change and remediation*

A number of recent surveys (e.g. Allman *et al.*, 2004; Demeritt and Langdon, 2004; LGA, 2004; SNIFFER, 2005) have shown that the local authority response to climate change is slowly developing, but is still highly variable. This was confirmed in the responses to the survey undertaken for this research, which sought to explore the particular adaptation measures being initiated in relation to land remediation.

In general terms, the survey revealed that just over half of local authorities have a dedicated climate change officer and a similar proportion consider themselves adequately informed of climate change impacts. Significantly fewer (42%) are confident about their knowledge of climate change adaptation practices. There is an encouraging sign that local politicians are becoming more aware of climate change issues, with 65% of respondents acknowledging that local authority members were giving it more priority, albeit from a low and variable base level. Despite this increase in interest, only 36% of our respondents' authorities had signed the Nottingham Declaration on Climate Change and 23% had an adopted climate change strategy. In terms of the provision of information on climate change, that originating from government agencies (including UKCIP) and regional networks is perceived as most reliable and widely used. This has implications for the effective dissemination of information on contaminated land remediation processes and appropriate adaptation measures.

As Figure 11.19 illustrates, land remediation is seen as a relatively important issue in relation to climate change impacts, even though it is given a lower priority than some more obvious issues such as flooding

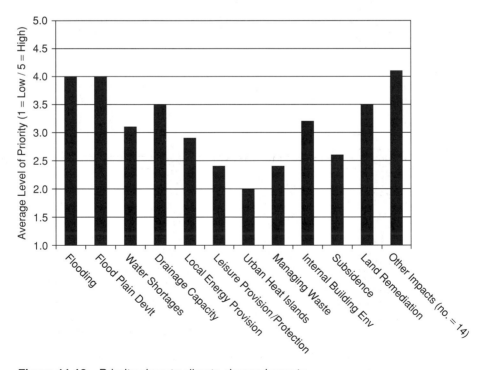

Figure 11.19   Priority given to climate change impacts.

and flood plain development. Notwithstanding this, taking that concern through into action has been less notable, with only about 10% of local authorities undertaking an appraisal of the robustness of contaminated land remediation measures and just 15% adopting measures to improve the robustness of past remediation works. In a more positive light, about a third of our respondents said that they were *considering* introducing measures to improve the robustness of past remediation (see Figure 11.20).

In a similar vein, many local authorities are considering changing specific mechanisms to assist in climate change adaptation and these are illustrated in Figure 11.21. They clearly show the potential role of the planning system

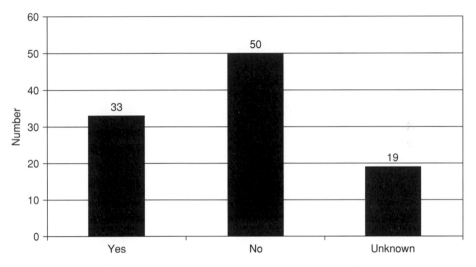

Figure 11.20   Local authorities who are considering measures to improve the robustness of past remediation.

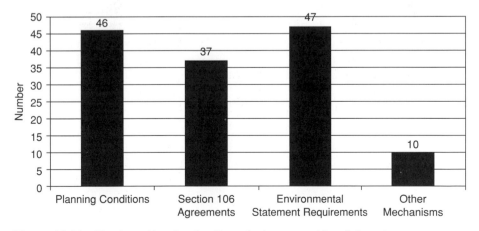

Figure 11.21   Number of local authorities who have considered changing mechanisms in relation to contaminated land remediation.

in making improvements to the land remediation process through the imposition of more stringent conditions on planning permissions, the use of legal agreements with developers or strengthening the requirements in environmental assessment. This opportunity for local authorities to (re)shape land remediation processes is in keeping with the government's vision of integrated 'spatial' planning (ODPM, 2004), although the restructuring is not without its tensions and problems (see Doak and Parker, 2005; Raco *et al.*, 2006).

11.6.3 General remarks

The review of policy and practice within the development sector and local authorities suggests that there is a growing awareness of the generalities and specific implications of climate change for land remediation policy and procedure. However, that awareness and action has risen from a very low base, and many developers and local authorities are still largely ignorant of the issues and inactive on developing strategies and mitigation/adaptation measures to deal with climate change. Although land remediation is given some priority in the (long!) list of relevant issues, most developers and local authorities are currently operating according to a 'business as usual' scenario. Overall the survey findings support the view that 'there is still a long way to go in delivering successful adaptation on the ground, and allowing adaptation to be incorporated into all major decisions' (Defra, 2005, p. 6).

11.7 Conclusions

From the published literature, some of which was presented in this chapter, and the findings from the experimental work and numerical modelling reported in sections 11.2 and 11.3 respectively, it is clear that certain climate change scenarios are expected to have significant impacts on current and future contaminated land and containment systems. Examples include severe physical damage to soil cover systems and stabilised/solidified soils and extensive soil water erosion and associated contaminant transport. These impacts will have major effects on the future management of contaminated and remediated sites and are expected to influence the way risk is managed on those sites and the design of future remediation strategies. A conceptual adaptation strategy has been developed and presented in section 11.4, highlighting four stages to be considered when addressing the impact of climate change in the current risk-based contaminated land management regulatory framework in the UK, together with a case study illustrating how to apply the developed methodology to a contaminated site.

Finally, the results of the surveys carried out on a range of stakeholders, namely the development industry and local authorities, reported in section 11.5, clearly demonstrate that these stakeholders are still largely unaware of the issues surrounding climate change and its impact on contaminated land management and redevelopment, and are hence not yet considering potential impacts of climate change and related evidence in their decision-making process. This has inherent dangers associated with it.

In conclusion, the authors argue that this chapter has identified three principal findings that have relevance for further research on climate change and sustainable urban brownfield development. First, it has shown that it is important for scientists and technical experts not only to develop new ways of monitoring and measuring the effects of change on physical processes and containment technologies but also to develop new modes of communication and knowledge transfer. The chapter has gone some way towards this end by developing models of risk assessment and analysis, but future work could examine new modes of working and develop new vocabularies and knowledge sets to encourage better communication. Second, the chapter has indicated that multidisciplinary research can provide new insights into the processes underpinning sustainable brownfield development, particularly in a context where political, social, economic and environmental influences are subject to significant change and variation over time. The authors have drawn on a variety of methods and ways of working to demonstrate the limited awareness that policy makers, developers and others have of the longer-term impacts of climate change on contaminated urban sites. Finally, the chapter has also shown that multiple understandings and interpretations of urban sustainability, risk and climate change exist among different stakeholders. The pre-eminence of short-term perspectives that prioritise profit-making and political popularity could undermine the longer-term sustainability of remediation and renewal investments. New ways of thinking about risk and planning are urgently needed if climates continue to change as predicted.

Acknowledgements

The authors would like to acknowledge the contribution made by the work package steering group members: Jan Hellings (BURA), Joanne Kwan (CIRIA), Richenda Connell and Courtney Blodgett (UKCIP). Tim Dixon would like to acknowledge the input of former colleagues (Jude Shephard, Sarah Kenney, Mike Waters and Yasmin Pocock) at the College of Estate Management, Reading, on the surveys that underpinned the 'developer response' part of this work package.

References

Albrecht, B.A. and Benson, C.H. (2001) Effect of desiccation on compacted natural clays. *Journal of Geotechnical and Geoenvironmental Engineering*, **127**, 67–75.

Alexander, M. (1995) How toxic are toxic chemicals in soil? *Environmental Science & Technology*, **29**, 2713–17.

Allman, L., Fleming, P. and Wallace, A. (2004) The progress of English and Welsh local authorities in addressing climate change. *Local Environment*, **9.3**, 271–83.

Bazzaz, F. and Sombroek, W. (1996) *Global Climate Change and Agricultural Production: Direct and Indirect Effects of Changing Hydrological, Pedological and Plant Physiological Processes.* John Wiley and Sons Ltd, Chichester, UK.

Benson, C.H. (1999) Final covers for waste containment systems: a North American perspective. *Proceedings of the XVII Conference of Geotechnics in Torino: Control and management of subsoil pollutants.* November. Turin, Italy.

Berger, K. (2000) Validation of the hydrologic evaluation of landfill performance (HELP) model for simulating the water balance of cover systems. *Environmental Geology*, **39**, 1261–74.

Berkhout, F., Hertin, J. and Gann, D. (2004) *Learning to Adapt: Organisational Adaptation to Climate Change Impacts.* Tyndall Centre Working Paper 47. Tyndall Centre for Climate Change Research, University of East Anglia, Norwich, UK.

Boes, N. and Al-Tabbaa, A. (2001) Long-term durability of in-situ auger-mixed stabilised/solidified contaminated made ground. *3rd BGA Geoenvironmental Engineering Conference: Geoenvironmental Impact Management*, Edinburgh, September, pp. 171–6.

BRE and The Concrete Centre (2005) *Concrete in Aggressive Ground.* BRE Special Digest 1 (BRE SD 1). BRE Bookshop, Watford, UK.

BRE, DTI, NHBC (National House Builders Council) and AGS (Association of Geotechnical & Geoenvironmental Specialists) (2004) *Cover Systems for Land Regeneration: Thickness of Cover Systems for Contaminated Land.* BRE Report 465 (BR 465), BRE Bookshop, Watford, UK.

Chadwick, O.A., Gavenda, R.T., Kelly, E.F., Ziegler, K., Olson, C.G., Elliott, W.C. *et al.* (2003) The impact of climate on the biogeochemical functioning of volcanic soils. *Chemical Geology*, **202**, 195–223.

Defra (2005) *Adaptation Policy Framework: A Consultation by the Department for Environment, Food and Rural Affairs.* Department for Environment, Food and Rural Affairs, London.

Defra (Department for Environment, Food and Rural Affairs) and EA (Environment Agency) (2002) *Contaminated Land Report 11* (CLR 11). Water Research Centre, Swindon, UK.

Demeritt, D. and Langdon, D. (2004) The UK climate change programme and communication with local authorities. *Global Environmental Change*, **14**, 325–36.

Ditsch, D.C. and Collins, M. (2000) Reclamation considerations for pasture and hay lands receiving sixty-six centimeters or more precipitation annually. In: *Reclamation of Drastically Disturbed Lands* (eds R.I. Barnhisel *et al.*). Monogr. 41, pp. 241–71. ASA, Madison, WI.

Doak, J. and Parker, G. (2005) Networked space? The challenge of meaningful community involvement in the new spatial planning. *Planning Practice and Research*, **20**, 23–40.

Foster, G.R. (2004) *User's Reference Guide for RUSLE2* (Draft). USDA-Agricultural Research Service, Washington, DC.

Foster, G.R., Toy, T.E. and Renard, K.G. (2003a) Comparison of the USLE, RUSLE1.06c, and RUSLE2 for Application to Highly Disturbed Lands. *Proceedings of the 1st Interagency Conference on Research in the Watersheds*, 154–60.

Foster, G.R., Yoder, D.C., Weesis, G.A., McCool, D.K., McGregor, K.C. and Bingner, R.L. (2003b) *User's Guide for RUSLE2* (Draft). USDA-Agricultural Research Service, Washington, DC.

Franzluebbers, A.J., Haney, R.L., Honeycutt, C.W., Arshad, M.A., Schomberg, H.H. and Hons, F.M. (2001) Climatic influences on active fractions of soil organic matter. *Soil Biology and Biochemistry*, **33**, 1103–11.

Hulme, M., Jenkins, G.J., Lu, X., Turnpenny, J.R., Mitchell, T.D., Jones, R.G. *et al.* (2002) *Climate Change Scenarios for the United Kingdom: The UKCIP02 Scientific Report.* Tyndall Centre for Climate Change Research, School of Environmental Sciences, University of East Anglia, Norwich, UK.

International Panel on Climate Change (IPCC) (2001) Climate change 2001: impacts, adaptation, and vulnerability. Contribution of Working Group II to the *Third Assessment Report of the Intergovernmental Panel on Climate Change.* Cambridge University Press, Cambridge.

Kenney, S., Pottinger, G., Plimmer, F. and Pocock, Y. (2007) *Flood Risk and Property Impacts on Commercial and Residential Stakeholders' Strategies.* College of Estate Management, Reading.

Korodjouma, O., Badiori, O., Ayemou, A. and Michel, S.P. (2006) Long-term effect of ploughing, and organic matter input on soil moisture characteristics of a Ferric Lixisol in Burkina Faso. *Soil and Tillage Research*, **88**, 217–24.

LGA (2004) *Sustainable Energy and Climate Change: A Survey of Local Authorities.* Local Government Association, London.

Moffat, A.J. and McNeill, J.D. (1994) *Reclaiming Disturbed Land for Forestry.* Forestry Commission Bulletin 110. HSMO, London.

Morgan, R.P.C. (2005) *Soil Erosion and Conservation*, third edition. Blackwell Publishing, Oxford UK.

Nelson, R. and Winter, S. (1982) *An Evolutionary Theory of Economic Change.* Harvard University Press, Cambridge, MA.

ODPM (2004) *Creating Local Development Frameworks: A Companion Guide to PPS 12.* Office of the Deputy Prime Minister, London.

Papatheodorou, E.M. (2004) Responses of soil chemical and biological variables to small and large scale changes in climate factors. *Pedobiologia*, **48**, 329–38.

Perera, A.S.R., Al-Tabbaa, A., Reid, J.M. and Johnson, D. (2005) State of practice report UK stabilisation/solidification treatment and remediation, Part V: Long-term performance and environmental impact. *Proceedings of the International Conference on Stabilisation/Solidification Treatment and Remediation*, 12–13 April, Balkema, Cambridge, UK, pp. 437–57.

Raco, M., Parker, G. and Doak, J. (2006) Reshaping spaces of local governance? Community strategies and the modernisation of local government in England. *Environment and Planning C: Government and Policy*, **24**, 475–96.

Renard, K.G. and Freimund, J.R. (1994) Using monthly precipitation data to estimate the *R*-factor in the revised USLE. *Journal of Hydrology*, **157**, 287–306.

Renard, K.G., Foster, G.R., Weesies, G.A., McCool, D.K. and Yoder, D.C. (1997) *Predicting Soil Erosion by Water: A Guide to Conservation Planning with the Revised Universal Soil Loss Equation (RUSLE).* Agriculture Handbook 703. USDA-Agricultural Research Service, South-West Watershed Research Center, Tucson, AZ.

SEERA (2006) *Climate Change Mitigation and Adaptation Implementation Plan for the Draft South-East Plan*. South-East of England Regional Assembly, Guildford.

Smit, B., Burton, I., Klein, R.J.T. and Wandel, J. (1999) An anatomy of adaptation to climate change and variability. *Climatic Change*, **45**, 223–51.

SNIFFER (2005) *A Survey of Scottish Local Authority Activities on Climate Change*. Scotland and Northern Ireland Forum for Environmental Research, Edinburgh.

Sowerby, A., Emmett, B., Beier, C., Tietema, A., Penuelas, J., Estiarte, M. *et al.* (2005) Microbial community changes in heathland soil communities along a geographical gradient: interaction with climate change manipulations. *Soil Biology and Biochemistry*, **37**, 1805–13.

Stern, N. (2006) *Stern Review on the Economics of Climate Change*. HM Treasury, London.

Toy, T.J. and Foster, G.R. (1998) *Guidelines for the Use of the RUSLE Equation, Version 1.06, on Mined Lands, Construction Sites, and Reclaimed Lands*. OTT, Western Regional Center, Office of Surface Mining, Denver, CO.

Twining, J.R., Zaw, M., Russell, R. and Wilde, K. (2004) Seasonal changes of redox potential and microbial activity in two agricultural soils in tropical Australia: some implication for soil-to-plant transfer of radionuclides. *Journal of Environmental Radioactivity*, **76**, 265–72.

UNCED (United Nations Conference on Environment and Development) (1992) *Agenda 21*. United Nations, New York.

Van Rompaey, A.J.J., Vieillefont, V., Jones, R.J.A., Montanarella, L., Verstraeten, G., Bazzoffi, P., Dostal, T., Krasa, J., De Vente, J. and Poesen, J. (2003) *Validation of soil erosion estimates at European scale*. EUR 20827 EN, Office for Official Publications of the European Communities, Luxembourg.

Wagner, J.F. and Schnatmeyer, C. (2002) Test field study of different cover sealing systems for industrial dumps and polluted sites. *Applied Clay Science*, **21**, 99–116.

12

Evaluating the Sustainability of Brownfield Redevelopment Projects

Kalliope Pediaditi, Walter Wehrmeyer and Kate Burningham

> The only way in which a human being can make some approach to knowing the whole of a subject is by hearing what can be said about it by persons of every variety of opinion, and studying all modes in which it can be looked at by every character of mind. No wise man ever acquired his wisdom in any mode but this; nor is it in the nature of human intellect to become wise in any other manner.
>
> John Stuart Mill, 'On Liberty'

12.1 Introduction

Preceding chapters of this book have explored the 'policy push' for brownfield redevelopment and its relation to the government's sustainability agenda (see Chapters 1–6). From the UK government perspective it can be argued that brownfield redevelopment is seen as inherently 'sustainable'. One key reason for this view is that brownfield redevelopment offers the potential to revitalise communities while simultaneously permitting the use of existing infrastructure and integrating development into the wider urban context (DETR, 1998). This is achieved by means of creating a more spatially integrated, mixed urban environment composed of resource-efficient and high-quality buildings (DETR, 1998). Furthermore, encouraging greater amounts of development on previously developed land helps to reduce pressure on greenfield sites as demand for housing is projected to grow from 20.2 million in 1996 to 24 million households by 2021 (ODPM, 1999). For these reasons, the government has stipulated that a minimum of 60% of new housing has to be built on brownfield sites.

However, the unequivocal equating of 'brownfield' with 'sustainability' is more problematic than it would at first appear. First, it is specious to assume that a redeveloped site is 'sustainable' simply because its old land use does not exist anymore, or that all traces of it have been removed. In addition, it asserts that the new land use must be 'more' sustainable than the old use, which, in the absence of methodologically sound sustainability indicators, must remain a matter of conjecture. Furthermore, Deakin and Edwards (1993) as well as Imrie and Thomas (1993) argued that few critical checks are undertaken to ensure that regeneration projects are sustainable. Housing on brownfield sites has often been of poor quality (Adams and Watkins, 2002), with buildings on brownfield regeneration projects (BRPs) having low environmental efficiency (Ball, 1999), in many cases inferior to that of much greenfield development. In addition, where redevelopment includes remediation, risks that arise from this process need to be evaluated, managed and placed in the context of the beneficial future land use.

Decision makers hence need to understand and improve the sustainability of brownfield developments across their life cycles. However, in order for them to do this we argue that they require better sustainability monitoring support tools. A large number of indicator systems exist to perform parts of this function; for example, SUE-MoT (2004) lists more than 600 sustainability indicator tools, few of which have found practical application (Mitchell, 1996; Innes and Booher, 2000; Deakin *et al.*, 2002; Bell and Morse, 2003). Therefore, rather than inventing another (underused) indicator system, we have developed the Redevelopment Assessment Framework (RAF) to dovetail, as much as possible, with existing decision-making and land-use policy parameters, with as much practitioners' influence as possible.

In this chapter we will offer a guide to the RAF, and report on the experience of its application on a brownfield development. The chapter is set out in three main parts. In the first, we review a number of existing sustainability assessment tools to examine their use and the barriers to their adoption. In the second part, we describe the RAF and offer a step-by-step guide for its application. In the final part, we report on a pilot of the RAF on a brownfield development project, which was evaluated by stakeholders in the brownfield development process.

12.2 Sustainability evaluation in brownfield projects

As indicated, there is currently a wide variety of sustainability indicators, most of which have been developed in an ad hoc way without adequate attention being given to how these would be embedded into the decision-making process. Innes and Booher (2000, p. 174) characterised the mushrooming of sustainability evaluation tools as follows:

[T]his movement is developing so quickly that little has as yet been published documenting, much less critically evaluating, these experiments
or assessing their impact. The internet is a much better source than the
library for finding out about much of this work, although its descriptions
are sketchy and reflect the image each group wants to offer.

It is as yet unclear which set of tools will prevail and little is known about
the extent of their use (SUE-MoT, 2004 and Deakin *et al.*, 2002). Therefore,
adopting an *a priori* approach, the essential characteristics of such a tool in
relation to brownfield redevelopment were established:

- *Holistic*: The environmental, social and economic aspects of BRPs must
  be assessed in an integrated manner that still allows the evaluation of
  trade-offs between these aspects.
- *Site- and project-specific*: The sustainability of the totality of the site,
  its conditions and the project must be assessed, as opposed to limiting
  the focus to single buildings, the organisation conducting the project or
  wider, regional concerns.
- *Long term*: The sustainability of a BRP needs to be measured throughout
  its land-use life cycle, spanning the planning and design phase, the construction and remediation phase and eventual operation phase.
- *Participatory*: Evaluation users as well as relevant stakeholders need to
  make their values and risk perceptions explicit as well as to develop
  their own sustainability indicators based on these.
- *Integrated within existing decision-making processes*: It is essential to
  dovetail new tools with existing (compulsory as well as voluntary) activities and policies.

(See Pediaditi *et al.* (2006a) for more details.)

As part of the SUBR:IM research to develop the RAF, the 27 most widely
cited and appropriate sustainability evaluation frameworks were examined
to establish whether they fulfilled the criteria set out above (Pediaditi *et al.*,
2006b). This was followed up with 41 semi-structured interviews of BRP
stakeholders (see Table 12.1) to establish current practice as well as potential barriers to uptake. In addition, questions were included on a large-scale
survey of volume housebuilders to identify the development industry's
current evaluation and monitoring practices, working closely with other
researchers in the SUBR:IM programme (see Chapter 5).

This review found that despite the plethora of existing sustainability
assessment and monitoring tools, there are none directly applicable to
BRP, despite the clear need for such a system. In particular, there is no tool
designed to holistically evaluate the sustainability of remediation and
reclamation schemes (Bardos *et al.*, 1999; Rudland and Jackson, 2004).

Table 12.1 Stakeholders interviewed and questions asked.

Stakeholders interviewed	Number (n = 41)	Main questions asked
Contaminated land relevant: EA, EH, NHBC, remediation contractor*	6	• How and to what extent can you influence the sustainability of a BRP?
Building control	1	• Do you and the processes you use monitor the sustainability of a BRP?
LA sustainability manager	2	
LA senior Development Control (DC) officers	2	• Do you and the processes you use assess the sustainability of BRP?
LA policy and regeneration managers	5	• What do you perceive to be the benefits of assessing and monitoring the sustainability of BRP?
Other LA internal planning consultees	2	• What are the barriers to the sustainability assessment and monitoring of BRP?
Private planning and sustainability consultants	3	
Architects and designers	3	• Recommendations for the assessment and monitoring of BRP?
Developers	10	
Sustainability assessment tool developers	5	* The above questions were put to contaminated land professionals with regard to contamination monitoring in addition to sustainability monitoring and assessment of BRP
Government policy EP Millennium Communities	2	

Although theoretical frameworks exist for remediation option selection (for example, Harris *et al.*, 1995), they have limited practical application as remediation decisions are currently based on Defra and EA (2004) model procedures for managing land contamination, which do not consider sustainability. Furthermore, these frameworks do not assess the sustainability of different options, nor do they allow an assessment of site-specific remediation processes[1] or post BRP completion.

There is, then, currently no tool which is capable of assessing the sustainability of a BRP throughout its land-use life cycle (Pediaditi *et al.*, 2005). The tools that currently exist are neither holistic nor long term, tending to focus on building performance and environmental issues either during construction or with respect to future land use. Yet defining and evaluating sustainability involves complex and typically value-based decisions, and is thus a social, participatory process in itself (Ukaga and Maser, 2004). However, the review of existing tools revealed that they tend to be top-down, leading to limited ownership by users, and typically fail to involve the public affected by the development proposals. This is more serious with respect to redevelopment projects as process transparency is a key criterion of 'good' participatory decision-making as well as risk communication (Wehrmeyer, 2001).

With regard to the integration of existing planning and decision-making processes into these tools, the review found that this is not being addressed sufficiently, if at all (see Pediaditi *et al.*, 2006b). That is, apart from the South-East England Development Agency (SEEDA) Sustainability Checklist, all are developed by consultancies or are copyright protected, thus making them expensive to use. It is for this reason that the RAF only uses existing tools which disclose their indicators and benchmarks on which the BRP evaluation would be based.

Further work was carried out to scope out these issues. A wider set of interviews with key stakeholders (including developers) as well as a survey of 987 developers (see Chapter 5 for more details), for example, showed low levels of use (and awareness) of existing indicator frameworks. These pointed to a set of profound barriers against the implementation of sustainability indicators (summarised in Table 12.2). The most significant of these barriers were procedural in nature, such as shortages of time, resources and expertise. Interviewees also felt that sustainability assessments tend to lack a structured process for their operation, even when making development control and planning application decisions. A broad cross section of interviewees identified problems of communication between, and even within, the development industry and the public sector. This lack of communication often led to extended project expenditure as well as delays. Many commented on the lack of integration of indicator tools within the planning and development process, expressing a desire for such tools to be integrated with existing processes such as Strategic Environmental Assessment (SEA), Environmental Impact Assessment (EIA), or monitoring and evaluation procedures specific to some local authorities. In general, developers were found not to be opposed to sustainability assessment and monitoring in principle. Indeed many welcomed the idea on the basis that it would provide a more equitable and predictable assessment of planning applications.

Table 12.2 Barriers to BRP sustainability evaluation.

Barriers to adoption	Procedural limitations	Tool limitations
• Lack of understanding of sustainability • Lack of market demand • Lack of enforcement/ resources and skills • Too many tools resulting in lack of confidence in them • 'Build and forget' development culture	• Lack of time • Lack of a structured process to follow • Lack of communication • Lack of ownership of the assessment process • Lack of integration of existing tools with planning processes e.g. planning application process, EIA, SEA, SA	• Scope of assessments limited to building performance • Scope of assessments mostly covering environmental issues • Lack of context-specific assessments • Lack of measurable benchmarks • Output approach to monitoring

Additionally, local authority officers suggested that their ability to require such assessments to be undertaken would be limited and sporadic without their integration into existing planning processes.

The review therefore concluded (a) that there exists the need for an effective and practical sustainability assessment framework for BRP; and (b) that such a framework should be simple, structured, cost-effective and integrated within existing planning and development processes. It must assess and monitor the environmental, social and economic implications of a site's redevelopment throughout its life cycle. The framework must be flexible and participative, and allow contextualisation and ensure that public perceptions of risk are taken into account as far as possible. These principles formed the design specifications upon which the RAF is based.

12.3 The Redevelopment Assessment Framework

The overall aim of the RAF is to inform stakeholders about the sustainability profile of a site across its life cycle in a way that is practical and integrated within existing BRP processes. It is a process to facilitate the development of site-specific sustainability indicators in a participatory manner. However, the RAF is not designed to make decisions about the viability of a project, its fit with existing policies or its general suitability for a site. In other words, the RAF process is a supplement to, not a substitute for, the planning application process. It is also not designed to compare different development proposals or assist in the design of these. The RAF is directed mainly at large or complex developments which require an EIA or Statement of Community Involvement (SCI). However, a balance needs to be struck between starting the RAF early, and having sufficient clarity and certainty about what the future site and its land use should look like.

To overcome local authority resource limitations, the RAF is undertaken by the developer/owner (here called the lead partner). As it is aimed at large developments, it is envisaged that the developer would hire a facilitation consultant to coordinate the RAF process. To ensure the process does not delay the planning application process, the RAF has been devised so that it can be undertaken in a minimum of two half-day stakeholder workshops and one meeting (which assumes adequate background research by the lead partner). In this respect, the lead partner (either the developer or the consultant) should have a 'basic understanding' of sustainability, be seen to be independent, and have demonstrated facilitation training and design skills.

To ensure that the recommendations that arise from the RAF process are implemented and post-development monitoring takes place, it uses S106[2] agreements and planning conditions which are determined in the initial planning phase of a BRP. As illustrated in Figure 12.1, the RAF consists of a simple procedure divided into six steps, through which site-specific indicators can be developed. The first three steps cover the preparatory stages undertaken by the lead partner and include information-gathering and stakeholder identification to enable the subsequent participatory development of indicators. Figure 12.1 shows the process in schematic form.

In order for its usefulness to be evaluated, the RAF was piloted on a case study site. The project covered a large mixed-use brownfield redevelopment in the Greater Manchester area consisting of approximately 520 residential units, a school and some employment units. The site was contaminated from its past use as a paper mill as well as from a nearby landfill site that was in the process of being restored. More recently the site had also included the development of a high school and associated playing fields. Where appropriate, references to outputs from the pilot study are made to illustrate outcomes or practical aspects of administering the RAF.

12.3.1 Step 1: team-building

The first step in the RAF process involves the lead partner identifying and selecting all relevant stakeholders involved in the BRP for an 'evaluation task force' (which should primarily include the evaluation users) (Patton, 1997). A list of potentially relevant stakeholders can be found in Box 12.1, which can be used in conjunction with the following questions (Environment Council, 2002, p. 6):

- Who is directly responsible for the decisions on the issues?
- Who holds positions of responsibility in stakeholding organisations?
- Who is influential in the area, community or organisation?
- Who will be affected by any decisions around the issue?
- Who will promote a decision – provided they are involved?
- Who will obstruct a decision – if they are not involved?
- Who has been involved in the site in the past?
- Who has not been involved up to now – but should have been?

This is an inherently subjective and sensitive process. Therefore the lead partner needs to consult with the development control officer to ensure that an equitable, appropriate, manageable and functioning set of representative stakeholders is selected. The list should then be circulated to all identified stakeholders as part of the invitation to participate, with the request to suggest other stakeholders that may have been omitted.

However, it must be stressed that there is no infallible method of selecting a representative group of stakeholders, which contains all relevant perspectives within a manageable group. This is because some sites are more complex, diverse or politically sensitive than others. However, facilitation

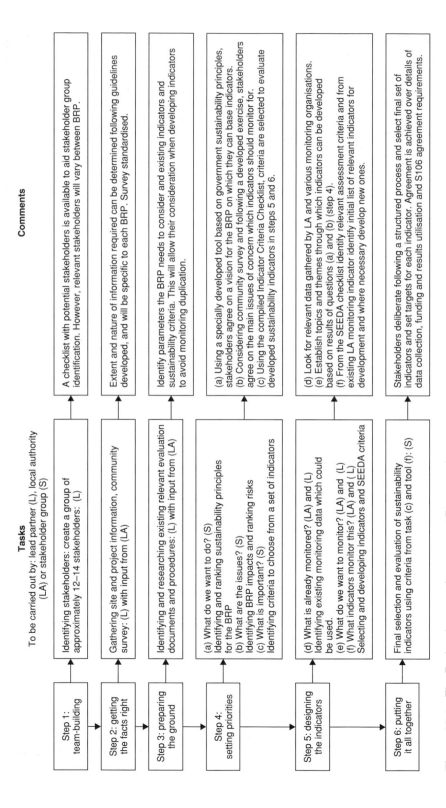

Figure 12.1 The Redevelopment Assessment Framework.

Box 12.1   Potential Stakeholders for the RAF.

- Building control
- Development Control (including regeneration officer)
- Sustainability officer LA21
- 'Environmental Health Officer' (incl. contaminated land officer)
- Health and Safety Executive (HSE)
- Highways
- Councillors
- National House Building Council (NHBC)
- National Health Service (NHS)
- Environment Agency (EA)
- Police
- Utility regulators
- Service providers
- Architect
- Engineer (incl. remediation consultant)
- Contractors
- Landowner
- Developer
- Investors
- Partners

- End users, occupants/residents
- Housing Associations
- Banks/financial institutions/insurers
- Aid/grant providers (e.g. English Partnerships)
- NGOs
- Central government departments
- Regional authorities (RDAs)
- Statutory and non-statutory consultees
- Residents' associations
- Community groups/pressure groups
- Individuals
- Business groups
- Building site operatives
- Visitors/workers on commercial sites
- Neighbours
- Purchasers or tenants
- Development surveyor (incl. planning consultant)
- Cost consultant
- Estate agent
- Community liaison officer
- Lawyer

Source: Pediaditi *et al.* (2006a)

---

guidance (Environment Council 2002; IEMA, 2002) suggests that in situations that require specific questions and detailed tasks to be undertaken in a limited time frame, small groups of 10–15 individuals are preferable. This is an artificial target, but the point remains that more consultees do not necessarily lead to better decision-making. As long as larger stakeholder groups such as local residents have effective representation (for example, through accepted and functioning local residents' associations), there is no reason why, for the majority of cases, the consultation group cannot be kept small. Essentially the RAF is an evaluation process which uses participatory methods, and this set-up clearly limits public participation to representation. The RAF incorporates methods for community representation as well as mechanisms for information exchange (Figure 12.2) which are elaborated throughout this chapter.

To ensure that the process is meaningful and its outcomes are sustainable, it is strongly recommended to include the following participants as a minimum:

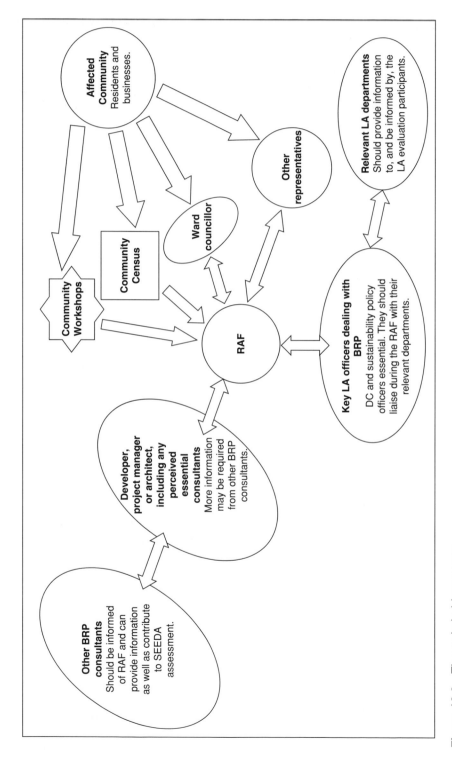

Figure 12.2 The stakeholder process.

- *Developer(s):* they are needed to fund the RAF, and their presence is necessary to ratify any decisions as well as enable the participation of private consultants. Given that it is 'their' project, they also have a moral duty/right to lead, let alone be involved.
- *Architect or project manager(s):* they are needed to provide insight into the nature of the development as well as to act upon changes which may emerge from the RAF process.
- *Councillor(s):* they are required to democratically represent local community views.
- *Sustainability or relevant policy officer(s):* at the minimum, they need to facilitate step 3, namely to identify relevant monitoring information and baselines, and to ensure that indicators will feed into policy.
- *Development Control Officer(s):* they lead the statutory consultation process and deal with the planning application, and can thus advise on stakeholder selection as well as coordinate the S106 agreements to ensure that monitoring takes place.

12.3.2 Step 2: getting the facts right

In order to make informed decisions about the likely impacts of a development and thus to select relevant monitoring indicators, step 2 involves two information-gathering tasks: first, the lead partner needs to collate information on the BRP proposal; second, the community needs to be consulted to identify and document the main concerns and aspirations of those most likely to be affected by the development.

12.3.2.1 Task 1: gathering relevant information
The list of potentially useful information about sites is vast. It is recommended here that the Environmental Impact Statement (EIS) review criteria (Tables 12.3 and 12.4) are used as guidance as to what information should be collated. The lead partner should produce a non-technical summary to circulate to all participating stakeholders.

The information requirements specified in Tables 12.3 and 12.4 do not add additional financial or time burdens for such sites, as this information would be required for large development planning applications anyway. Therefore the RAF uses existing information for the purpose of developing indicators.

12.3.2.2 Task 2: consulting the community
The second task in the second stage of the RAF involves consulting the community with regard to their views on the proposed development, focusing strongly on their aspirations and sustainability principles for their area. This information should guide the evaluation task force in step 4

Table 12.3   Sample Information to describe the development.

Criteria*	
Principal features of the project	• Explains the purposes and objectives of the development. • Indicates the nature and status of the decision(s) for which the information has been prepared. • Gives the estimated duration of the construction, operational and where appropriate, decommissioning phase and the programme within these objectives. • Provides a description of the development comprising information on the site, design and size of the development. • Provides information with regard to the influx of people, number of jobs, resulting from the project. • Identifies the impact of the development on services, e.g. public transport, schools, health care. • Provides diagrams, plans or maps and photographs to aid the description of the development. • Describes the methods of construction. • Describes the nature and methods of production or other types of activity involved in the operation of the project. • Describes any additional services (water, electricity, emergency services, etc.) and developments required as a consequence of the project. • Describes the project's potential for accidents, hazards and emergencies.
Land requirements	• Defines the land area taken up by the development and/or construction site and any associated arrangements, auxiliary facilities and landscaping areas and shows their location clearly on a map. For a linear project, describes the land corridor, vertical and horizontal alignment and need for tunnelling and earthworks. • Describes the uses to which this land will be put and demarcates the different land use areas. • Describes the reinstatement and after-use of land taken during construction.
Project inputs	• Describes the nature and quantities of materials needed during the construction and operation phases. • Estimates the number of workers and visitors entering the project site during both construction and operation. • Describes their access to the site and likely means of transport. • Indicates the means of transporting materials and products to and from the site during construction and operation and the number of movements involved.
Residues and emissions	• Estimates the types and quantities of waste water, energy (noise, vibration, light, heat radiation, etc.) and residual materials generated during construction and operation of the project, and rate at which these will be produced. • Indicates how these wastes and residual materials are expected to be handled/treated prior to release/disposal and the routes by which they will eventually be disposed of to the environment. • Identifies any special hazardous wastes which will be produced and describes the methods for their disposal as regards their likely main environmental impacts.

* Criteria have been added to describe the socio-economic aspects of the development.
Source: adapted from Weston (2000a)

Table 12.4 Sample Information to describe the development's environment.*

Description of the area occupied by and surrounding the project	• Indicates the area expected to be significantly affected by the various aspects of the project with the aid of suitable maps. Explains the time over which these impacts are likely to occur. • Describes the land uses on the site(s) and in surrounding areas. • Describes the area with regard to unemployment, crime and considers the effect the development is likely to have on the area. • Identifies whether existing services and facilities e.g. schools, recreational, retail, have the capacity to accommodate development impacts. • Defines the affected environment broadly enough to include any potentially significant effects occurring away from the immediate areas of construction and operation. These may be caused by, for example, the dispersal of pollutants, infrastructural requirements of the project, traffic.
Baseline conditions	• Identifies and describes the components of the affected environment potentially affected by the project. • Uses existing technical data sources including records and studies carried out for environmental agencies and for special interest groups. • Reviews local regional and national plans and policies and other data collected as necessary[1]. Where the proposal does not conform to these plans and policies the departure is justified.

* It is recommended that baseline conditions relevant to all policies are reviewed based on Annual Monitoring Reports, Sustainability Appraisals and SEA reports. It is recommended that baseline conditions relevant to all policies are reviewed based on Annual Monitoring Reports, Sustainability Appraisals and SEA reports.
Source: adapted from Weston (2000b) with criteria added to describe the socio-economic aspects of the development

when developing sustainability indicators to focus on the priority issues of the community.

At a minimum, the consultation requires a questionnaire (see Pediaditi, 2006), which should be sent out to all residents in the catchment area (as specified in the relevant regulations on planning application consultation). The census, as opposed to a (stratified) sample, is essential to give everybody the opportunity to participate. The lead partner is responsible for funding and carrying out the survey, but this should be done in close consultation with the relevant local authority. This is a cost-effective and straightforward process as addresses are logged electronically on all local authority geographical information systems (GIS). However, consideration should be given to whether the survey should be extended to a wider area.

The survey should cover standardised specific questions for all BRP. It should be devised so that its results can feed into step 4 easily. Briefly, it should identify the major impacts of the scheme and prioritise the sustainability objectives undertaken in the step 4 workshop (see below). This allows the stakeholders in the workshop to compare the differences between their perceptions and aspirations for the site and those of the community. Thus, it is proposed that in future RAF applications the survey questions are not modified but are used as a standard format.[3] We suggest that the questionnaire, as a minimum, should cover the following sections:

(a) the significance of the development on specific impact categories (for example through an EIA scoping study) supplemented with a qualitative question which justifies the score provided
(b) open-ended questions to identify the main concerns about the development and aspirations for the area
(c) a ranking exercise of general sustainability principles, based on government sustainability principles and modified from Dair and Williams (2004) (see below)
(d) a Likert scale question to gauge their perceived risks from the development
(e) further Likert scale questions about their satisfaction with the consultation they have received to that point as well as the extent to which they feel positively or negatively affected by the proposed development

This survey, however, should not be mistaken for a two-way communication and does not replace community input into the planning application process. Best practice community consultation literature proposes the use of a variety of methods, including community workshops, open days and local press publicity (Sanoff, 2000; SNIFFER, 1999). Therefore RAF best practice includes separate workshops which are carried out in combination with the survey in order to obtain community input (see above). It is suggested that community workshops follow the same format as those conducted in step 4 (see below) to ensure that the results can feed into subsequent steps.

Since the implementation of the Planning and Compulsory Purchase Act 2004, developers are required to provide Statements of Community Involvement (SCI) to demonstrate that they have consulted with the community on their proposals at a pre-application phase. The RAF is designed to be incorporated as part of the SCI, again minimising the costs involved in applying it as it dovetails as much as possible with existing requirements. This implies that the SCI can be used as an enforcement mechanism by local authorities to require the RAF or similar processes.

The process output of this step 2 is therefore twofold: a non-technical summary with information regarding the site and the proposed development, as well as a report representing the views of the community. Both documents should then be circulated by the lead partner to all the RAF stakeholder participants for their consideration prior to the step 4 workshop (see below).

12.3.3 Step 3: preparing the ground

Many evaluation procedures, frameworks and guidelines exist which are directly or indirectly relevant to a BRP, predominantly in the design and planning phase of a project: SEA, Local Development Framework (LDF) indicators, community strategy indicators, local authority sustainability checklists and funders' sustainability criteria are of particular relevance (Table 12.5). In step 3, the lead partner needs actively to monitor the sustainability profile of the BRP to ensure that the outcomes will be covered by the sustainability indicators the RAF aims to produce. This step helps to avoid data duplication and demonstrates the large and complex volume of information that needs to be collected in the planning process anyway. Table 12.5 shows existing guidelines and policies and how data requirements in these may be used in the RAF.

Table 12.5 serves only as a guide as not all sources are relevant to all projects and the specific requirements for individual BRPs may vary. For instance, a government-sponsored development requires an EIA, which means there are two sources of sustainability-related monitoring: the EIA demands post-monitoring and government funding typically requires meeting additional sustainability criteria. Here, RAF could address them collectively, minimising duplication and overlap. Future policies may well add to Table 12.5, in particular the forthcoming Code for Sustainable Buildings which should also be considered as part of these assessment requirements.

However, there are also some problems with the use of such external indicators. To begin with, as RICS (2003, p. 1) states, 'the public sector holds potentially valuable information but again not in a format that is always conducive to facilitate analysis'. In addition, not all such indicator systems are likely to prevail in the future (or have a sufficiently long record back to show consistency). It is also at times unclear who collects the data, with different departments within local authorities being responsible for collecting different sets of data and, until recently, no centralised system for storing and accessing of information. Different data collectors may have different reasons to collect that data, which may make it biased and thus difficult to use or difficult to compare across local authorities. Equally, the origin, policy context and extent to which they have been developed in a participatory mode matters for their usefulness, acceptability and meaning for the RAF. The lead partner should therefore communicate with the local authority officer(s) responsible for these reports. However, once these processes have been established and all local authorities have collated their monitoring data, step 3 should be a simple matter of downloading these publicly available documents from the local authority website.

Table 12.5 Sources of BRP monitoring requirements, indicators and baselines.

	Monitoring/assessment tools or potential sources of data	How to use/things to consider	Limitations	Case study use
Development sustainability monitoring and assessment requirements.	**Project funders' sustainability criteria or monitoring requirements** (e.g. Millennium Communities fund sustainability criteria.)	Does the project receive external or government funding? If so, are monitoring obligations (benchmarks etc.) attached to the grant? Does it require the disclosure of confidential information?	Funders' monitoring is usually output and financially based which may not be suitable for publicising. Some funders require/use sustainability statements/objectives, which the RAF can provide if it incorporates these in Step 4	N/A
	LA development sustainability checklist.	Does the LA have its own checklist? Benchmarks included? Are the topics changeable? Involve relevant LA officer.	Some LA have checklist; others don't. They vary a lot in structure, requirements and relevance. Some lack benchmarks and read more like wish lists.	The case study LA did not have a sustainability checklist.
	Developers' or partners' own sustainability indicators	Does the developer or partner's company have sustainability? Are they relevant here?	Some of the developer indicators are very general, and focus on the company, not the site.	The developer did not have a sustainability policy/indicators.
	EIA post-monitoring requirements	Does the development require an EIA? What post-monitoring requirements have the different consultants specified for each significant effect?	Timing is an issue. These requirements are known towards the end of the process, and may have to be obtained late in the RAF process. In some cases performance benchmarks are not specified.	EIA long-term monitoring was included in monitoring strategy. RAF ensured consistency, specified benchmarks.
	Other planning application assessments e.g. traffic impacts	Are assessments required for the planning application? Long-term monitoring?	As above.	All assessments were included within the EIA.
	Other? (Code for Sust. Buildings)	N/A	Still to be developed.	N/A
Existing sustainability indicator data and baselines of potential use	**Community Strategy Indicators**	Does the LA have a community/LA21 strategy? Are its indicators consistent? – do they have baselines?	From experience, the quality in particular of the indicators varied. They often report on the LA performance instead of the site.	The group agreed that they were not relevant and lacked baselines.
	SEA & SA LDF indicators	Has an SEA or SA scoping report been carried out for the LDF or a relevant area plan? What are the indicators/are they consistent? (discuss with policy officer)	LA are currently in the process of collating existing monitoring information for the purpose of SEA and SA. However, from discussions with policy officers inconsistencies in data collection were an issue.	SEA and SA indicators were collated for select on in Step 5. Policy officer indicated inconsistencies.
	Annual LDF monitoring report indicators	Has a LDF annual monitoring report been prepared? What are the indicators/which ones are consistent?	As above. Similar if not identical indicators are being utilised for AMR as for SEA and SA. They can change to reflect specific policies (lack of consistency) and are subject to political pressures.	AMR was in process of preparation. However, the indicators were almost identical to SEA and SA indicators.
	Best Value performance indicators	Are the indicators relevant to the state of the environment? Do they only focus on LA performance? If so they shouldn't be used.	Best Value performance indicators often focus purely on LA performance in delivering services and lack focus on baseline conditions – relevance?	BVPIs focused on LA performance and were considered non-applicable.

12.3.4 Step 4: setting priorities

Sustainable development is a value-laden concept which is difficult to define, operationalise and evaluate (see Chapter 13). Yet before we can develop sustainability indicators for a particular site, relevant stakeholders need to define what they interpret sustainability to mean and entail. Therefore step 4 is designed to define the 'sustainability' objectives of the BRP as well as the aspirations and concerns stakeholders have for the proposed development.

In step 4 the identified group of relevant stakeholders (see step 1) is brought together in a workshop where participants consider the results of step 2 (background information and community consultation) and undertake three tasks collectively (Figure 12.1). Although this is a large work programme, experience has shown that it can be done in a minimum of one half-day workshop.[4] The workshop is divided into three sessions to reflect on three tasks, which are described in turn.

Prior to commencing the workshop, agreement needs to be sought regarding the acceptability of the RAF process (in contrast to the wider planning application process), the purpose of the workshop in the RAF context, as well as some ground rules common to participatory workshops.

12.3.4.1 Task 1: identifying a vision, concerns, and benefits

In session 1, the evaluation task force should be split up into small groups, of typically three to six participants each. Participants are then asked to state on 'Post-it' notes their main individual short- and long-term concerns, and their visions for the site and proposed development. These should then be grouped by participants into themes such as design visions, employment visions, environmental visions and so forth. This part is essentially a meta-plan operating at a high strategic level (Environment Council, 2002). Using a Carousel technique (Environment Council, 2002), brief presentations are made on each theme with comments or additions made to these.

Following this exercise, each group makes a presentation on the pertinent topics: that is, on the main concerns, the main benefits and the main visions for the project emerging from the exercise. It is only at this point that the results of the community consultation should be presented as doing so earlier may unduly influence the group. The following discussion is needed to consider possible differences between the community and those of the group. This task should lead to some agreement on the main benefits and concerns, which are then to be monitored.

12.3.4.2 Task 2: prioritising sustainability objectives

In session 2, a prioritisation or voting exercise on general sustainability objectives, such as those presented in Table 12.6, should be undertaken.

Table 12.6   Some sustainability objectives.

Social objectives	Environmental objectives	Economic objectives
• To provide adequate local services to serve the site • To provide a safe environment for people to work and live in • To provide housing to meet needs • Integrate the development within the locality • To provide good accessibility for all • To share benefits and burdens of the development equitably	• To minimise the use of resources • To minimise pollution during construction and use • To remediate existing contamination • To protect biodiversity and the natural environment • To protect the landscape • To protect heritage and historic buildings	• To enable businesses to be efficient and competitive • To provide employment opportunities • To promote the local economy • To provide transport infrastructure to meet business needs • To support local business diversity • To raise land prices and yields

Source: adapted from Dair and Williams (2004)

Posters with economic, social and environmental sustainability objectives should be presented and participants provided with ways to express their priorities for each of these objectives. These could be expressed in the form of coloured dots, voting crosses or, less appropriately, a show of hands. Participants should prioritise between objectives within each objective category to reduce the opportunity for trade-offs between the social, environmental and economic objectives of sustainability, and to ensure an equitable weighting between these three.

A presentation of the results of the same prioritisation exercise undertaken by community consultation respondents and sustainability objectives of the local authority Community Strategy should follow. This allows participants to clearly see possible differences between their sustainability priorities and those of the community. After discussing these differences, an agreement needs to be reached over the main sustainability objectives of the development. At the end of this task, participants can combine the themes of task 1 with the agreed priority sustainability objectives identified in task 2 to produce a list of site-specific sustainability objectives for which they feel indicators should be developed. Box 12.2 shows some objectives developed for the pilot study.

12.3.4.3   Task 3: agree on the nature of the evaluation (procedural issues)

Practical aspects should be addressed with regard to the nature and function of the final indicators. Task 3 consists of an exercise where participants are asked to consider an indicator selection criteria checklist (see Appendix to this chapter). A discussion should follow to develop agreement about the

> **Box 12.2   Example sustainability objectives developed for case study.**
>
> (1) Improved image and integration of the area in terms of architecture, design and social aspect as well as the combination of all.
> (2) A safe environment for people to work and live in.
> (3) Improved education in terms of academic achievement and infrastructure and design.
> (4) Improved local economy, particularly with regard to small businesses and the creation of quality employment opportunities.
> (5) Improved mix between housing and businesses as well as types of housing. The need to create a new housing balance – a property ladder enabling people to stay in the area.
> (6) Improved biodiversity in terms of habitat creation and water management.
> (7) Improved accessibility (traffic management and transport links).
> (8) Ensure safety with regard to contamination.

nature of the indicators. In particular, agreement should be sought on the following questions:

- Who should manage the monitoring process?
- Who should collect the data?
- Who should utilise the results?

These procedural issues will affect the nature of the indicators (or future sets) and make them more prescribed, which, in turn, will increase their practicality, feasibility and ultimately their utility. This task completes step 4, and the facilitator should disseminate the results to the group.

12.3.5 *Step 5: Designing the indicators*

In this step, all the information collated so far should be considered so that an initial set of long-term sustainability indicators can be developed, as well as a selection of sustainability criteria for the assessment of the development proposals. Due to the limited time realistically available for the RAF process, this step can be undertaken in a meeting between the lead partner, his or her planning consultant and relevant development control and policy officers. However, where time and resources permit, the exercise could be undertaken utilising electronic consultation techniques with all participants for comments prior to the meeting. Likewise, a further meeting may be seen as appropriate. In each case, accountability and the need for information sharing with the wider stakeholder group remains with whoever attends that meeting.

Step 5 answers three distinct questions:

(a) What is already monitored?
(b) What do we want to monitor?
(c) What indicators monitor this?

These questions require the following material:

- the step 3 list of relevant indicators and monitoring requirements collated using the checklist outlined above (Table 12.5)
- the updated potential EIA post-monitoring requirements (where relevant)
- the stakeholder-ratified step 4 results report, which contains agreed priority sustainability objectives for which indicators should be developed
- the complete SEEDA or, where available, relevant regional development agency development sustainability checklist
- where remediation has been identified as a priority sustainability objective, RESCUE's remediation sustainability criteria (RESCUE, 2005)

The meeting is usefully divided into two sessions: the first consists of identifying development proposal sustainability assessment criteria; the second deals with the long-term monitoring indicator development. Each session is now described.

12.3.5.1 Session 1: selecting BRP sustainability assessment criteria

In session 1 the thematic topics or priority sustainability objectives identified in step 4 (cf. Box 12.2) requiring development evaluation and monitoring are put to the group for consideration throughout the meeting. The SEEDA development sustainability checklist, which contains a number of predefined sustainability criteria, is put to the group for the selection of criteria which are relevant to the specific BRP's sustainability objectives. Where relevant, the RESCUE contaminated land criteria should be put forward for consideration of the contamination sustainability objective, as the SEEDA checklist does not cover this issue in sufficient detail. Although the SEEDA checklist contains pre-specified criteria, it is provided for five reasons. First, it considers the development as a whole rather than purely building performance. Second, it provides benchmarks relevant to policy and government guidance. Third, it addresses holistically environmental, social and economic issues. Fourth, it requires a justification of the attributed benchmark performance. Finally, the criteria is transparent and use of the checklist is free.

Additionally, the SEEDA checklist is in the process of being launched throughout the English regions and is aimed for use by local authorities, providing a standardised basis for which sustainability evaluations could be undertaken. Although the SEEDA checklist has an interactive website where a development can be scored and overall performance results provided based on pre-set criteria weightings, the web-based element of this tool is not used through the RAF. This weighted web-based method is not endorsed by the authors (for its lack of transparency) and therefore a paper version of the checklist is used for participants to choose criteria which are context-specific.

In summary, session 1 involves the identification of relevant development sustainability assessment criteria from the SEEDA and RESCUE checklist for consideration by the whole group in step 6. It should be highlighted that at this step 5 meeting and at the step 6 workshop there is the flexibility and opportunity to add additional criteria where considered relevant.

12.3.5.2 Session 2: developing long-term BRP sustainability indicators

Session 2 develops indicators to monitor the long-term sustainability of the development. Again, step 4 sustainability objectives should be put forward for consideration: that is, the question 'What do we want to monitor?' together with the list of existing indicators identified in step 3, which deal with the question of 'What is already monitored?'. Session 2 usefully starts by identifying which of the step 3 indicators are relevant by theme to the sustainability objectives, so asking the question 'What indicators monitor this?'. Participants are then asked to examine in more detail the chosen indicators and identify whether they are relevant in scale, timing and so forth, and whether additional indicators are required.

The result is a report listing the sustainability assessment criteria and long-term indicators developed. This report should be presented to the whole stakeholder group for further deliberation (step 6). It should provide specific questions for stakeholders to consider (see Boxes 12.3 and 12.4 for examples of questions) regarding each criterion and indicator. Individual consultations on the report should be provided to the facilitator prior to the commencement of the workshop. The facilitator should collate the consultations and present comments at the step 6 workshop.

12.3.6 *Step 6: putting it all together*

In step 6, the stakeholder group meets for the final half-day workshop to review the proposed indicators and assessment criteria and to (hopefully) agree on a final set and a monitoring strategy. All stakeholders at the time of the workshop should have had the opportunity to provide their individual feedback. To achieve this, step 6 is divided into three structured and facilitated sessions, which are described below.

Box 12.3   Stakeholder task and questions (session 1 of step 6).

Task for stakeholders:
Review long-term indicators presented by answering the following questions in the task boxes. DA = Disagree, N = Neutral, A = Agree.

Indicator review criteria	DA	N	A	Comments
Does this indicator appropriately inform the sustainability objective stated?				
Is this indicator useful i.e. provides information which can be used in decision-making?				
Is it relevant to the sustainability objective stated?				
Is it cost-effective?				
Any other comments?				

Indicator characteristics review	DA	N	A	Comments
Is the *monitoring task* specified relevant and appropriate for obtaining representative information on the stated indicator?				
Do you think the data collection *timing* is appropriate?				
Do you think the *sample is* representative?				
Do you agree with the stated *benchmarks?*				
Do you agree with the *additional information* collected or do you think there should be more?				
Do you have any other comments?				

Box 12.4   Stakeholder task and questions (session 2 of step 6).

Task for stakeholders:
Review the sustainability criteria presented by answering the following questions presented in the task boxes. DA = Disagree, N = Neutral, A = Agree

Indicator review criteria	DA	N	A	Comments
Is it relevant to the sustainability objective stated?				
Does this criterion appropriately assess the sustainability objective stated?				
Do you consider the criterion's stated benchmark is appropriate regarding the proposed development and locality				
During which phase should these criteria be used to assess the sustainability of the development? (please circle one)	Outline planning application			Detailed application
Do you have any other comments?				

12.3.6.1 Agreeing on sustainability criteria and indicators

In sessions 1 and 2 a combined Nominal Group Technique (NGT) (Environment Council, 2002) and Carousel are recommended: the workshop stakeholders are divided into groups of between three and six and seated at different tables. Each group should have a balance of private and public sector participants. Each table requires a stationary participant (dubbed 'the station master') who is selected based on his or her knowledge of the particular sustainability objective. For example, if the sustainability objective is 'effective transport management', the station master should ideally be either a local authority highways officer or the transport private consultant. The station master's role is to facilitate the answering of the questions (Boxes 12.3 and 12.4) with the rotating group participants (Figure 12.3) and then to collate and present the conclusions to the whole group at the end of each session.

At each rotation the new group, guided by the station master, needs to answer the questions listed in Boxes 12.3 and 12.4 for each of the presented

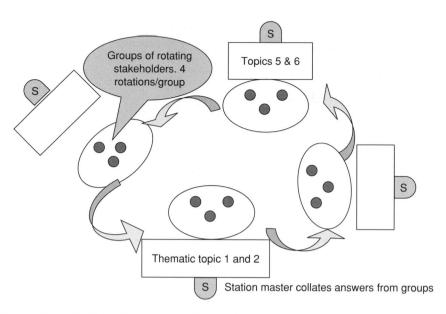

Figure 12.3 Participation technique for sessions 1 and 2 of step 6.

long-term indicators and criteria, add new ones if appropriate, and make a final selection. The rotations ensure that all stakeholders, except the station masters, have an opportunity to deliberate on all indicators for all sustainability objectives. Station masters can change between sessions 1 and 2. Keeping within agreed time limits is critical and difficult. After the rotations, station masters make a short presentation on the collective results, which is followed by a discussion that focuses particularly on any remaining areas of disagreement.

12.3.6.2 Session 3: agreeing on the procedural issues

Finally in session 3, a simple NGT exercise can be carried out which asks participants to answer in groups the questions in Box 12.5, followed by discussion. Each participant, however, has a questionnaire with the same questions to answer individually. Session 3 is sensitive in nature as it deals with funding and resource issues, so it is appropriate to enable participants to express their views in writing, and confidentiality is desired to ensure that dominant participants do not (over-)influence decisions.

12.3.7 Implementing the RAF

This section reports on a pilot study of the RAF on a site in the north of England and shows how the results of the process could be implemented. As the main aim of the RAF is to produce results that are useful, it is vitally

---

Box 12.5   Questions for session 3 of step 6.

Task for stakeholders:
Answer the following questions.

Question 1
Please select preferred option:
1. The developer should be responsible for the preparation of the monitoring report as well as the surveys and for obtaining and collating information from the LA.
2. The developer should put some money aside for the surveys and monitoring frameworks, yet it is the LA's responsibility to analyse the data and write the monitoring reports (this can also be undertaken by obtaining consultant help).

Question 2
Should the results of the assessment of the development based on these criteria be reported in a sustainability assessment to be handed to LA for consideration with the EIA and planning application?

Question 3
How public should the results of the assessment and monitoring surveys be made? Should they be made available to the local community or should they be used purely to inform LA, regional and other relevant government bodies?

---

important that its potential use, and the information produced, is specified. Table 12.7 makes recommendations for how the various results of the RAF, step by step, can be utilised and dovetailed into existing processes. It should be noted that what we present here are only guidelines and may not be relevant to all BRP.

12.3.8 Enforcing and using the RAF Results – the GM case study

This section explains how the RAF was utilised on an actual brownfield development. The pilot covered all six steps of the RAF and resulted in a S106 agreement enshrining future sustainability monitoring. The developer agreed to fund the monitoring, but the local authority assumed responsibility for the identification of the consultant who would carry out the monitoring and review the results. The community survey (undertaken in RAF step 2) formed part of the Statement of Community Involvement for the development, with further consultation sought by the developer. Additionally, information from this community survey was included within the socio-economic impact assessment (see Pediaditi, 2006).

When participants were faced with the results of the community survey and the identified sustainability objectives for the development (step 4), it was agreed that a socio-economic impact assessment should be included

Table 12.7   The main RAF results.

RAF outputs	Potential application/use
Community survey	• As part of requirements for Statement of Community Involvement • Results of survey could feed into a social impact assessment.
Sustainability assessment criteria	Results of the sustainability assessment based on these criteria could be handed in as: • a stand alone supplementary document to the application form; or • part of an EIA.
Long-term sustainability monitoring indicators	• The indicators need to be enforced through S106 agreements. • They can be detailed in a separate document or if relevant within the EIA post-monitoring section. • Results of monitoring can be reviewed by policy officer to inform policy reviews as well as other participants; they should also be made available to the public. • Should potential funders' monitoring indicators be included in monitoring strategy, they should be provided with the results report.
Site and development information.	• This information is required to support the planning application; or • The Site and Project description sections of an EIA.
Step 4 identified main concerns and benefits	• Feed into EIA scoping report, identification and agreement on significant impacts requiring a detailed assessment.

within the EIA. As a result of step 4, agreement was achieved about the significant impact areas which the EIA would have to address by carrying out detailed impact assessments (scoping opinion agreement). This illustrates that, in practice, the development and planning process is not as linear as may be implied by the literature and may in fact represent an iterative process of re-examining issues based on new evidence. For example, an EIA scoping decision may not have been provided until detailed effect assessments are already underway. This can lead to delays or wasted resources from carrying out assessments that are not necessarily required. When interviewed, the private planning consultant of the development stated:

> [T]here is a tendency in EIA to carry out detailed assessments of all effects and not only of significant ones, in fear that additional assessments will be required later on in the planning application process which would result in project delays.

The RAF allows for this by enabling a joint consensus on the significant impact areas which would otherwise require a detailed investigation early on in the application process.

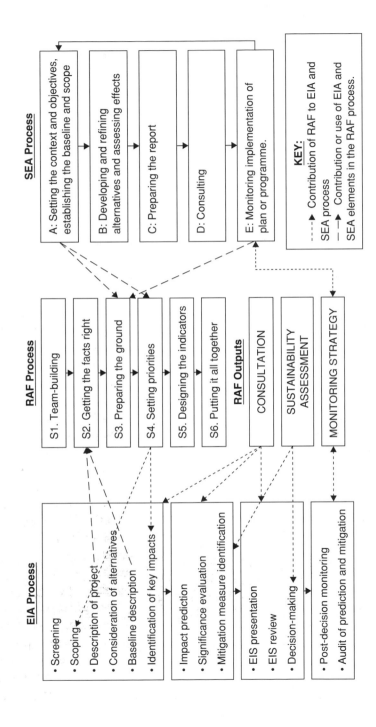

Figure 12.4 Links between the SEA, EIA and RAF processes.

The site and development information, required as part of step 2 of the development, was obtained by the EIA consultants and architect and covered the information collated for the EIA project and site description. However, it was found that clauses for more site socio-economic information should supplement the description of the environment, as recommended in Tables 12.3 and 12.4.

Having identified and agreed on a set of SEEDA sustainability assessment criteria and performance benchmarks (step 6), these were provided to the relevant consultants (for example, transport criteria to the transport consultant, design criteria to the architect). They then needed to establish the potential performance of the proposed development, using the SEEDA benchmarks. By design, the SEEDA sustainability criteria require those carrying out the assessment to justify the performance scores they allocate for each criterion. This process thus enhances transparency and provides an additional layer of scrutiny of the development's performance results, although the web-based incarnation provides less transparency. Results from all relevant consultants were then collated and introduced as a distinct sustainability assessment chapter within the Environmental Impact Statement. The performance results for each of the eight sustainability objectives were collated and presented in the form of bar charts, clearly indicating the overall development performance. This is recommended to enable EIA reviewers (e.g. councillors) to undertake a rapid appraisal, to be provided with the detailed information of the assessment.

The long-term indicators agreed upon in step 6 were detailed as part of the post-monitoring requirements of the EIS in a separate section, to support the S106 agreement covering all post-monitoring requirements. This method follows that recommended by Tinker *et al.* (2005) and the results obtained from initial interviews, which proposed measures such as EIA post-monitoring requirements, could be omitted from S106 agreements as they are dispersed throughout the EIS.

Pilot participants agreed that the results would be made available to the public through standard local authority publicity channels such as websites and bulletins. There was consensus that the results of the indicators, once obtained, should be circulated to all involved participants and be reviewed in detail by the policy officer to ensure compliance with existing policies and overarching local authority indicator trends. Because the BRP long-term indicators are essentially based on local authority Sustainability Appraisal (SA) and SEA indicators, they can use the same baseline. It is therefore possible, from the results obtained, to determine whether the development is performing better or worse than the borough as a whole with regard to a specific issue (for example, perception of crime), although the differences in the definition of sustainability and site-specific aspirations may make such comparison in some cases much harder. It may in due course be determined that the specific policies that shaped the BRP will have had the desired outcome.

Other pilot participants stated that they use the monitoring information to improve practices and ascertain whether improvements or changes need to be made for future or different developments. For example, the local authority required the development to achieve a high Secure by Design standard, with the aim of reducing the perception of crime, which was identified as an issue in step 4. However, by monitoring long-term residents' perception of crime, it can be ascertained whether the Secure by Design standards actually helped improve perceptions of crime and thus should be enforced for all developments, or if in fact they have had no effect and thus alternative measures would need to be identified. Thus, through this process, an information feedback loop is created, providing information which decision makers require.

12.3.8.1 Evaluating the RAF and its future prospects

An essential part of this project involved the piloting of the RAF on an actual BRP, with an evaluation of its effectiveness by its users. To learn most from this experience, questionnaires and interviews were conducted with all pilot participants to document the experience of utilising the RAF.

Overall, participants were very positive about the RAF process, with the local authority stating that they would apply the process to future major development applications. The developer stated that it found the process useful as it allowed it to make improvements to the site design which would otherwise have caused difficulties later on; it did not delay the planning application and was overall a very cost-effective and efficient process. However, the main perceived benefit was the increased communication and collaboration fostered by the RAF process which allowed an exploration of the development issues in a structured way. This also incorporated community views – sourced by the community survey – which was seen as a major advantage even though it requires initial additional work. Importantly, local authority officers confirmed that the RAF would be compatible with the planning process, with the developer and planning consultant considering it a useful tool for EIA.

Having a facilitator to coordinate the workshops was found to be essential, and participants appreciated the use of the SEEDA benchmarks, stating that it was good not to have to reinvent the wheel and also to have the flexibility to introduce additional criteria where needed. It was especially the case regarding the contamination criteria proposed by RESCUE (2005), which was considered less helpful and was in fact replaced by site-specific criteria based on the site investigation and risk assessment.

In addition, it was found that using the SEA and community strategy indicators ensures relevance and information feedback to planning policy, something which currently does not exist and which local authority participants found particularly helpful. All stakeholders appreciated having feedback on the development. In particular the policy officer felt that owing to the compatibility of developed indicators with existing local authority indicators, the results could feed into future policy. All participants stated that the time and resources allocated were reasonable and well spent and that the initiation of the RAF early in the development decision-making process was appropriate. From a sustainability perspective, participants also emphasised the benefits of actually considering the meaning of this contested concept in the context of a specific site.

In general, participants felt that it would be useful to see a wider adoption of the RAF. They pointed out, however, that the process could only realistically be applied to large-scale developments because of the time and resource implications involved. Furthermore, both planning consultants and local authority officers stated that the RAF would only be more widely adopted if this was required by government guidance or policy. However, in developments where there was no public sector involvement, there might be little incentive for developers to carry out such a process. Furthermore, the local authority participants stated that the RAF would need to be required by all local authorities otherwise it could be classified an unreasonable burden should the case go to planning appeal.

However, it should be emphasised that the RAF, despite its clear advantages, also has limitations. In particular, it is no substitute for formal public consultation. The RAF uses participatory methods to improve communication and understanding regarding the sustainability issues of BRP. However, the purpose of such participation is not to decide whether the development should go ahead or not (and in what shape), but merely how its sustainability can be evaluated in the long run. Likewise, the RAF cannot assess or monitor the sustainability of remediation strategies, as it reviews the whole-life sustainability profile of the site, not the remediation technology applied during the engineering phase. Equally, the RAF is not a substitute for risk assessment, EIA, SEA or EMS. These are regulated formal processes which are well established, albeit with their limitations, and the RAF should be perceived as a complement to these rather than a substitute.

Finally, much has been made in this chapter of the participatory nature of the RAF. It must be stressed that public participation is substantially limited in the RAF when it comes to large-group stakeholders, and although the survey was perceived by participants as an effective way to obtain public views it should not be used in isolation. In addition, even though it worked in the pilot study, it should be recognised that two half-day workshops are not a long time, and that for more complex or larger developments longer workshops are likely to be required.

The RAF could only have worked with the explicit acknowledgement that complex and perhaps disputable trade-offs have been made, in particular between theoretical best practice and practical feasibility. This has affected the role of 'the public'[5] as well as the technical accuracy of the indicators. However, the indicators that were selected are somewhat different from those the 'experts' would have chosen, and at no stage did the result appear to be divergent from what participants knew about the views of 'the public' in this matter. In fact, the results seem to be closer to the lay person's perspective than the experts', indicating that the power balance within the RAF deliberations was not to the disadvantage of 'the public'. In other words, if the ends justify the means (i.e. from a utilitarian perspective), the efforts to produce a manageable and efficient deliberative process, at least in the pilot case, did not seem to disadvantage the most 'vulnerable' group.

12.4 Conclusions

To conclude, the RAF is a process which has been designed to enable the functional and relevant use of existing indicators to assess the sustainability of redevelopment projects. What sustainable development means for a site is likely to remain a contentious question. In addition, defining indicators by which this can be measured is complex in itself. This chapter has shown that too many existing indicator tools perform poorly, failing to provide either a meaningful definition of sustainability across a site's life cycle, or an appropriate and practical process by which they can be implemented. This chapter has in effect turned the discussion upside down; by departing from the dominant top-down approach, it proposes a process by which some aspects of a site's sustainability can be evaluated in a participatory manner. This also includes fundamental questions about the meaning of sustainability and the relative priorities those affected by the development attribute to the site and to sustainability. The proposed process appears complex precisely because it tries to minimise work and time delays by making use, as much as possible, of existing processes and information sources in a way that should not preclude site-specific indicators. The experience of the pilot in Greater Manchester has shown that, rather than complicating things, it has simplified the indicator development process. It has also allowed a collaborative spirit to emerge between stakeholders who have engaged in a constructive dialogue throughout.

It is always difficult to generalise on the basis of one sample, but this process has appeared effective and has received support from practitioners for wider application. Participants identified a number of benefits of the RAF, including the greater communication and understanding of the different sustainability issues it provided. However, it is recognised that despite the compatibility of the RAF to existing BRP and planning processes, it remains a voluntary process which can be ignored by others. Wider adoption of the RAF will only be possible through its incorporation into central government guidance, subsequently feeding into regional and local planning policies. The RAF should be initiated at the pre-application phase of a development as it is at this point that decisions are made which will affect its future sustainability. The RAF offers the potential to establish wider dialogue about sustainability on BRPs and to help form a greater local consensus on what this actually may mean in practice. Furthermore, it offers a means of better integrating the diverse policies on sustainability that relate to brownfield developments. Finally, it has the advantage of generating better information about the sustainability profile of a site.

Even with the RAF, sustainability assessment of a site remains problematic as there are many specific issues that deserve more research. We believe this is the first model to do this in a practical, applied and meaningful way and do so across the life cycle of the site. In this light, the positive response from practitioners towards the RAF is encouraging as it is to acknowledge academic research designed to develop a practical tool for the complex, multi- and interdisciplinary world (see Chapter 2) of sustainable urban regeneration.

12A.1 Appendix

The following was presented to case study participants on the first workshop in session 3. Participants found it too hard to make the selection and proposed an open discussion talking through each of the criteria and their relevance to the relevant BRP monitoring strategy.

Box 12A.1 BRP monitoring strategy.

Please read through the criteria and their specific elements circling a **maximum of 2** elements for each criterion. Then rank the criteria according to the level of importance you think they should play when deciding which indicators to choose. Start by putting 1 for your most important criterion, 2 for your second most important, etc. in the boxes provided.

CRITERIA	SPECIFIC ELEMENTS
Development and use of indicators both by local government and public	• Be relevant to local government but also to the ordinary citizen • Relevant to ordinary citizens as well as to local government and easy to understand • Ensure participation of decision makers to secure a firm link to adopted policies and resulting action

	• Clearly assigning responsibility and providing ongoing support in the decision-making process
Indicators should enable the setting of targets, thresholds and trends	• Lead to the setting of targets or thresholds • Linked to setting targets for action • The identification of targets and trends that allow progress towards or away from sustainability to be determined • Comparing indicator values to targets, reference values, ranges, thresholds or direction of trends as appropriate • Develop a capacity for repeated measurement to determine trends • Show trends over reasonable timescales
Indicators should be integrating	• Have a relationship to other sets of indicators • Be both individually and collectively meaningful • Integrating • Relation to other indicators: as well as being meaningful on its own does the indicator have a collective meaning? • Linkage: do they link environmental, economic and social issues?
Indicators should be scientific and measurable	• Measurable (implies that it must be a quantitative indicator) • Relevant to the issues of concern and scientifically defensible • Measurable • Expressed in a way that makes sense (percentage rate, per capita, absolute value) • Stable and reliable: compiled using a systematic and fair method? • Make explicit all judgements, assumptions and uncertainties in data and interpretations • Valid: do they measure something that is related to the state of the system?
Indicators should be sensitive, iterative, adaptive and responsive to change	• Sensitive (must readily change as circumstances change) • Likely to change from year to year and more importantly, open to being changed as a result of local action • Sensitive to change across space and social groups • Sensitive to change over time • Responsive: do they respond quickly and measurably to changes? • Proactive: do they act as a warning rather than measure an existing state? • Long range: do they focus on the long term? • Adopt a time horizon long enough to capture both human and ecosystem timescales, thus responding to current short-term decision-making needs as well as those of future generations • Build on historic and current conditions to anticipate future conditions; where we want to go, where we could go

(Continued)

	• Be iterative, adaptive and responsive to change and uncertainty because systems are complex and change frequently • Adjust goals, frameworks and indicators as new insights are gained
Indicators should be simple to understand and have educational value	• Be clear, easy to understand and educate as well as inform • Understandable and if appropriate resonant • Understandable: simple enough to be interpreted by lay persons? • Be designed to address the needs of the audience and set of users • Aim from the outset for simplicity in structure and use of clear and plain language • Promote development of collective learning and feedback to decision-making
Indicators should be able to influence policy, services and lifestyles	• Provoke change in policies, services, lifestyles, etc. • Policy relevance: relevance to public or corporate policy? • Ensure participation of decision makers to secure a firm link to adopted policies and resulting action
Indicators should be based on broad participation	• Developed with the input from multiple stakeholders in the community • Community involvement: were they developed and acceptable by the stakeholders of the system of concern? • Make the methods and data that are used accessible to all • Obtain broad representation of key grass roots, professional, technical and social groups including youth, women and indigenous people to ensure recognition of diverse and changing values
Indicators should consider both the local and global scale	• Have a reasoned relationship to sustainability at both global and local level • Linked to sustainability, ideally both locally and globally • Act locally, think globally: do they promote sustainability at the expense of others? • Define a space of study large enough to include not only local but also long-distance impacts on people and ecosystems
Indicators should be context-specific	• Reflect local circumstances • Resonance: would the audience empathise with the indicator? • Representative: as a group, they cover the important dimensions of the focus area • Be designed to address the needs of the audience and set of users • Supporting development of local assessment capacity

Indicators should be comparable	• Comparability: is the indication capable of comparison with other values reported elsewhere? • Standardising measurement whenever possible to permit comparison
Data should be made publicly available	• Providing institutional capacity for data collection, maintenance and documentation • Measurable either by the local authority or by a body that can make the data available
Indicator practicality and procedural issues	• Usable (practicable) • Available (it must be relatively straightforward to collect the necessary data for the indicator) • Cost-effective (it should not be a very expensive task to access the necessary data) • Be based on relatively easy to collect information • Available and timely: can the data be collected on an annual basis? • Stable and reliable: compiled using a systematic and fair method? • A limited number of indicators or indicator combinations to provide a clearer signal of progress • Draw from indicators and other tools that are stimulating and serve to engage decision makers • Aim from the outset for simplicity in structure and use of clear and plain language.

Notes

1. Apart from RESCUE (2005), a checklist for the assessment of BRP funding applications which include assessment criteria of remediation processes.
2. Section 106 of the Town and Country Planning Act 1990 allows a local planning authority (LPA) to enter into a legally binding agreement or planning obligation with a land developer over a related issue. The obligation is sometimes termed a Section 106 (or S106) agreement. Such agreements can cover almost any relevant issue and can include sums of money. Possible examples of S106 agreements could be that the developer will transfer ownership of an area of woodland to an LPA with a suitable fee to cover its future maintenance, or the local authority will restrict the development of an area of land, or permit only specified operations to be carried out on it in the future (sourced from www.idea-knowledge.gov.uk).
3. Introductory information to questionnaires should be modified to describe the individual BRP.
4. Of course, more time would be better and some projects that are more complex, more politically sensitive or simply larger in scope and impact may well require more time. This chapter outlines the RAF process as a functioning minimum.
5. Although it is by no means clear who is included in this often-used phrase.

References

Adams, D. and Watkins, C. (2002) *Greenfields, Brownfields and Housing Development.* Blackwell Science, Oxford.

Ball, M. (1999) Chasing a snail: innovation and housebuilding firms' strategies. *Housing Studies*, **14** (1), 9–22.

Bardos, R.P., Nathanail, C.P. and Weenk, A. (1999) *Assessing the Wider Environmental Value of Remediating Land Contamination, A Review.* R&D Technical Report P238. Environment Agency, Swindon.

Bell, S. and Morse, S. (2003) *Measuring Sustainability: Learning from Doing.* Earthscan, London.

Dair, C. and Williams, K. (2004) *Sustainable Land Re-Use: The Influence of Different Stakeholders in Achieving Brownfield Development in England.* Brownfield Research Paper, No.1. Centre for Sustainable Development, Oxford Brookes University, Headington.

Deakin, M., Huovila, P., Rao, S., Sunikka, M. and Vreeker, R. (2002) The assessment of sustainable urban development. *Building Research and Information*, **30** (2), 95–108.

Deakin, N. and Edwards, J. (1993) *The Enterprise Culture and the Inner City.* Routledge, London.

Defra and EA (2004) *Model Procedures for the Management of Land Contamination, Contaminated Land Report 11.* www.environment-agency.gov.uk/commondata/105385/model_procedures_881483.pdf

DETR (1998) *Planning for Communities of the Future.* The Stationery Office, London.

Environment Council (2002) *Dialogue for Sustainability: Facilitation Skills and Principles.* Environmental Council C1/V1, London.

Harris, M.R., Herbert, S.M. and Smith, M.A. (1995) *Remedial Treatment for Contaminated Land: Classification and Selection of Remedial Methods, IV, SP104.* CIRIA, London.

IEMA (2002) *Perspectives: Guidelines on Participation in Environmental Decision-making.* Institute of Environmental Management and Assessment, Lincoln.

Imrie, R. and Thomas, H. (1993) The limits of property led regeneration. *Environment and Planning C: Government and Policy*, **11**, 87–102.

Innes, J.E. and Booher, D.E. (2000) Indicators for sustainable communities: a strategy building on complexity theory and distributed intelligence. *Planning Theory and Practice*, **1** (2), 173–86.

Mitchell, G. (1996) Problems and fundamentals of sustainable development indicators. *Sustainable Development*, **4**, 1–11.

ODPM (1999) *Projections of Households in England 2021* (19 October 1999) www.odpm.gov.uk/stellent/groups/odpm_housing/document/page/odpm_house_604206.hcsp

Patton, M.Q. (1997) *Utilization-Focused Evaluation: The New Century Text*, 3rd edn. Sage, Thousand Oaks, CA and London.

Pediaditi, K. (2006) *Evaluating the sustainability of brownfield redevelopment projects: the Redevelopment Assessment Framework (RAF).* PhD thesis, Surrey University.

Pediaditi, K., Wehrmeyer, W. and Chenoweth, J. (2006a) Developing sustainability indicators for brownfield redevelopment projects. *Engineering Sustainability*, **159** (March), 3–10.

Pediaditi, K., Wehrmeyer, W. and Burningham, K. (2006b) Evaluating brownfield redevelopment projects: a review of existing sustainability indicator tools and their adoption by the UK development industry. In: *Brownfields III, Prevention, Assessment, Rehabilitation and Development of Brownfield Sites* (eds C.A. Brebbia and U. Mander), pp. 51–62. WIT Press, Billerica MA.

Pediaditi, K., Wehrmeyer, W. and Chenoweth, J. (2005) Monitoring the sustainability of brownfield redevelopment projects: the Redevelopment Assessment Framework (RAF). *Contaminated Land and Reclamation*, **13** (2), 173–83.

RESCUE (2005) *Development of an Analytical Sustainability Framework for the Context of Brownfield Regeneration in France, Germany, Poland and Wales.* Workpackage 1. www.rescue-europe.com

RICS Foundation (2003) Benchmarking urban regeneration. *FiBRE Findings in Built and Rural Environments*, December. Royal Institution of Chartered Surveyors, London.

Rudland, D.J. and Jackson, S.D. (2004) *Selection of Remedial Treatments for Contaminated Land: A Guide to Good Practice.* CIRIA C622, London.

Sanoff, H. (2000) *Community Participation Methods in Design and Planning.* John Wiley and Sons, New York, USA.

SNIFFER (1999) *Communicating Understanding of Contaminated Land Risks*, SR97(11)F. Stirling.

SUE-MoT (2004) http://www.sue-mot.org.uk

Tinker, L., Cobb, D., Bond, A. and Cashmore, M. (2005) Impact mitigation in environmental impact assessment: paper promises or the basis of consent conditions. *Impact Assessment and Project Appraisal*, **23** (4), 265–80.

Ukaga, O. and Maser, C. (2004) *Evaluating Sustainable Development: Giving People a Voice in their Destiny.* Stylus Publishing, Virginia.

Wehrmeyer, W. (2001) A guide to communicating contaminated land risk. *Land Contamination and Reclamation*, **9** (1), 21–8.

Weston, J. (2000a) EIA, decision-making theory and screening and scoping in UK practice. *Journal of Environmental Planning and Management*, **43** (2), 185–203.

Weston, J. (2000b) Reviewing environmental statements: new demands for the UK's EIA procedures. *Planning Practice and Research*, **15** (1/2), 135–42.

13

Is Brown the New Green?

Philip Catney, David N. Lerner, Tim Dixon and Mike Raco

13.1 Introduction

The objective of the SUBR:IM consortium was to develop cross-disciplinary research to address the various technical, policy and socio-economic issues that brownfields present (see Chapter 2). The preceding chapters have sought to demonstrate the complexities of brownfield regeneration, in both social scientific and technical terms. In this chapter we will summarise and synthesise the key messages and findings that have emerged from the contributions to this volume and assess their significance for wider debates over environmental sustainability, remediation and urban redevelopment. The chapter is organised into three parts. We first examine the concept of sustainability and its ambiguities. We then look at how it has been used by the various contributors to this volume. The third section highlights wider lessons about interdisciplinary and integrated research drawn from the experience of SUBR:IM.

13.2 Sustainable brownfield regeneration

Brownfields lie at the nexus of a variety of socio-economic and technical issues. Understanding their complex dynamics requires theories, concepts and methods that have traditionally resided within discrete academic disciplines. As Raco and Dixon (Chapter 2) noted, various recent trends have emerged to push cross-disciplinary research back up the academic agenda. One of the key factors over the last 20 years has been the rise of the concept of sustainable development. This concept has acted as a powerful centripetal force, requiring greater cross-disciplinary working and the formation of more holistic research agendas. One of the main aims of this volume, and SUBR:IM more generally, has been to undertake research into the key facets of a sustainable brownfield agenda. We have used the term frequently, and in the first part of the book it was discussed in a 'regeneration' context. But what do we mean by 'sustainable', and how has the term been used in the various contributions to this book? It is to this question that we now turn.

13.2.1 Sustainable development

The concept of sustainable development was popularised by the 1987 Brundtland Report (formally known as the World Commission on the Environment and Development) and endorsed by political leaders from across the globe at the Rio Earth Summit in 1992. Over the course of the 1990s, the concept of sustainable development became embedded in the language of policy makers and academics to the point where it has been described as a new meta-narrative (Meadowcroft, 2000, p. 370) or a 'neo-renaissance idea' (O'Riordan and Voisey, 1997, p. 4). It is an attempt to resolve the traditional tension in environmental politics between striving for economic growth and protecting the environment (see Meadows *et al.*, 1972). The discourse of sustainable development suggests that governments and their citizens can seek to promote economic growth, but that they must take greater responsibility for protecting the (global) environment from further damage. Furthermore, and particularly relevant for the purposes of brownfield regeneration, it suggests that we need to have greater regard for how future generations can enjoy similar resources and opportunities to the ones we presently enjoy. Hence, one of the key challenges for sustainable development is to integrate plans that enable stable growth to complement social inclusion and the protection of the environment and its natural resources (DETR, 1999). It is by means of this triad – economic, social and environmental – that various activities are commonly judged to be 'sustainable'.

Meadowcroft (2000, p. 374) argues that much of what governments have done in the name of sustainable development constitutes a repackaging of traditional environmental policies (for example, pollution control). However, this does not imply that we should dismiss sustainable development as an empty concept that has no core. Rather, governments have interpreted sustainable development in the light of current activities and have developed new initiatives to accompany these efforts:

> There has been reform to structures and procedures, designed to integrate environmental problem-solving into the workings of the main branches of the public administration. It has accepted – at least in principle – that environmental policy cannot be operated as a post hoc corrective to normal (that is non-environmental) decision processes; rather the environmental

dimension should be factored-in from the outset. (Meadowcroft, 2000, pp. 344–5)

The principles of sustainable development must, therefore, become an integral factor in the development of a broader spectrum of public policies than previously, and not shunted into the sidelines as just an optional consideration. With this understanding, economic policy cannot be seen as isolated from environmental or social factors; rather, these realms should be seen as inextricably linked. Whether they actually are taken to be so is, of course, dependent on the national context and the extent to which the concept permeates the style of national policy-making (see Lafferty and Meadowcroft, 2000). Young (2000) argued that New Labour came to office with a more sophisticated appreciation of what sustainable development is and widened this to incorporate a greater social dimension. This can be seen in the government's initial consultation paper, *A Better Quality of Life* (DETR, 1999), which set four broad objectives for a sustainable development strategy: maintenance of high and stable levels of economic growth and employment; social progress that recognises the needs of everyone; effective protection of the environment; and the prudent use of natural resources.

The recycling of brownfield and contaminated sites has come to be seen as a *sine qua non* for, among other things, the sustainability of the UK economy and the housing market. Freeing up urban areas for development is central to the Labour government's objectives for developing an urban renaissance in Britain's cities and towns. For example, the Urban White Paper, *Our Towns and Cities: The Future* (DETR, 2000, sec. 2.30), stated:

Clearly any strategy to reduce the adverse environmental impact of how we live today has to focus on urban areas both because they are the source of much of the problem and because the scope for measures such as increased re-cycling are so much greater. But how we plan for the future shape and design of our towns and cities can contribute more directly in minimising the use of green field land by re-using derelict and contaminated land. How we develop can reduce car use, increase the use of public transport and have a positive impact on the local and global environment.

The current emphasis on redeveloping brownfield land instead of green-field sites has arisen in part because of the UK's high population densities, particularly in south-east England, and the perceived need to protect the UK's rural heritage and to create more sustainable compact urban areas. In order to realise this, the current UK government has made changes to the principles underpinning the British land-use planning system with a

Table 13.1 The sustainability of brownfield redevelopment.

Environmental benefits	Social benefits	Economic benefits
Reduction of development pressure on greenfield sites	Renewal of urban areas	Attraction of domestic and foreign investment
Protection of public health and safety	Elimination of stigma attached to communities residing in affected areas	Development of remediation/decontamination technologies
Protection and recycling of soil resources	Reduction of community fear (ill health, environmental damage and reduction in property values)	Increasing land values in inner city areas
Protection of groundwater resources		

Source: Adapted from Sousa (2001)

greater emphasis on increasing brownfield development at the expense of greenfield sites. It has set a national target for regional/local planning authorities that 60% of all development should be on brownfield land by 2008, the effect of which has been to increase pressure for development of contaminated sites. The target for development on brownfield sites was achieved by 2001, though this was at a time when aggregate housing completions across England had declined to historically low levels (English Partnerships, 2003, p. 6). An interpretation of the sustainability aspects of brownfield redevelopment is presented in Table 13.1 (see also Dixon, Chapter 5; Raco *et al.*, Chapter 6).

Alongside this more 'directive' approach, the government has identified areas of growth in areas such as the Thames Gateway and Milton Keynes in the South East where considerable brownfield and contaminated land exists alongside areas of economic deprivation and housing shortages. In addition, it has given its main regeneration body, English Partnerships, a lead role in identifying and supporting development activities that will bring these sites back into productive use. It has also made the redevelopment of brownfield sites a key measurement of the success of the regional development agencies (RDAs) that the government created in the late 1990s for English regions and the Greater London area. The redevelopment of brownfield land in areas like the Thames Gateway is at the heart of the government's agenda for tackling housing shortages in London and the South East and creating 'sustainable communities'. In February 2003, the Labour government produced *Sustainable Communities: Building for the Future*, which set out the government's plans for a £22 billion initiative aimed at tackling housing shortages in the South East and low demand in northern areas (see Dixon, Chapter 5). However, as Raco *et al.* (Chapter 6; see also

Raco, 2005) noted (see below), there is good reason to believe that the government avoided the hard (environmental) issues that would be required for the programme to be described as 'sustainable' in 'sustainable development' terms.

13.2.2 An elusive concept?

While the general principles of sustainable development seem clear, it is often criticised for its ambiguity. This has given rise to a variety of discourses that interpret sustainability in different ways within and across nations (see, for example, Torgerson, 1993; Dobson, 1998; Lafferty and Meadowcroft, 2000; Rydin, 2003). As Dixon (Chapter 5) noted, 'sustainability' is not static in terms of either temporal or spatial perspective; it requires activities and actions to balance short- and long-term effects over generations and for the multiple levels on which such impacts occur to be appreciated – for example, the global, national, regional, local and site levels.

One of the early difficulties that SUBR:IM faced was providing a satisfactory definition of 'sustainability' and 'sustainable development' that could anchor the proposed cross-disciplinary research agenda. While there was general agreement on the importance of the triad of economic–social–environmental dimensions, there was substantially less agreement on how the term should be put into operation in practice. But why was this?

Part of the answer to this question lies in the nature of the term. It needs to be noted that sustainable development is not a phenomenon: that is, something observable that exists in its own right. It is a concept, and the words used are open to multiple interpretations (Wittgenstein, 1958, p. 118). It is what philosophers refer to as an 'essentially contested concept'. This perspective asserts that when we examine different uses of certain terms, we find that there is no one definition that 'could be set up as its generally accepted and therefore correct or standard use' (Gallie, 1964, p. 157). Moreover, it asserts that certain terms fulfil a variety of functions and are not resolvable by argument, although they are often sustained by them (Gallie, 1964, p. 158). But how are we to recognise and identify such concepts? Gallie (1964, p. 161) outlined five conditions that are necessary for a concept to be defined in this way:

- It must signify some form of valued achievement.
- This achievement should be internally complex yet it must be judged as a whole.
- Any explanation must cite the various contributions that its parts or features make (including potential rival descriptions).
- The recognised achievement must make allowance for significant modification in changing circumstances, and such modifications cannot be prescribed or predicted in advance.

- Each party must acknowledge that the use of a concept may be contested by other parties and therefore recognise that its use may vary according to how the concept is being employed.

Gallie goes on to place concepts such as art, democracy, religion, science and social justice beneath this rubric. Each of these can be recognised in their general form, but they can assume a variety of guises depending on the context in which they are articulated. This does not, however, render them meaningless. For Jacobs (1999, p. 25), such concepts have two levels of meaning. On the first level, concepts such as 'democracy' can have a multitude of definitions but still retain core features that define and operationalise them and that other terms cannot connote. Such concepts are constituted by a set of 'core ideas' that give them their distinctive form. Jacobs (1999, pp. 26–7) argues that the discourse of sustainable development is structured around six such 'core ideas':

(1) *environment-economy integration*: allowing for the integration of economic development and environmental protection in planning and implementation
(2) *futurity*: ensuring that current activities do not over-burden future generations
(3) *environmental protection*: a pledge to impose greater restrictions on pollution levels and environmental destruction, and to encourage the more efficient use of resources
(4) *equity*: ensuring that the basic needs of the poor of the present generation are met (as well as equity between generations)
(5) *quality of life*: acknowledging that human well-being requires more than just financial well-being
(6) *participation*: attaining sustainable development requires the participation of all groups or 'stakeholders'

While the 'internal' or 'core' ideas that structure the discourse of sustainability can be articulated with some degree of precision at one level, how they are to be operationalised in practice remains vague. This is Jacobs's second level of meaning. It is at this level that such concepts are contested: for example, few would argue against environmental protection being a core element that defines sustainability; however, the *degree* of protection afforded is debatable (Jacobs, 1999, p. 31). It is at this level that we see rival conceptions of sustainability (based on interests, beliefs and/or disciplinary focus) emerge that seek to shape the precise form that action takes. In the context of their discussion on greenspace, for example, Moffat and Hutchings (Chapter 7) demonstrated some of these varying interpretations of the concept of sustainability and how they can affect how academics research sustainability issues:

For engineers, the main concerns will be the effectiveness and longevity of pollution-control measures . . . For silviculturalists and horticulturists, the most important issue might be the longevity of the vegetation planted. For community leaders, it might be focused on the utilisation of the greenspace by the public and the degree of social cohesion engendered by it. For the local authority, it might be the balance between the enhancement of the value of the properties adjoining the greenspace and the cost of running it. And developers will inevitably be most interested in supporting site sustainability design criteria so long as budgets and profits aren't threatened.

In the next section we return to some of the key findings of the research presented in this volume and, through an exploration of several of sustainable development's 'core ideas' outlined above, examine what they tell us about the sustainability of current practices associated with brownfield redevelopment.

13.3 Sustainability in action

13.3.1 The role of government and policy

A key issue that emerged from the preceding chapters is the role that government at all levels can and should play in promoting brownfield regeneration. As Raco *et al.* (Chapter 6) noted, the potential role that local authorities could fulfil in supporting brownfield regeneration in their areas is often overlooked. Indeed, they are actually criticised for 'interfering' with local property markets through the planning process. The authors highlighted the importance of public sector support in the regeneration of the Salford Quays development. Salford City Council played a key role in setting out the development 'visions' for the scheme, directly by the purchase of land for the original developments and by cultivating the confidence of the developer community in the scheme.

Dixon's research suggested that, with limited gap funding now available for development, further public sector funding and improved grant regimes will be needed to tackle 'hardcore' sites. It also argued that there is a clear need for government and related agencies to ensure infrastructure is in place prior to development. The importance of this was demonstrated in the case study of the Paddington Basin development by Doak and Karadimitriou (Chapter 4) where the creation of infrastructure kick-started regeneration networks in the area. Dixon also suggested that the absence of full government funding and support may make the introduction of a planning gain supplement (or equivalent) inevitable. He advocated a key role for English

Partnerships in providing local infrastructure and serviced sites. In addition, Dixon recommended that government bodies such as the Environment Agency take on a greater role in publicising and disseminating information about the alternatives to landfill more widely, as his findings indicated that small housebuilding firms remain unaware of other options for remediation of contaminated sites. He further suggested that the Environment Agency should help develop 'realistic' risk guidelines for clean-up.

While the public sector can play a positive role in facilitating sustainability in brownfield redevelopment, its structures and policies can often erect unintentional barriers to this. Dixon's investigation (Chapter 5) of private developers showed that they are increasingly coming to terms with brownfield risks, but that they claim policy issues are a major obstacle. For example, while government policy appears to have been successful in pushing greater amounts of development onto brownfield sites, Dixon suggested that conflicting policy aims, such as the EU Landfill Directive, may threaten the continued success of the agenda. Furthermore, private developers suggested that contamination and waste legislation and guidance need to be streamlined and rationalised. Dixon reports that there appeared to be support for the introduction of a single remediation permit system and a review of the Soil Guideline Values to ensure that a 'sensible' balance is created between safety and risk to public health.

Another barrier cited is the complex structure of governance that exists in this policy area. It is argued that clearer designation of responsibilities between the various governmental agencies at national, regional, local and site levels is required to prevent substantial planning delays and overwhelming bureaucracy (especially in institutionally and policy-dense areas such as the Thames Gateway). This proposal is supported by Raco *et al.* (Chapter 6) who suggest that the institutional roles and responsibilities of the various organisations involved in brownfield governance need to be clarified so that local programmes can be more clearly embedded in a wider set of development practices. They suggested that development practice often becomes confused between different agencies, each with their own targets, resources and priorities.

13.3.2 *Preparing for the future*

One of the critical challenges for societies over the next century will be how they adapt to climate change. In Chapter 11, Al-Tabbaa *et al.* examined the potential consequences that climate change may pose to pollution containment systems and the extent to which policy actors and practitioners have incorporated these into their policies, plans and practices. They showed how long-term changes in national and global temperatures will have significant consequences for the robustness of remediation technologies, such as containment systems; they suggest, for example, that severe physical damage will be sustained by soil cover systems and stabilised/solidified soils. It is their view that such impacts will have major effects on the future management of contaminated and remediated sites, influencing how risk is managed and the design of future remediation strategies.

They also highlighted that while there is a growing consciousness of the implications of climate change for contamination remediation, the ensuing action has arisen from a very low base; many developers and local authorities remain largely unaware of the issues and are inactive on developing strategies and adaptation measures. Al-Tabbaa *et al.* (Chapter 11) argued that new ways of monitoring and measuring the effects of climate change on physical processes and containment technologies are required, and that this needs to be supported by more multidisciplinary research, which can provide further insights into the processes underpinning sustainable brownfield development. In particular, this form of research is important in that it seeks to examine how political, social, economic and environmental influences are subject to significant change and variation over time. It can also unpack the multiple understandings and interpretations of key brownfield concepts such as urban sustainability, risk and climate change that exist among different stakeholders.

Moffat and Hutchings (Chapter 7) suggested that greenspace provides significant opportunities that can help adaptation to climate change. They argued that, with further increases in urban populations predicted alongside higher temperatures, urban dwellers will require more greenspace for recreational purposes as well as to help provide shade. Another opportunity for greenspace could be the way it can help mitigate future urban flooding, and Moffat and Hutchings suggested that areas of greenspace could be developed within towns and cities to reduce local rainwater run-off and provide areas where surface water can be allowed to flood, rather than flooding residential or industrial sectors. However, they also noted that water provision will be affected by climate change, particularly in the south and east of England. Hence they suggest that the design of greenspace should increasingly consider future climate change scenarios so that planned vegetation can be supported in the light of summer water demand.

13.3.3 Environment–economy integration and environmental protection

One key finding that emerged from SUBR:IM's research was that the extent to which environmental protection has been integrated with the pursuit of economic development in policy and practice remains uncertain. Raco *et al.* (Chapter 6) argued that a conspicuous feature of the government's guidance for creating 'sustainable communities' has been the lack of emphasis on 'green' policies and practices. They noted that the only environmental components included concern the reduction of car travel and access to open space for local residents. Similarly, Moffat and Hutchings suggested that while the environmental benefits of greenspace are specifically acknowledged in the government's report on greening in the Thames Gateway report, the key drivers for greenspace are to enhance the 'liveability' of existing and newly created domestic and work places, with environmental drivers being relevant insofar as they have a utility for humans.

However, SUBR:IM researchers have explored the benefits of utilising technical solutions to promote better forms of remediation. Moffat and Hutchings point out that sustainably remediating brownfield sites is highly complex because of the heterogeneous nature of ground conditions, contaminants present and potential multiple pathways to receptors. Remediation of contaminated sites in the UK has focused on cleaning up sites to generic levels (that is, the standard of remediation is the same for different end-usage) with the particular formation of the final landscape being considered in isolation from the remedial process. The authors contended that the integration of remediation and greening systems can provide more robust, reliable, cost-effective and sustainable design methods for the future. They outlined research findings that showed that utilising different types of vegetation as part of a greenspace strategy on brownfield sites could prove a less costly overall remediation strategy for developers while providing a superior standard of restored functional greenspace for eventual users and stakeholders.

In a similar vein, Ouki *et al.* in Chapter 8 pointed to the potential that novel composts and additives may hold for treating various contaminants on brownfield sites. They presented findings from empirical research into using novel sustainable remediation techniques that will rely on the use of waste-produced materials (composts) combined with naturally occurring minerals (clays, zeolites) in order to enhance the biodegradation and immobilisation capability of the compost. They suggested that this technique could play an important role in mitigating at least some of the threats posed by contaminated sites. In addition, they argued that utilising composts would also facilitate plant growth and provide soil conditioning and nutrients to a wide variety of vegetation.

While the utilisation of vegetation and composts can achieve considerable palliative results, they are often inappropriate for application on various severely contaminated sites on their own. Often physical containment systems are the only viable and cost-effective means of isolating contamination and cutting its pathways.

Chapters 9 and 10 explored two different techniques that are commonly employed in remediation: capping, and stabilisation and solidification (S/S). In Chapter 9, Al-Tabbaa *et al.* showed that little research has been undertaken into assessing and comparing the sustainability of individual remediation methods employed in the UK. Like Moffat and Hutchings, the authors argued that the risk-based and site-specific approach adopted in the UK often results in the possibilities for improving these methods being overlooked as decision makers and technical experts are concerned with achieving the core objectives for each site and neglecting wider societal implications. They suggested that remediation techniques can be judged on sustainability in five key dimensions:

- Future benefits outweigh cost of remediation
- Environmental impact of the implementation of the remediation process is less than the impact of leaving the land untreated
- Environmental impact of bringing about the remediation process is minimal and measurable
- The timescale over which the environmental consequences occur, and hence inter-generational risk, is part of the decision-making process
- The decision-making process includes an appropriate level of engagement of all stakeholders

Al-Tabbaa *et al.*'s analysis of the effect of different factors on robustness and durability of remediation techniques suggests that improvements can be made in the sustainability of these methods, thereby producing enhancements in terms of lower remediation costs, and/or lower environmental impact.

Chapter 10 examined the durability and sustainability of remediation technologies directed towards acid tar lagoons in the UK, particularly capping technologies. Through a case study of an existing acid tar site, Talbot *et al.* sought to examine potential remedial scenarios to mitigate the contamination hazards posed. This chapter did not simply view the challenges of such sites through a technical prism; it also sought to analyse the socioeconomic and risk communication challenges that such sites pose (see also Catney *et al.*, Chapter 3). It examined how the public authorities dealing with the site sought to mediate the competing viewpoints about how to deal with the site (see below).

13.3.4   Social equity

One of the 'core ideas' of sustainable development is that any development should meet the basic needs of the poor of the present generation (as well as ensuring equity between generations). Raco *et al.* (Chapter 6) argued that future major brownfield developments need to pay more attention to fostering social and economic links between those living on the site and residents outside. Their findings suggest that local communities adjacent to brownfield developments feel that the focus of such schemes is to attract middle- to high-income workers and consumers to the area and to drive established communities away through processes of commercial and residential gentrification. In addition, they found that there exist major difficulties in connecting local residents to the emerging employment opportunities offered by major brownfield developments. The jobs that they can fill tend to be relatively low-skilled positions, for example as cleaners or security guards. While their two case studies of Salford Quays and Paddington Basin showed that some efforts have been made to encourage local employment linkages, the results have not been a resounding success.

Raco *et al.* found that brownfield developments can engender a sense of pride in local residents, despite their generally limited feelings of 'ownership' over the schemes. However, these positive feelings are qualified by feelings of exclusion that are created by a lack of employment opportunities and poor-quality housing and urban environments. These feelings of apathy and disconnection are further amplified by a sense that these developments are not for them. Dixon's research (Chapter 5) supports this view, suggesting that area-based initiatives focused solely around property development were more likely to fail in their aims. He argued that, instead, people-based initiatives, such as re-skilling, are needed to complement these developments so as to enable local people and businesses to thrive.

13.3.5 Participation

Another theme that these chapters unearthed concerned the democratic performance and the incorporation of stakeholders affected by contamination remediation decision-making and broader development plans. In terms of the sustainable development paradigm, democratic and holistic decision-making is a crucial factor in winning broader public acceptance of governmental decisions (Meadowcroft, 1997). It is widely acknowledged that decision-making processes need to arrive at negotiated solutions with actors and agencies from across the various tiers of government and the different spheres of society for actions to be legitimate, effective and sustainable. As well as actors from the public and private sectors, this can incorporate non-governmental organisations and other organisations from civil society that represent environmental groups and community action groups.

As noted above, developing governance and policy structures that can aid brownfield recycling is clearly an important aspect of achieving greater sustainability in land use. Yet relatively little is known about the attitudes of the two key groups outside government critical to the effective implementation of brownfield regeneration: institutional investors and private developers. Chapter 4 sought to outline a sophisticated theoretical framework through which the interactions of various actors, but particularly institutional investors, can be conceptualised. Doak and Karadimitriou's research showed that institutional investors, while often having limited direct involvement in brownfield governance, can play an important role in structuring the network negotiations and agreements that take place. However, institutional investors have a tendency to be risk-averse and so avoid investing in brownfield regeneration projects because the imprint of traditional ways of operating reinforces the view that profits cannot be made from such projects. Yet Doak and Karadimitriou report that this might be changing. They suggest that a select but growing group of investors and developers are joining brownfield development networks as they become aware of the potentially high returns on these investments.

Dixon's research (Chapter 5) on the private development industry confirmed that it has started to come to terms with the challenges posed by the brownfield agenda. He suggests that this change has been not just the result of national government policy, but a realisation that it is a potentially profitable activity in a relatively buoyant property market. Although the findings of his national survey indicated that contamination was still seen as an important challenge, infrastructure, density and governance issues were considered more important obstacles to development (see above).

In terms of implementation of sustainability principles, Dixon noted that it is widely held that the property and construction industry has been slow to react to the challenge. He suggests that in some cases the development industry seems to have been merely playing 'lip service' to the concept. One reason for this, he suggests, is the contested nature of the concept, which causes confusion over what parts of a development need to be sustainable (for example, the land remediation process, the planning process, the buildings themselves, or the resulting community?). Indeed, he suggests that the development industry is playing an increasingly influential role in the 'sustainability' agenda, taking the initiative and defining sustainability on its own terms. However, this may lead to wide variations in how the term is applied across the sector, and he advocates the formulation of an agreed, industry-wide definition, although he does concede that there remains some scepticism about this within the industry.

In addition, Raco *et al.* (Chapter 6) observed that major brownfield redevelopment projects still involve little direct participation from local communities or environmental groups that may challenge the normative assumptions underpinning such developments. They argued that there exists a contested 'politics of time' concerning a development, with different groups having very different understandings of exactly how and when the development should deliver on its objectives: for example, this could include what different actors see as the proper timescales for a course of action to be sustainable and why; timescales could be interpreted through such things as the commercial development cycle, human lifetime cycles, profit-making/payback cycles, action deadlines, electoral cycles, memory spans of actors and institutions, ecological timescales (the time it takes for trees to grow). These have profound effects on (1) the perception of what is sustainable (for example, how long it will take for developers to make a profit); (2) the capacity of actors to undertake a particular course of action on a site (for example, do we know enough yet to take action?); (3) the durability of a particular course of action (will a particular technology last?). Also of importance are the ways in which geological, politico-social (e.g. electoral), economic and ecological cycles can link spatial and temporal scales (see Meadowcroft, 2002).

In their two case study areas, Raco *et al.* reported mixed evidence for the extent to which local decision-making processes incorporate 'outside' perspectives. They observed that it was not just local communities who felt excluded from important decisions; existing firms can feel undervalued and excluded from developments taking place around them. Raco *et al.* argue that the success of brownfield redevelopment projects is dependent on new forms of 'active' and 'sustainable' citizenship being fostered through participatory brownfield decision-making, enabling local communities to play a 'stewardship' role over such developments (see also Dobson, 2003; Dobson and Bell, 2006).

There are, however, some encouraging signs that local community actors are being integrated into more recent brownfield developments. For example, Doak and Karadimitriou (Chapter 4) found that on the New Islington Millennium Communities development in east Manchester local residents became a key 'sounding-board' for proposals being developed for the area and had a key role in selecting and briefing the social housing developer required to build the replacement accommodation in the area. Moreover, local residents changed the development agenda by insisting that the social housing for the development should be located within central residential plots and not pushed to the edges of the development. They also rejected the high-density vision contained in government guidance and demanded individual 'houses'.

Research from other chapters showed that the extent to which lay communities are involved in remediation decisions is questionable. For example, Catney *et al.*'s examination (Chapter 3) of risk communication on two contaminated sites suggested that poor implementation of the principles of open risk communication may be prevalent. They argued that open and democratic risk management processes are necessary because they generate higher levels of social trust than do closed ones. Their two case studies identified key differences in organisational culture of the two councils, which resulted in one council adopting an 'open' approach to communication while the other operated a 'closed' approach. Their findings, derived from a large-scale questionnaire of residents' perceptions of risk and quantitative research, reinforced the message that risk perceptions, attitudes and trust are closely interconnected.

Yet participatory decision-making can lead to considerable delays. Talbot *et al.*'s cross-disciplinary examination (Chapter 10) of an acid tar lagoon found that despite a quantified risk being identified and a carefully constructed regulatory system being in place to protect human health, inaction occurred. They suggest that this was due to a misalignment in stakeholders' interests on the site. They found that the 'internal' stakeholders to the consultation process were unable to agree on whether the quantitative risk assessment accurately established the risks to human health. This led to regulatory paralysis, with the local authority preferring to direct funding to sites that it considered to be of more immediate concern.

Such diverse stakeholder interests and concerns can run the risk of producing the kind of paralysis exhibited in the case outlined by Talbot *et al.* In order to avoid this, and to promote 'bottom-up' means of identifying indicators for monitoring the sustainability of brownfield redevelopment projects, Pediaditi *et al.* (Chapter 12) developed the Redevelopment Assessment Framework (RAF). The RAF is a process that enables the more effective use of existing indicators to assess the sustainability of redevelopment projects. It is a participatory process which addresses central questions about the meaning of sustainability and the relative priorities of a wider variety of actors affected by proposed developments. Pediaditi *et al.*'s pilot of the RAF on a site in Greater Manchester demonstrated that it was indeed effective at simplifying the indicator development process. Importantly, it promoted a constructive dialogue between the various stakeholders through which sustainability issues could be discussed effectively.

13.4 Constructing cross-disciplinary research: lessons from the SUBR:IM experience

As Raco and Dixon noted in Chapter 2, there are various ways in which cross-disciplinary research can be characterised (inter-, multi- and transdisciplinarity). These are often considered as challenges to existing research paradigms and disciplinary structures, helping to promote new methodological approaches. In this section we return to this theme to discuss possible ways of overcoming some of the barriers to cross-disciplinary working identified from the experience of SUBR:IM.

It is first, however, important to note that increased interdisciplinarity may not necessarily lead to the end of disciplinary structures as we know them. Indeed Barnett (1994, p. 126) claimed that interdisciplinarity 'paid compliments to disciplines: it did not seek to displace them but to build forms of integration between them'. This suggests that researchers need not be wary of cross-disciplinary ventures; in fact, such activities can help researchers gain greater insights into their own discipline as well as others. However, it needs to be appreciated that researchers wishing to engage in cross-disciplinary working are often confronted with the potential for information overload and entry barriers to certain specialities, including the lack of necessary background knowledge to be able to operate effectively (Wilson, 1996).

The joining of engineers and natural scientists with social scientific researchers can also benefit their traditional audiences as it broadens their perspectives by introducing them to the relevance of other disciplines in the understanding of particular issues (Wear, 1999, p. 301). However, the diversity of an audience can produce tensions in terms of the manner in which research findings are communicated, the particular concerns that take priority in the analysis and the style of analysis adopted (for example, between understanding phenomena and their critical analysis) (Wear, 1999, p. 299). Audiences for SUBR:IM events were small but heterogeneous, often bringing together academics from different disciplinary fields; policy actors from various levels of government; environmental activists; and private sector consultants and contractors. These audiences were generally small precisely because they were only interested in research in areas that were related to their own expertise or their interests, or that they could use in the course of their work. The diversity of knowledge and professional backgrounds of these groups created difficulties for researchers attempting to communicate their research findings in knowing at what level these should be pitched. For example, should the social scientist assume that the audience understands policy networks theory? Should the engineer take for granted that the audience understands the technology behind MRI scanners? This can, however, exaggerate the knowledge barriers that exist as sub-disciplines within disciplines may encounter similar obstacles to communicating research. For instance, in social science there is often considered to be a gulf of understanding between quantitative and qualitative researchers (see Smith, 1983).

This leads us to consider the substantial linguistic barriers to effective cross-disciplinary working. Wear (1999, pp. 299, 301) argued that language is an important barrier to effective interdisciplinary research

> because scientists speak in dialects that are specialized to their disciplines . . . The fundamental challenge to interdisciplinary communication is the different ways we see the world, that is our constitutive metaphors. The greater the divergence between these foundations, the more difficult it is for communication to be effective.

These linguistic difficulties arose over a number of different concepts including essentially contested terms such as 'sustainability' (see above), 'quality' and so forth. These terms are often imbued with subtly different meanings according to academic disciplines and sub-disciplines. Crow *et al.* (1992, pp. 749–50) observed:

> The dilemma of negotiating an interdisciplinary approach to produce a coherent narrative forced us to consider again the disciplinary paradigms influencing each of us. Words acted as signals, alerting us to the languages of our respective disciplines and to the different meanings embedded in language use.

This reinforces the point made earlier that cross-disciplinary research can improve our knowledge of our own disciplines as well as of other disciplines. To address such concerns, Crow *et al.* (1992, pp. 748–9) suggested that a common language is essential for effective interdisciplinary working. In a similar vein, Wear (1999, p. 301) suggested that in order to avoid some of the linguistic and methodological pitfalls that hamper interdisciplinarity, research needs to be constructed from the 'ground up', with the models developed, the methods adopted and the data sought and manipulated being clearly defined and developed by the researchers involved, rather than reaching for various 'off the shelf' methods from discrete disciplines. This way, the basic premises behind the terminology adopted and the framing theories and methodologies utilised by differing disciplines are exposed to critical dialogue. Similarly, Crow *et al.* (1992, p. 750) stress that conceptual and linguistic barriers need not be stumbling blocks to collaborative research if there is some 'goodness of fit' between the varying disciplinary interpretations of particular concepts.

As stated above, the chapters of this volume demonstrated a multiplicity of interpretations of how to conceptualise 'sustainability' and how to analyse it in practice. These interpretations arose from different interests, beliefs, disciplinary focuses and theoretical interests. Yet this need not necessarily be considered a limitation to cross-disciplinary research:

> [T]he theoretical fragmentation and eclecticism that marks the domain of environmental political economy provides no real cause for complaint. Human interactions with the environment are not separate from other social behaviours. So why should theories that contribute to understanding alternative dimensions of social interaction not contribute to making sense of environmental issues? Moreover, in political and economic terms 'the environment' is not one thing. Rather it is a fractured and multi-dimensional mosaic that touches social life at many points. So perhaps our theories are similarly destined to remain

fragmented and partial, as well as borrowed and contested. (Meadowcroft, 2005, p. 483)

One lesson that has arisen from the SUBR:IM experience is that successful cross-disciplinary research requires significant time and resources from the beginning. This view is supported by Turner and Carpenter (1999, p. 275), who noted that interdisciplinary research requires longer time frames for success than more traditional forms of academic inquiry, which are more discipline-based and hence already operate according to some form of common idiom. As Raco and Dixon (Chapter 2) observed, limited time frames and resources in the early days of the Sustainable Urban Environments (SUE) programme, in addition to the particular way in which the funding council apportioned funds (allocating funding to principal investigators before they had developed a properly integrated research programme), reduced the scope for establishing such a framework from the beginning of SUBR:IM. It was only well into the project that cross-disciplinary working started to emerge as individual researchers began to trust other researchers and form working relationships. This was fostered through the use of cross-project initiatives, such as a researchers' day. However, by this time the project plans were already in place and were not up for revision to suit the interdisciplinary agenda.

In short, trust is an essential factor in constructing inter-, multi- and transdisciplinary research (Turner and Carpenter, 1999, p. 275; Bruhn, 2000). One of the main barriers to effective cross-disciplinary working can be resistance from academics themselves. Without establishing confidence between researchers, establishing cross-disciplinary research is impossible.

13.5 Conclusions

Achieving sustainable development is now, in theory at least, at the heart of the government's agenda for an array of policy areas, including urban regeneration. Within this area, increasing focus is being afforded to the recycling of brownfield land, with various planning and economic development policies being implemented to move development away from greenfields and onto brownfields. The importance of brownfield redevelopment has recently been heightened by increases in urban populations due in part to the growth of inward migration to the UK, specifically to south-east England, and London's successful bid for the Olympics, with its concentration in the Lea Valley area, notable for the amount of brownfield land that it contains.

However, while the government has become attuned to the necessity of pushing urban development away from greenfields and onto brownfields, we suggest that it has given little thought to the multifaceted social and technical issues which challenge this relatively new direction in policy. These include, among other things, the technical difficulties of remediating sites for future use, the durability of remediation technologies in the face of climate change, ensuring that quality is maintained throughout the site investigation and remediation processes, the resistance of property markets that could remain sceptical of the benefits of brownfield redevelopment, and more generally the difficulty in connecting the benefits of redevelopment to deprived local communities in order to create truly sustainable communities. This view carries some resonance with that expressed by Gans and Weisz (2004) in their work on 'extreme sites':

> A brownfield is a latent condition, as yet unreclaimed, unbuilt. Too often the approach to its remediation is driven by economics and litigation rather than by a conceptual framework of landscape, urbanism and culture.

Finally, we suggest that future policy and research on brownfield redevelopment needs to consider three issues. First, there remain technical issues and scope for innovation to increase certainty and reduce costs in brownfield regeneration, particularly with the onset of climate change. Second, further research is required into the financing of brownfield regeneration and how brownfields are to be more effectively integrated within broader issues related to cities, society and sustainability. Lastly, there is scope for research agendas to focus on individual brownfields as part of a larger and longer-term process of urban development.

It is clear then that brownfield regeneration is a key component of a new 'green agenda' in the UK and indeed in other developed countries. In this sense, 'brown is the new green'. In the future it is likely that the term 'brownfield' will become increasingly mainstreamed within the context of a much wider 'sustainable urban regeneration' agenda. After all, some 60% of the world's population will be urbanised by 2030, and London's population is expected to be some 8.6 million in 2015 compared with 7.7 million in 2007. It will therefore be critical to continue to find ways of regenerating our urban areas, and to develop and provide a built environment that is sustained over a longer time frame and provides liveable spaces. The real question is can we develop sustainable cities that really will stand the test of time? Only time will tell.

References

Barnett, R. (1994) *Limits of Competence: Knowledge, Higher Education and Society.* Open University Press, Buckingham.

Bruhn, J.G. (2000) Interdisciplinary research: a philosophy, art form, artifact or antidote? *Integrative Physiological and Behavioral Science*, **35** (1), 58–66.

Crow, G.M., Levine, L. and Nager, N. (1992) Are three heads better than one? Reflections on doing collaborative research. *American Educational Research Journal*, **29** (4), 737–53.

DETR (1999) *A Better Quality of Life: A Strategy for Sustainable Development for the UK*. Department for Environment, Transport and the Regions, London.

DETR (2000) *Our Towns and Cities: The Future – Delivering an Urban Renaissance*. Department for Environment, Transport and the Regions, London.

Dobson, A. (1998) *Justice and the Environment: Conceptions of Environmental Sustainability and Dimensions of Social Justice*. Oxford University Press, Oxford.

Dobson, A. (2003) *Citizenship and the Environment*. Oxford University Press, Oxford.

Dobson, A. and Bell, D. (eds) (2006) *Environmental Citizenship*. MIT Press, Cambridge, MA.

English Partnerships (2003) *Towards a National Brownfield Strategy*. English Partnerships, London.

Gallie, W.B. (1964) *Philosophy and the Historical Understanding*. Chatto & Windus, London.

Gans, D. and Weisz, C. (eds) (2004) *Extreme Sites: The Greening of Brownfield*. Wiley-Academy, London.

Jacobs, M. (1999) Sustainable development as a contested concept. In: *Fairness and Futurity: Essays on Environmental Sustainability and Social Justice* (ed. A. Dobson), pp. 21–45. Oxford University Press, Oxford.

Lafferty, W.M. and Meadowcroft, J. (eds) (2000) *Implementing Sustainable Development: Strategies and Initiatives in High Consumption Societies*. Oxford University Press, Oxford.

Meadowcroft, J. (1997) Planning, democracy and the challenge of sustainable development. *International Political Science Review*, **18**, 167–90.

Meadowcroft, J. (2000) Sustainable development: a new(ish) idea for a new century? *Political Studies*, **48** (2), 370–87.

Meadowcroft, J. (2002) Politics and scale: some implications for environmental governance. *Landscape and Urban Planning*, **61**, 169–79.

Meadowcroft, J. (2005) Environmental political economy: technology transitions and the State. *New Political Economy*, **10** (4), 479–98.

Meadows, D.H., Meadows, D.L., Randers, J. and Behrens, W.W. (1972) *The Limits to Growth*. Earth Island, London.

O'Riordan, T. and Voisey, H. (1997) The political economy of sustainable development. *Environmental Politics*, **6** (1), 1–23.

Raco, M. (2005) Sustainable development, rolled-out Neo-Liberalism and sustainable communities. *Antipode*, **37** (2), 324–46.

Rydin, Y. (2003) *Conflict, Consensus, and Rationality in Environmental Planning: An Institutional Discourse Approach*. Oxford University Press, Oxford.

Smith, J.K. (1983) Quantitative versus qualitative research: an attempt to clarify the issue. *Educational Researcher*, **12** (3), 6–13.

de Sousa, C. (2001) Contaminated sites: the Candian situation in an international context. *The Journal of Environmental Management*, **62**, 131–54.

Torgerson, D. (1993) The uncertain quest for sustainability: public discourses and the politics of environmentalism. In: *Greening Environmental Policy: The Politics of a Sustainable Future* (eds F. Fischer and M. Black). Paul Chapman Publishing, London.

Turner, M.G. and Carpenter, S.R. (1999) Tips and traps in interdisciplinary research. *Ecosystems*, **2**, 275–6.

Wear, D.N. (1999) Challenges to interdisciplinary discourse. *Ecosystems*, **2**, 299–301.

Wilson, P. (1996) Interdisciplinary research and information overload. *Library Trends*, **45** (2), 192–203.

Wittgenstein, L. (1958) *Philosophical Investigations*. Basil Blackwell, Oxford.

World Commission on Environment and Development (1987) *Our Common Future*. Oxford University Press, Oxford.

Young, S. (2000) New Labour and the environment. In: *New Labour in Power* (eds D. Coates and P. Lawler), pp. 149–68. Manchester University Press, Manchester.

Index